AGILE LEADERSHIP FOR INDUSTRY 4.0

An Indispensable Approach for the Digital Era

AGILE LEADERSHIP FOR INDUSTRY 4.0

An Indispensable Approach for the Digital Era

Edited by

Tanusree Chakraborty

Tahir Mumtaz Awan

Ashok Natarajan

Muhammad Kamran

AAP | APPLE ACADEMIC PRESS

First edition published 2023

Apple Academic Press Inc.
1265 Goldenrod Circle, NE,
Palm Bay, FL 32905 USA

760 Laurentian Drive, Unit 19,
Burlington, ON L7N 0A4, CANADA

CRC Press
6000 Broken Sound Parkway NW,
Suite 300, Boca Raton, FL 33487-2742 USA

4 Park Square, Milton Park,
Abingdon, Oxon, OX14 4RN UK

Library and Archives Canada Cataloguing in Publication

Title: Agile leadership for industry 4.0 : an indispensable approach for the digital era / edited by Tanusree Chakraborty, Tahir Mumtaz Awan, Ashok Natarajan, Muhammad Kamran.
Names: Chakraborty, Tanusree (Faculty member at the Centre for Management Studies) editor. | Awan, Tahir Mumtaz, editor. | Natarajan, Ashok, editor. | Kamran, Muhammad (Lecturer in management sciences), editor.
Description: First edition. | Includes bibliographical references and index.
Identifiers: Canadiana (print) 20230146333 | Canadiana (ebook) 20230146384 | ISBN 9781774911877 (hardcover) | ISBN 9781774911884 (softcover) | ISBN 9781003314615 (ebook)
Subjects: LCSH: Leadership. | LCSH: Industry 4.0. | LCSH: Personnel management. | LCSH: Industrial management.
Classification: LCC HD57.7 .A35 2023 | DDC 658.4/092—dc23

Library of Congress Cataloging-in-Publication Data

CIP data on file with US Library of Congress

ISBN: 978-1-77491-187-7 (hbk)
ISBN: 978-1-77491-188-4 (pbk)
ISBN: 978-1-00331-461-5 (ebk)

About the Editors

Tanusree Chakraborty

Tanusree Chakraborty, PhD, is currently a faculty member at the Centre for Management Studies, Administrative Staff College of India, Hyderabad, in India. With 17 years of overall work experience and 16 years of academic experience, Dr. Chakraborty has been associated with several reputable business schools and universities in India. She is presently working with Administrative Staff College of India, Hyderabad, in organizational behavior/human resource management. She is a former research fellow of the Indian Council of Medical Research and has many publications to her credit in Scopus-indexed and ABDC-listed journals, as well as in University Grants Commission-approved journals. She has presented research papers at several conferences and is an empaneled corporate trainer with training consultancies and government productivity councils. She has conducted over 150 corporate training programs in the areas of behavioral science and HR and psychometrics. She has been an active member in clubs and committees of the institutes she has worked for. She has been editor of several journals and newsletters and a reviewer for many journals. She is currently working on multiple edited books with reputed publishers. Dr. Chakraborty holds a PhD in Applied Psychology from Calcutta University, India; an MA (Applied Psychology) with a specialization in Organizational Psychology; a Postgraduate Diploma in Human Resource Management; and MBA (Human Resources).

Tahir Mumtaz Awan

Tahir Mumtaz Awan, PhD (Marketing), is currently the Head of China Study Centre, COMSATS University Islamabad, Pakistan. He is also a Senior Research Fellow at the China Brand Centre of the Academy of Entrepreneurship, Sun Yat-sen University, Guangzhou, China. He trains students in the SKOLKOVO Executive MBA program at the Moscow School of Management and at Shenzhen, China, annually, and is a regular member of the global business exposure visits team of the school. He was a United Nations Systems Staff College Fellow in Big Data for Sustainable Development. He was received research productivity awards by the Higher Education Commission of Pakistan and Common Ground Publishing of the University of Illinois, Urbana-Champaign, USA. He has delivered multiple talks, training workshops, and keynotes about the China–Pakistan Economic Corridor (CPEC), the Belt and Road Initiative (BRI), marketing in developing countries, big data, cryptocurrencies, etc. He has written more than 40 academic articles and edited five books and serves various journals as a member of editorial teams.

Dr. Awan's aim is to help improve the people-to-people relationships among Pakistani and Chinese citizens, to encourage youth development for better placement opportunities, to change the stereotype of "Made-in-China" products, and ultimately to make Pakistan a better consumer market in the world. To realize this aim, he is currently providing hands-on experience to Pakistani youth in learning about China and helping them to become the ultimate choice of Chinese firms working in Pakistan. He also runs a Pak-China Youth Club from his China Study Centre.

Ashok Natarajan

Ashok Natarajan, PhD, was formerly at the College of Business Administration, California State Polytechnic University, Pomona, CA, USA. He has over 15 years of experience in the academic arena in addition to more than five years of experience as a practising manager, both in the USA and India. In addition, he has been a consultant on several business valuation projects. His area of specialization is accounting and finance, and he has published several research papers in the top accounting and finance journals. His current research interests include accounting conservatism, business valuation, earnings management, and industrial organization. In addition, he works in the areas of corporate governance and corporate social responsibility (and the role of accounting and finance in these two areas) and corporate strategy. His other interests include curriculum development for management education, leadership in business schools, academic mentoring, as well as the relationship between research output and academic quality. His teaching portfolio includes several courses in accounting, finance, corporate strategy, and allied areas. Dr. Natarajan has won numerous awards for his teaching and has been recognized for his service and mentoring roles.

Muhammad Kamran

Muhammad Kamran, PhD, has more than 15 years of teaching and administrative experience in academia. He has taught for more than eight years at different Polish universities. He has participated in national and international conferences that include Harvard University, USA. He is currently serving Wyższa Szkoła Gospodarki Euroregionalnej im. Alcide De Gasperi, Jozefow, Poland, as Assistant Professor. In addition to this, he is well equipped with both quantitative and qualitative research methods and data analysis techniques. Dr. Kamran earned a Doctoral of Philosophy (PhD) degree in Management and Quality Sciences from Faculty of Management, University of Warsaw, Poland.

Contents

Contributors

Wajid Shakeel Ahmed
Department of Management Sciences, COMSATS University Islamabad, Islamabad, Pakistan;
E-mail: wajid_shakeel@comsats.edu.pk

Ali Eren Alper
Faculty of Economics and Administrative Sciences, Nigde, Turkey

Findik Özlem Alper
Faculty of Economics and Administrative Sciences, Nigde, Turkey; E-mail: alierenalper@gmail.com

Ashish Arya
FMS, Gurukula Kangri (Deemed to be University), Haridwar, India

Tahir Mumtaz Awan
Department of Management Sciences, COMSATS University Islamabad, Islamabad, Pakistan;
E-mail: tahir_mumtaz@comsats.edu.pk

Trisha Barua
Deloitte India (Offices of the US), Hyderabad, Telengana 500081, India;
E-mail: trisha.barua6@gmail.com

Anup Baul
Department of Management & Humanities, NIT, Arunachal Pradesh, India;
E-mail: anupmax.max@gmail.com

Swaha Bhattacharya
Calcutta University, Kolkata, India

Tanusree Chakraborty
Administrative Staff College of India, Hyderabad, Telangana 500034, India

Satish Chinchorkar
MIT World Peace University, Pune, India; E-mail: chinchorkar@gmail.com

Aditya Dhiman
FMS, Gurukula Kangri (Deemed to be University), Haridwar, India;
E-mail: adityadhiman7@gmail.com

Naveen Farooq
Lecturer, Social Sciences Department, Fatima Jinnah University, Rawalpindi, Pakistan

Tayyba Fatima
Department of Management Sciences, COMSATS University Islamabad, Islamabad, Pakistan

L. Gandhi
SDM Institute for Management Development (SDMIMD), Mysore, Karnataka, India;
E-mail: gandhi@sdmimd.ac.in

Madhurima Ganguly
Heritage Business School, Kolkata, India; E-mail: mganguly10@gmail.com

Esha Garg
Bennett University, Greater Noida, India

Radha Iyer
Alkesh Dinesh Mody Institute for Financial and Management Studies, Mumbai, Maharashtra 400098, India; E-mail: radha.mohan258@gmail.com

Jawed Kalangade
Alkesh Dinesh Mody Institute for Financial and Management Studies, Mumbai, Maharashtra 400098, India

Muhammad Kamran
Assistant Professor, Wyższa Szkoła Gospodarki Euroregionalnej im. Alcide De Gasperi, Jozefow, Poland. E-mail: muhammad.kamran@wsge.edu.pl
ORCID: 0000-0002-5154-0504

Sarita Kar
Department of Humanities and Social Science IIT (ISM), Dhanbad, India

Kamaljit Kaur
Institute of Information Technology and Management, Janakpuri, Delhi, India

Palvinder Kaur
University of Delhi, Delhi, India. E-mail: royalpalvi16@gmail.com

Myra Khalid
Department of Management Sciences, COMSATS University Islamabad, Islamabad, Pakistan

Hamayun Khan
Finance Specialist, The Siena School, Silver Spring, MD, USA

Mahesh Luthia
Chetana's Institute of Management and Research, Mumbai, Maharashtra 400051, India; E-mail: mahesh.luthia@cimr.in

Pankaj Madan
FMS, Gurukula Kangri (Deemed to be University), Haridwar, India

Geeta Mahajan
Institute of Information Technology and Management, Janakpuri, Delhi, India

Sohaib Jalil Malik
Department of Management Sciences, COMSATS University Islamabad, Islamabad, Pakistan

Nandita Mishra
Chetana's Institute of Management and Research, Mumbai, Maharashtra 400051, India; E-mail: nanditamishra06@gmail.com

Soumita Mitra
Department of Humanities and Social Science IIT (ISM), Dhanbad, India; E-mail: soumita.rupa@gmail.com

Tusharika Mukherjee
Work and Organizational Psychologist, Justus Liebig University, Giessen, Germany; E-mail: tusharikamukherjee@gmail.com

Anand Nadar
Bennett University, Greater Noida, India

Ashok Natarajan
Formerly at California State Polytechnic University, Pomona, California, USA

Gurcem Ozayturk
Faculty of Economics and Administrative Sciences, Nigde, Turkey

Ipsita C. Patranabis
Techno International New Town, Kolkata, India

Nalanda Roy
Amity University, Kolkata, AIPAS, India

Ramandeep Saini
Director Principal, Chandigarh Business School of Administration, Landran, India;
E-mail: ramandeepsaini2000@gmail.com

Anubhuti Saxena
Bennett University, Greater Noida, India; E-mail: anubhuti.saxena@bennett.edu.in

Mihir Kumar Shome
Department of Management & Humanities, NIT, Arunachal Pradesh, India

Smita Shukla
Alkesh Dinesh Mody Institute for Financial and Management Studies, University of Mumbai, Mumbai, India.

Manvi Sodani
Amity University, Kolkata, AIPAS, India

Malabika Tripathi
Amity University, Kolkata, AIPAS, India; E-mail: malabikatripathi1@gmail.com

Hafsah Zahur
National University of Modern Languages (Management Sciences), Islamabad 440000, Pakistan;
E-mail: hafsazahoor6585@gmail.com

Abbreviations

AI	artificial intelligence
AMCS	Automatic Milk Collection Systems
AWRM	Agile Wheel Reference Model
BCDR	business continuity and disaster recovery
BI	business intelligence
BIR	British Industrial Revolution
BPM	business process management
COR	conservation of resources
CPS	cyber-physical systems
CR	composite reliability
COVID	coronavirus
DR	disaster recovery
DSS	decision support system
EFA	exploratory factor analysis
EI	emotional intelligence
ERP	enterprise resource planning
HR	human resource
HRA	human relation approach
HRM	human resources management
HRIS	Human Resource Information System
ICT	information and communication technology
IIoT	industrial internet of things
ILC	institutionalization of leadership culture
IM	impression management
IoD	Internet of Data
IoP	Internet of People
IoS	Internet of Services
IoT	internet of things
IQ	intelligent quotient
IT	information technology
ISFJ	introvert, sensing, feeling, and judging
KMO	Kaiser-Meyer-Olkin
ML	machine learning
M2M	Machine to Machine

NCW	National Commission for Women
OA	organizational agility
OEMs	original equipment manufacturers
OTSH	organizational tolerance of sexual harassment
PHM	prognostic and health management
PMI	Project Management Institute
PSEB	Pakistan Software Export Board
QWL	quality of work life
RA	role ambiguity
RAMI4.0	Reference Architecture Model
RCS	resilient control system
SAFe	Scaled Agile Framework
SEM	structural equation modeling
SHe-box	Sexual harassment electronic box
SHW	Sexual Harassment at Workplace
SHR 4.0	Smart Human Resources 4.0
SIR	Second Industrial Revolution
SKU	stock keeping units
SME's	small- and medium-sized enterprises
TAA	transparency, accountability, and action
TAT	trait activation theory
TEL	Tennessee Electronic Library
TSN	Time-Sensitive Network
VUCA	volatility, uncertainty, complexity, and ambiguity
VR	virtual reality

Acknowledgment

At the outset, we would like to extend our heartfelt gratitude and acknowledgment to everyone who was a part of this edited book. There were many people who were in different ways associated with the editing process and have contributed to the successful compilation of the book.

Almighty, who has always been supportive and gracious in our journey, we thank you wholeheartedly. Without your intention and enlightenment, the accomplishment of this job would not have been possible. There were several ups and downs, more particularly because of the pandemic; but we could go on with our work with sustained efforts being blessed by you.

We would also like to thank our friends, family, colleagues, and different other people who have played important roles in getting chapters, circulating the call for chapters, connecting us to authors, and being with us with their tremendous support to complete the project.

We thank all the academicians who have helped us in reviewing chapters, guided authors to do quality revisions of their chapters, and enlightened us to the final production of the book manuscript.

Dear contributors, we thank you all so much for the chapter contributions, continued quality work, and supporting us with on-time revisions and modifications of chapters as and when required.

Thank you all for giving us this opportunity to achieve our purpose today by being able to complete the book successfully.

We would be remiss in not thanking our publishers Apple Academic Press for giving this opportunity to undertake this task and for guiding and helping in every phase of the process.

Thank you.

Editors

Preface

The term Fourth Industrial Revolution, or Industry 4.0, was coined by a group of German scientists in 2013 who were trying to develop a high-tech strategy for the German government to counter the cheap manufacturing advantages of developing countries. However, the term has been seized upon by many as a means of describing a new paradigm in describing technologies that integrate software, hardware, and biology (cyberphysical systems). The World Economic Forum has been one of the biggest proponents of Industry 4.0 and has recommended it as a solution to rebuild the global economy on a sustainable basis following the COVID-19 pandemic.

A global crisis tests all of us, but it especially tests the character and mettle of our leaders. A global health crisis, such as the COVID-19 pandemic, helped us identify those leaders whose leadership skills are forged in the crisis, as a pandemic of this nature was hardly ever envisaged. It is in this context that the concept of "agile leadership" comes to the fore. With the advent of Industry 4.0, the world is seen more as a volatile, uncertain, complex, and ambiguous (VUCA) environment, and leaders have to be agile and be prepared to lead in the digital transformation of their organization. Tools and methods of an agile leader include agile process models, servant leadership, delegation, motivation, change management, social systems, innovation, communication, and mental fitness.

This book strives to bring forward agile leadership during the pandemic and Industry 4.0 and the issues and problems therein. There are three broad themes in this book. We first review some theories of agile leadership and show how technology and the COVID-19 pandemic have affected such theories. We highlight the role of virtual leadership in an age of working from home. We present case studies that provide valuable lessons for the agile leader, and finally, we examine the role of the human resource functions in providing agile leadership to the organization.

The book will prove to be of interest for the academic as well as the practicing manager alike in acting as a practical and theoretical guide in explaining and articulating agile leadership. It is our hope that learning organizations find these chapters instructive in helping them evolve, along with their leadership.

We would like to take this opportunity to thank our diverse and international review board members for their reviews. They have spared their valuable time and effort to help us complete this project, and we acknowledge the tremendous debt that we owe them, especially in this pandemic situation. We also solicit feedback and suggestions from our readers that will help us further improve the quality of the book.

—Editors

Introduction

Economic historians have long debated the term "Industrial Revolution," with some even doubting the term that perhaps conflates economic progress events with political events. Nevertheless, the term industrial revolution has survived as a means of demarcating important economic events and how these affect humanity. Of late, economists and management writers have tied the term to technological changes that have occurred over time and how these technological innovations, to use the term that is very popular nowadays, "disrupt" the existing scheme of things and create a new economic reality for the society at large. Thus, there seems to an agreement now that we have observed three such technological shifts in the past and naming them as three industrial revolutions; the first one denoting a shift from human or water power to steam power; the second from steam to electric or oil power; and the third to the introduction of computer power. These three industrial revolutions also denoted changes in society itself, moving from an agrarian society to a society of mass production, and then to semiautomated processes of production. In 2011, the German Government applied an industrial revolution terminology to what they saw happening right then; the integration of information and communication technologies to industry and labeled it "Industry 4.0," or the "Fourth Industrial Revolution."

This industrial revolution builds on the third industrial revolution that already had computer technology embedded in it by expanding the process to include network connectivity making use of internet communication capabilities. These lead to production automation, with cyber-physical production systems, leading to smart factories utilizing artificial intelligence and machine learning. This digitization of manufacturing is predicted to change the ways goods are manufactured and distributed, and how products are refined and put into service.

This concept has captured the imagination of many economists and management thinkers. Firstly, the fourth industrial revolution is not something that is part of history, but a contemporary event. Secondly, the scope and scale of the breakthroughs that are happening in this revolution seem to have no historical precedent. The revolution itself seems to be unfolding at an exponential instead of a linear rate and threatens to disrupt, or has already disrupted every sector of the economy. It is believed that this will lead to a transformation of the society itself. Very similar to the revolutions that came before it, this Industry 4.0 has the potential to increase income levels on a global basis, and improve the quality of life for populations around the world. On the other hand it is also posited that Industry 4.0 will result in greater inequality especially as it threatens to disrupt established labor markets.

Given the major impact that 'industry 4.0' will have on businesses, it is important to see how existing organizations will evolve in the future and be sustainable. Disruption from agile innovative competitors can threaten well established incumbents. An environment that changes quickly, with a constant threat of new disruptive technology being introduced, along with an emphasis on digitization and with the evolving democratization of information, together with competition for talent on a global basis—these have led to the evolution of an "agile organization," where the paradigm is that of an organization as a living organism. The agile organization is both nimble and stable at the same time, enabling the organization to adapt quickly to new challenges and opportunities, and at the same time having a framework that can sustain these dynamics. Academic studies reveal that an organization that has undergone an agile transformation tend to also perform well on a financial basis, with a better chance of being in the top quartile of organizational health, enabling them to be able to sustain their superior performance over time.

The concept of an agile organization has captured the imagination of business leaders and many organizations have started the process of transforming themselves into agile organizations. Surveys have shown that 75% of business leaders have committed themselves to make their organizations agile, labeling it as one of their top priorities, and about 2/5ths of them stating that they have already started this process. The COVID-19 pandemic has accelerated the desire of organization to transform themselves into agile ones, in an environment where employees work from home, and are left largely to their own supervision, one of the trademarks of an agile organization.

While business thinkers have come up with numerous factors that make an organization agile, a common theme that runs through these descriptions is the role of the leader of these organizations, since agile leadership is seen as the key factor that enables organizations to become agile. The transformation of a leader to develop the agile outlook and create a new organizational mindset is seen as the crucial element. The characteristics of an agile leader include the capability of transforming themselves into agile leaders, an ability to be able to transform their teams to work in this new paradigm, and the willingness and the ability to build the framework for their organizations to transform themselves into agile organizations.

It is in this context that this book was undertaken. This book attempts to study the role of agile leaders in the Fourth Industrial Revolution. We have invited academics to submit their research on various themes of agile leadership in this age of the Fourth Industrial Revolution. We have included chapters that look at various aspects of the interaction between agile leadership and Industry 4.0. These chapters include thought pieces on the technological aspects of Industry 4.0, while a few chapters consider the impact of Industry 4.0 and agile leadership on the Human Resources function and the challenge of managing human resources in a post-COVID era. Some chapters look at how agile leadership has to be "virtual" in the Fourth Industrial Revolution, while other chapters consider how employee engagement and empathy affect this agile leadership. The book also contains chapters that provide practical examples of agile leadership in various situations, such as in handling women's issues, branding, school leadership, and rural women empowerment. These chapters will give the practicing manager and the academic thinker a wide survey of the entire field and how the new paradigm has evolved over time, and provide many useful insights to the reader.

—Editors

CHAPTER 1

Role of the Internet in Connecting Agile Industries: Industry 4.0 Scenario

SATISH CHINCHORKAR

MIT World Peace University, Pune, India
E-mail: chinchorkar@gmail.com

ABSTRACT

Constantly increasing greed of business to increase the production at lower prices with fewer efforts is the origin of industrial revolutions. Mechanization was a characteristic of the first industrial revolution in the year 1765, attributed to the invention of the steam engine. The second industrial revolution in the year 1870 was related to sources of energy caused by the invention of electricity, gas, and oil. Computers initiated the third industrial revolution in the year 1969 in the field of electronics and telecommunications. Now we are at Industry 4.0, the revolution of connecting industries with virtual reality exploring the use of the Internet.

The Internet and specifically wireless networks make it possible for everyone to have permanent access to the network always and anywhere. The Internet is a pool of "Computers" that share information. Over some time the word "Computer" got replaced by "Devices," which can be a smartphone, Personal Digital Assistant PDA, robots, or television. Further, the Internet is being explored beyond just "sharing" the information or communication but to "execute" information remotely. The list of devices connected is appended with machines, robots, and smart equipment that can operate remotely; also the range of devices is being added with remotely operated machines, equipment, and robots that are covered under the emerging topic

Agile Leadership for Industry 4.0: An Indispensable Approach for the Digital Era, Tanusree Chakraborty, Tahir Mumtaz Awan, Ashok Natarajan, and Muhammad Kamran (Eds.)

Internet-of-Things (IoT) as a network of physical "objects"—embedded with sensors, software, and peripherals.

The concept of integrating the devices through the Internet is further explored to create the connected "Network of Industries" which can be called Industry 4.0 the fourth industrial revolution. In this revolution, the challenges are shifted from "technical" to "process." That means the future of the Internet is not about technology development, but how the process of change (i.e., agility) and evolution itself will be adopted and managed. Adoption of the new technologies and processes needs a paradigm shift of leadership mindsets from "transactional management" to "transformational leadership," for which Agile Leadership is significant.

1.1 INTRODUCTION

Industry 4.0 is also known as the fourth industrial revolution that includes the areas usually classified as industrial applications, the scope of which is extended further to the sectors such as smart cities. To understand the significance of this fourth industrial revolution, it is worth denoting the history of earlier industrial revolutions. With the start of mechanization the first industrial revolution arose, like steam power and water power. The second subsequent industrial revolution is related to the mass production and assembly lines concepts using the power of electricity. The third industrial revolution is all about the applications of electronics, IT systems, and automation, which eventually further lead to the fourth industrial revolution that is related to cyber-physical systems.

Industry 4.0 is all about the growing trend toward digitization automation and data exchange in technology and processes within the manufacturing industry, for which the Internet and Information and Communication Technology (ICT) framework is a basis. This includes the emerging technologies such as the Internet of Things (IoT) for remote operations, the industrial internet of things (IIoT) for technical communication, Cyber-physical systems (CPS) for machine and human interface, smart manufacture with advanced techniques, smart factories with infrastructure, cloud computing for optimizing IT cost with flexibility, cognitive computing to simulate the human thought process in complex situations, and artificial intelligence to automate the tasks.

The manufacturing system is being created by industrial automation with such types of technologies, whereby wireless connectivity and sensors are used to monitor, visualize, and manage an entire production process that improves the machine's capabilities in factories.

The use of the internet and wireless connectivity to integrate the factories with the augmentation of machines will be significantly advanced with the technologies such as 5G. This will provide quicker response times, facilitating real-time communication between the systems, which are very much needed. The fourth industrial revolution consists of digital technologies that can create virtual versions of real-world installations, processes, and applications. Such a cost-effective and well-tested automation facilitates interconnectivity between processes, information transparency, and technical assistance for decentralized and rapid decisions. Connecting the industries is a need of a business that is getting addressed by the constant growth of Internet technology with a focus on effective implementation and utilization of the same. Agile leadership assists to create an appropriate context for own organization. Agile leadership, which is a combination of transformational leadership style and digital leadership style, is significant to drive the change to take advantage of digital technology involved in Industry 4.0.

1.2 HISTORICAL BACKGROUND

A literature review of historical papers is chosen as the appropriate methodological approach for the achievement of this article's objectives.

While establishing the association between Industry 4.0 and Agile Firms (Şen and İrge, 2020) the organizational agility is being explored. The application of Industry 4.0 in the pilot sectors like domestic applications, machine systems, automotive, and chemical sectors is analyzed. According to them an Agile Leadership includes operational, partnership, and customer agility as a dynamic capability, explained with the conceptual model for implementing Agility. The agility capabilities are broadly defined as responsiveness, competency, flexibility, and speed; an agile leader is expected to have these capabilities. The components of Agile Firms as agile strategies, agile processes, agile linkages, and agile people are explained with an Agile Wheel Reference Model (AWRM) and major activities involved in it.

Thomas (Ooi, 2016) of Intel Internet of Things Group rightly narrated the Industry 4.0 from three major aspects that are worth outlining the various aspects of the topicas follows:

- Industry 4.0 from viewpoint of Data
  - There will be massive "raw" data

    - Data will be converted invaluable "oil"
    - Usage of better algorithms will be there

- Industry 4.0 from viewpoint of Software

    - The software will be reconfigurable
    - Virtualization will be part of a software

- Industry 4.0 from viewpoint of Speed

    - Usage of an advanced network like 5G will be there
    - The focus will be on timely delivery
    - Synchronization of man, machine, and software will be there

This article gives a fair outline of the terminology and aspects of "Industry 4.0" frequently used in the topic.

The German word for the fourth industrial revolution, also known as Industry 4.0 (I4.0) or Industrial Internet (Schuh et al., 2016), shows the close relationship between Industry and the Internet. This addresses the interconnection of machines, people, and products. The "collaborative productivity" approach can be measured with two innovative indicators such as "return on engineering" and "return on production." The framework design explained consists of two dimensions: Dimension-1 consists of "product" and "process" whereas Dimension-2 consists of "orientation," "data," "interaction," and "resources." Various requirements of Design for Industry 4.0 for the value chain in product optimization are identified and explored. Different types of interactions between providers, smart products, and customers are elaborated with eight key methods to implement Industry 4.0.

Industry 4.0 can be further better illustrated with various dimensions (Fonseca, 2018), where it is defined as advanced digitization and integration of manufacturing and logistics processes in which the internet and smart objects like machines and instruments are used to integrate the real or physical world and virtual or digital world by application of appropriate Information and Communication Technology (ICT). In the article the author further listed the specific benefits of such integration as shorter operation time, rapid delivery, reduced time to market, improved quality; provide options for customization and customer involvement and loyalty. The environment for Industry 4.0 is broadly classified into four major classes for better understanding as follows

- Internet of Things: that includes smart mobility and smart communication grid or infrastructure
- Internet of Data: consist of smart buildings and smart homes

- Internet of People: covers social web and business web on which people communicate
- Internet of Services: such as smart logistics needed for any business

Accordingly, automation, integration, configurable equipment, analytics, innovation, servitization, and smart devices are various dimensions defined for Industry 4.0.

In the paper How to Define Industry 4.0 (Erboz, 2017), the evolving customer trends are explained that caused the technical enhancements, and are further illustrated by stating that Industry 4.0 is a strategic approach for digitization in manufacturing that includes highly automated industries through human–machine interactions. Accordingly, the nine pillars of this fourth industrial revolution explain the scope as follows.

a. Big Data and Analytics: Consist of large, diverse, unstructured, and complex datasets used in organizations for making the strategies and decision-making.
b. Autonomous Robots: Used in the areas like production, logistics, and distribution that include complex and repetitive activities.
c. Simulation: Basically for creating a sustainable manufacturing environment.
d. Horizontal and Vertical Integration: Vertical—establishing the flexible and reconfigurable integration inside the factory. Horizontal—Integration of partners in the Supply Chain.
e. Industrial IoT: Basically to collect the data from physical objects by the Internet.
f. Cloud Computing: To obtain the Supply Chain Integration by facilitating the visualization of resources.
g. Additive Manufacturing: The application of advanced manufacturing technologies such as 3D printing enables customized products as per the requirements of the customer.
h. Augmented Reality: This is an interactive technology that integrates the virtual world with its users.
i. Cyber Security: Collaborative Risk Management to eliminate the harmful impact of any destructive incidences.

Umachandran et al. (2019) defined the fourth industrial revolution as a new form of manufacturing changing traditional manufacturing systems, competencies, complexities, and opportunities. They further differentiated the earlier three revolutions (from the 18th, 19th, and 20th centuries) from the fourth revolution as decentralized, automated, and controlled independently.

They also focused on earlier revolutions as the first revolution is about a production where the use of water, steam with heavy manpower is involved, whereas the second revolution is also related to production where the use of electricity is there with a technique like division of labor being deployed. According to them, the third revolution is related to electronics and information technology where the focus is on automation. Fourth revolution that is ongoing is the use of the Internet there for achieving the Cyber-Physical System (CPS) with smart mobile devices. They further elaborated the CPS as transformative technologies integrating the physical assets with computational capabilities. They invented unique "5C" architecture for CPS that includes five levels as Connection level, Conversion (data into information) level, Cyber level, Cognition level, and configurable level. This 5C model is further elaborated from virtualization, decentralization, open networks building, and servitization point of view.

The impact of the fourth industrial revolution specifically on the supply chain (Ghadge et al., 2020) is described with drivers and barriers for adapting Industrial Revolution 4.0 classified in terms of business dimensions as Organizational, Legal and Ethical, Strategic, and Technological. The implementation framework explained for Supply Chain can be extended appropriately for other business areas. The name "Supply Chain" indicates the sequence or series of links, which are individual industries. The success of a business completely depends upon the connectivity or integration of these links.

To understand the context of the fourth revolution it is worth noting that the complexity of business operations is increasing as the industrial revolution progresses with technological advancements (Koleva, 2018). The "open" production systems shall become flexible and agile "self-regulation" systems that will integrate and synchronize all the actors involved in the value chain. It is a paradigm shift for both business management and production processes. Industry 4.0 is a collection of connected digital technology solutions that includes automation, integration, and real-time data exchange within all the production processes. It is an industrial as well as technological transformation process followed by the development of scientific and production practices.

Before connecting the industries using the Internet creating the "smart" environment within the production system is a primary objective of Industry 4.0 (Crnjac et al., 2017) that includes changes specifically in manufacturing operations and information technology. The basis of Industry 4.0 is "smart factory" that gives business benefits such as virtual world as it can produce a simulation of a variety of cases for optimizing products, processes, and entire supply chain constantly, which is otherwise difficult to achieve in traditional industries. The focus of Industry 4.0 is on creating intelligent products, processes, and procedures. Use of the Internet while implementing Industry 4.0 is in the following three major categories

- Internet of Things (IoT): Using which every physical object is equipped with IPv6 address
- Internet of Services (IoS): in which service-based value-added processes are used
- Internet of Data (IoD): Integration of product and production data is obtained using big-data

Eight priority areas are being identified for implementation of Industry 4.0 by connecting the industries as follows:

1. Resources and processes: Process improvement with skilled resources
2. The use of property: Effective utilization of all resources
3. Operations: Improvement in the speed and productivity with reduction in wastage
4. Inventories: Optimization of inventory with effective controls
5. Quality: Compliance with quality that minimizes the cost
6. Demand-Supply Balance: Understanding and catering to the customer's needs
7. Time to market: Identification of appropriate time for the right opportunities
8. Service and Maintenance: Timely service and preventive maintenance that improves the efficiency

Most significant changes while a transition from Industry 3.0 to Industry 4.0 is described are as follows

- The manufacturing process to the product life cycle
- Lean manufacturing to smart manufacturing
- Decisions by experience to decisions by information
- Focus on saving money to focus on creating new revenue streams

AI (Artificial Intelligence) for current and future use of decision-making in Cyber-Physical Systems (CPS) is explained by a hierarchical cascading framework with a conclusion that there will be increasing integration of connected smart devices such as IoT in CPS (Radanliev et al., 2020). The five-level CPS architecture is explained that includes 5C as follows:

- Configure: Resilient Control System (RCS) for actions to avoid
- Cognition: Decision Support System (DSS) for prioritizing and optimizing the decisions
- Cyber-CPS: for self-comparison
- Conversation: Prognostic and Health Management (PHM) for self-awareness
- Connection: Condition-based monitoring for conditional monitoring

3C (Client, Company, and Competitor) model is explored for the execution and implementation of Industry 4.0 in the following levels

- Level-1: Project Management, where activities of resources planning are involved
- Level-2: Process Management, improvement in the practices and the quality
- Level-3: Management of Tools and Information Technology
- Level-4: Organization Management includes the provision of required infrastructure
- Level-5: Resources Management that consists of planning and deployment of skills

Vaidya et al. (2018), in their article regarding glimpses of Industry 4.0, explained the transition from mass production to customized production. Industry connects using the internet by achieving the system integration at all three levels as horizontal system integration that includes value creation network, whereas vertical system integration consists of manufacturing systems and end-to-end system integration covers the entire product life cycle. These all three levels of integration are feasible using the Internet that enables smart, efficient, effective, individualized, and customized operations at a reasonable cost.

Major communication in the production chain in I4.0 system is obtained with Time Sensitive Network (TSN) such as Ethernet-based Industrial Internet of Things (IIOT) using specific standards like IEEE 802.1 for the R-T enhancement of the existing public Internet (Zezulka et al., 2018). In this technical paper, it is stated that OPC UA (Open Platform Communication Unified Architecture) will be a common communication standard for the I4.0 factory of the future, which is an Open Software Interface and communication protocol specifically for the automation and information systems in I4.0 applications.

Various options for implementing the I4.0 specifically in the discrete manufacturing sector (Arm et al., 2018) include Asset Administration Shell and Consumer Provider Model based on OPC UA and Industrial CPS (Cyber-Physical Systems) using Internet achieves the business objectives such as curbing the product life cycle, increasing the product diversity, improving the productivity and quality with resources optimization, and finally the customer service enhancements.

While considering the fourth industrial revolution that is all about connecting the industries using the Internet, it might be worth denoting a brief history of the Internet (Leiner et al., 2009) because of its historic import. Since the events and activities related to the initial development of the Internet would be valuable contribution. According to them, Internet explored the potential opportunities in the area of computers and communications in a big way. The Internet is having better integration capabilities than all other inventions in the past. The Internet is having a huge capability to collaborate, interact, and communicate with a number of individuals and specifically to their computers and machines irrespective of geographic locations. This paper explains the technical evolution of the Internet from packet switching technology and the ARPANET and after having series of transitions it reached such a mature state that the industries would like to explore the Internet for connecting the business world with commercialization. In the paper, it is predicted that the future of the Internet will not be decided by innovations in technology, but it will be predicted based on the evolution of processes to adopt the Internet technology.

The Digital Leadership role is significant in the context of digital transformation needed for the creation of the Industry 4.0 workplace. In their paper, Mihardjo et al., (2019) assessed the role of digital leadership after surveying 88 senior leaders from the Indonesia telecom industry. According to them, digital leadership had direct and indirect impacts on customer experience while applying the innovation models related to digital technology that influences Industry 4.0.

Future leaders and values required in the Industry 4.0 workplace are analyzed after a survey of 371 relevant participants (Črešnar and Nedelko, 2020). According to them the power of achievement, openness to change, and conservation values are significant for future leadership. However, values needed in Industry 4.0 workplace like benevolence, universalism, and general self-transcendence have a negative impact on leadership. The paper concludes that there is a clear mismatch between values of future leaders and the values needed for the Industry 4.0 workplace. They classified future leaders from three types of generations as Generation-X (born before the year 1984), Generation-Y (born between years 1984–1996), and Generation-Z (born after the year 1997). According to them from Generation-Y and Generation-Z future leaders will take the leadership positions for upcoming emerging technologies required for workplaces in Industry 4.0. Generation-X believes more in values like stability, hard work, and security, whereas Generation-Y and Generation-Z (those who were born with technology) give more importance to self-enhancement, value money, promotions, progress at work-place, and fame of power. The transition from personal values to attitudes to behavior is needed for agile leadership specifically from traditional line hierarchy to a network organization.

1.3 INDUSTRIAL REVOLUTIONS

Perpetual growing hunger of business to grow the production at lesser expenses with less labor is the beginning of industrial revolutions. Jindal et al. (2018) in their article about the future of IoT, classified the industrial revolutions year-wise and feature-wise as follows:

- First Industrial Revolution: Year-1800—Related to the introduction of water and steam as a power resource in the industrial sector
- Second Industrial Revolution: Year-1900—Associated with electrically powered mass production and concept of division of labor
- Third Industrial Revolution: Year-2000—Is all about electronics and information technology with automation in manufacturing
- Fourth Industrial Revolution: Ongoing—Cyber-Physical System dealing with increasing productivity and complexity in the industry

Since IoT is the backbone of the fourth industrial revolution, they correlated the industrial revolutions with ICT development stages as follows.

- Web: Related to hypermedia where documents are interlinked with www (World Wide Web)
- Web 1.0: Associated with multimedia where multimedia are interlinked using technologies such as Java, XML, and UML
- Web 2.0: Is all about Social Media where people are interlinked with web services
- Web 3.0: Where Social Media is further explored to interlink the enterprises using application technology
- Web AB: Known as Cyber-Physical System in which the systems are integrated using technologies such as Internet of Things (IoT), Internet of People (IoP), and Internet of Services (IoS).

According to history, the British Industrial Revolution (BIR) initiated the factory (Kapás, 2008), then a multidivisional form (M-form) appeared as a Second Industrial Revolution (SIR), whereas in Third Revolution Project-Based firms come into existence.

History had shown that how because of ignorance of digitization the Kodak and Nokia empires are collapsed. Therefore companies need to be alert to identify and adopt the enhancements in traditional business models to survive and stay competitive. Industry 4.0 is a term that first came into existence in 2011 at the Hanover fair in Germany and was subsequently considered the "focal part of the high-tech strategy of German Government" (Kainer, 2017). Industry 4.0 is the comprehensive transformation of entire traditional production through the assimilation of digital technology and the internet with conventional industry. Evolution of business models is specific to Small and Medium-sized Enterprises (SME's) toward a transition to Industry 4.0 analyzed by classification as micro-, small-, and medium-sized enterprises based on the number of employees working in that company.

Evolution from steam power to electrical energy to automating production is reached to a level where Industry 4.0 revolution is considered into three stages as follows.

- Stage-I: Cyber-Physical System
- Stage-II: Cyber-Physical-Production System
- Stage-III: Industry 4.0

Accordingly, the evaluation grid is prepared for strategic level, business model layer, and process layer with the

- Value Proposition: this includes offerings to customer
- Value Chain: includes the creation of proposition
- Revenue Model: for financial viability

1.4 INDUSTRY 4.0

The Original Equipment Manufacturers (OEMs) and allied industries that fall behind this technology innovation graph could get swallowed by larger players or need to shut down the operations. Suppliers with traditional technologies will find that their assets and skill-sets will not have any demand. Therefore, they all are forced to remain relevant and respond to the rapid technology development to survive.

Players looking for strategic partnership or targeting toward acquisitions and mergers will not have any choice but to adopt these digital transformations and accordingly invest profoundly in relevant technology such as smart factories, mobile devices, robotics, artificial intelligence, process automation, digital supply networks, predictive maintenance, block-chain technology, and big data with data science. This fourth industrial revolution is going to impact the product and service offerings, business models, manufacturing processes, and ultimately customer experiences.

The scope of Industry 4.0 is beyond just manufacturing activities (Rojko, 2017). In traditional industry, the topics are related to production planning and control, ERP, and data management, whereas the topics for Industry 4.0 covers Cyber-Physical Systems, IoT, Smart Factory, Big Data, Cloud, and Cyber Security. The differences in the topic illustrate the scope and outline of Industry 4.0 in which smart factories use digital to physical conversion with effective usage of big data and cloud computing to make the manufacturing systems reconfigurable. The Three-dimensional Reference Architecture Model (RAMI4.0) addresses the standard issues that Industry 4.0 is facing. This model gives product life cycle in terms of location/functional/hierarchy levels (IEC 62264 and IEC 61512) and Value System Layers (IEC 62890). In the model, six dimensions of industries are considered for transformation as follows.

1. Strategy and Organization: This includes policies and management support
2. Smart Factory: The advanced infrastructural issues considered here
3. Smart Operations: Manufacturing processes like additive manufacturing are included here
4. Smart Product: Product with virtual identity and RFID
5. Data-Driven Services: Fact-based intelligent automated services include the use of AI
6. Human Resources: Skilled resources competent to plan, control, and manage Industry 4.0

Ultimately, industry 4.0 is certainly going to affect the world of manufacturing with some specific points such as

- Machine to Machine (M2M) Communication is machines that talk to each other
- Crash the process cycle time where radically time-saving will be achieved
- Improved safety in terms of physical as well as cyber-security

- Optimizing the cost by eliminating the waste and improving the productivity
- Enhancement human skills and standards by giving monotonous jobs to smart factories

Many different forms of industry design have been developed (Schuh et al., 2016) for the product life cycle comprising the recycling stage. Major types of industry designs based on either product requirements or development processes are as follows.

- Design for Quality
- Design for Manufacturing and Assembly
- Design for Serviceability
- Design for Environment

However, Industry 4.0 designs considering both product requirements and the development process also.

As per the research predictions by 2020 around 50.1 billion of IoT-connected objects will be there; this prediction introduced the challenges in product development which is addressed in the two-dimensional framework design suggested for Industry 4.0 that consists of

Dimension-I

1. Product: to be produced from industry
2. Process: to be adopted

Dimension-II

1. Product orientation: paradigm shift about concepts about the products
2. Data Management: introduction of complexity and information security issues
3. Interactions: communication like machine to machine becomes challenging
4. Resources: skilled and competent resources required for the new Industry Design

The design is based on diversified approaches in which the factors such as product development, production processes, product usage, and disposal are considered. Three types of interactions, that is, internal interactions, product interactions, and external interactions are reflected in the Industry 4.0 model.

1.5 IMPACT ON BUSINESS

The fourth Industrial Revolution allows for digital transformation for automated and autonomous manufacturing with integrated systems that can cooperate for scaling up productivity and efficiency. Specifically, the cyber-physical systems communicate and cooperate with human staff to create an integrated real-time data exchange and automation process for Industry 4.0 manufacturing.

The most apparent impact of Industry 4.0 is expected in the industrial manufacturing and management, logistics, and business process management areas (Ghadge et al., 2020). Digital innovations in businesses have made remarkable changes in the way businesses are done (Chakraborty and Saha, 2018).

Traditional business shall encounter some major barriers while transforming to Industry 4.0 (Fonseca, 2018) as follows

1. Difficulty in planning and coordinating the activities across different organizational units
2. Lack of courage to push through a radical transformation from traditional to Industry 4.0
3. Shortage of necessary talent and skills required
4. Increased risks of cyber-security
5. Absence of clarity in a business case
6. Concerns about data ownership
7. Uncertainty about make-or-buy decisions
8. Challenges while integrating the systems and data

A framework for analyzing a specific impact of fourth industrial revolution on a company's competitive environment (Nagy et al., 2018); the factors analyzed using Porter's Model are as follows

1. The quantity and influence of a company's competitive rivals
2. Potential new market entrants
3. Suppliers
4. Customers, and
5. Substitute products influence a company's profitability.

Industry 4.0 is going to be beneficial for businesses across most of the sectors in terms of efficiency, agility, innovation, better customer experience, cost efficiency, and increased revenues.

There is a huge potential of Industry 4.0 to ensure CE (Sustainable Economy) or more sustainable production methods (Werthmann and Blunck, 2017). Smart products and advanced technologies could produce huge environmental, economic, and social benefits and thereby contribute to CE. End-to-end engineering is being achieved by vertical and horizontal integration of the continuous flow of data. Huge unstructured data generated by smart machines and equipment (Big Data) will have a high volume of interactions between these cyber-physical objects that shall be stored in the cloud and connected using the internet. Technology like IoT in smart factory can use the smart data from the big data created and consumed by the Cyber-Physical-System (CPS). A closed-loop design of production patterns in the production and consumption model within the economic system is adopted. An economy that is currently locked in a linear model of production and consumption can be released by Industry 4.0.

1.6 BUSINESS AGILITY AND INDUSTRY 4.0

A speed and stability are essential factors in this digital age to adopt the internal and external changes, which is made possible by the emerging technologies in Industry 4.0. Agility in Business involves rapid, continuous, and systematic adoption, and evolution of values, culture, and capabilities through innovations to gain competitive advantages. It is essential to adjust the market changes as well as internal changes, essential for sustainable business. Emerging technologies in Industry 4.0 contribute substantially toward obtaining the delivery agility, product innovations, and organizational adaptability which is needed for business agility.

According to Götz (2019), Industry 4.0 induced agility and new skills needed to remain sustainable. According to him, the global value chain is getting transformed into the adoptive networks of enterprises by the fourth industrial revolution. Increased flexibility to reorganize the business structure and acquire the related competencies is required to remain competitive. Accordingly, the new skills required for changeable, reconfigurable, and virtual production consist of working on this new intelligent, integrated, digitalized, and automated advanced architecture. For this attitude of employees need to be adoptive; the agile leaders make it possible with skills such as the ability to extract and use the big data using cloud technology, understand potential threats, and explore opportunities of digitalization, prompt and better responses to changes, openness and adaptability to new opportunities, quick learning, understanding lean production, work in flexible, reconfigurable organization structure with decentralization of power, autonomy, and empowerment.

A leader had to be agile to think out of box and act across industry boundaries by thinking end-to-end. An agile leader should be able to grasp the technical knowledge available within and outside the organization and make use of it. He needs to be capable of adapting to constant changes along with taking the team on the journey toward Industry 4.0 transformation. Organizations are more agile after a disaster, so the role of human resource professionals in managing after effects of a crisis is very crucial (Chakraborty and Saha, 2017). A good example of Business Analysis in supply chain disruptions in a manufacturing firm illustrates an agile business that implements the ERP solution to fetch the appropriate data and generate dynamic reports to predict the disruptions and allow to manage the same. Application of Business Intelligence (BI), a technical process to analyze data and produce actionable information, is another example of technology making the business agile. Digitization in agile working structure using techniques like Scrum and KANBAN helps to make the business agile.

Implementation of emerging technologies and diverse business practices of the fourth industrial revolution have a major impact on business processes, agile leaders are required to adopt these radical technical changes while keeping the values together in new circumstances.

1.7 ROADMAP AHEAD

The fourth Industrial Revolution allows for digital transformation for automated and autonomous manufacturing with integrated systems that can cooperate for scaling up productivity and efficiency. Specifically, the cyber-physical systems interact with collaborators and human staff to create an integrated real-time data exchange and fully automation process needed for Industry 4.0 manufacturing.

Due to the digital transformation transition to digitalization, most of the business processes of the value chain should be integrated with IT (Hitpass and Astudillo, 2019) using media like the internet. Business Process Management (BPM) is a systematic methodology that consists of identification, correlation, documentation, design, development, implementation, measurement, and control stages of a business process. Similarly, electronic commerce (e-Commerce) activities consist of purchase, sales, marketing, and distribution of products and services. These (BPM and e-Commerce) two major components of the business are greatly impacted specifically by the digitization of processes and enhancement in internet technology which cannot be further managed in isolation. The impact of Industry 4.0 will cause the migration of manufacturing and production from a centralized to decentralized paradigm. Hence having strong interconnection of machinery and systems becomes a prerequisite for implementing Industry 4.0. New trends that Industry 4.0 shall lead can be as follows:

- Decentralization of processes
- Integration of e-Commerce
- Raise in automation
- Improvement and quality and productivity

Agile leaders need to shape the team's trust in new technology, make them aware of the data security issues, remove their fear about potential job loss, and mobilize them for adopting the changing needs of job requirements and skills. Because the success of applications of Industry 4.0 depends on other factors besides knowledge, such as communication and coordination within and outside the organization that agile leadership does. In the future organizations should focus on development of such agile leaders, leaders ready to adopt new skills and talents needed for Industry 4.0 transformation. Organizations in the present context would require a positive organizational culture and encouraging leadership (Chakraborty and Ganguly, 2019) will need positive approaches to understand employees by appreciative questioning (Chakraborty and Mishra, 2019; Chakraborty et al., 2019).

1.8 FINDINGS AND CONCLUSION

First, this study finds that the upcoming fourth industrial revolution includes connecting the industries using the Internet. For this transition, appropriate changes in the business processes are recommended.

Second, for effective implementation of the enhanced processes applicable up-skilling of human resources is suggested.

Third, while applying the emerging technologies for Industry 4.0 change of mindset from transactional management to transformational leadership is recommended; this is possible if an Agile Leadership is adopted.

By starting small and moving rapidly organizations can generate success stories with value and Industry 4.0 needs to articulate its value to their management, investors, employees, and customers.

The impact will be substantially high when industries are getting integrated (rather than an individual is getting connected) via media like the Internet. With the widespread availability of the internet (anytime and anywhere) and readiness of smart devices the technical issues are condensed in comparison to issues related to processes of implementation and sustenance of Industry 4.0.

As we have seen that advancements in technology such as the development of technical devices/sensors, information technology, and communication platforms like the internet are the obvious major players of the fourth Industrial Revolution characterized by digitization and interconnections of machines, products, people, value chains rather entire business models. However to create the "Network of Industries," which is the main purpose of Industry 4.0 such utter technological advancements are not sufficient. To integrate the technological advancements with business and people involved there is a strong need for "process improvement" incorporating and addressing all the issues and requirements which is a challenge. Agile leadership will become tomorrow's game changes for Industry 4.0 as they navigate the organization through emerging technologies by including the disruptive changes and motivate the employees to do the same. Agile leaders are more than just conventional or digital leaders. They demonstrate their entrepreneurial, strategic, agile, visionary, and technical role that is needed to implement the enhanced processes needed for Industry 4.0. A strong need for agile leadership is essential to create a collaborative, multicultural, and multidisciplinary organizational environment necessary for Industry 4.0 transformation and sustenance.

KEYWORDS

- **industrial revolution**
- **Industry 4.0**
- **Agile**
- **leadership**
- **digitization**
- **IoT**
- **machine learning**
- **virtual reality**

REFERENCES

Arm, J.; Zezulka, F.; Bradac, Z.; Marcon, P.; Kaczmarczyk, V.; Benesl, T.; Schroeder, T. Implementing Industry 4.0 in Discrete Manufacturing: Options and Drawbacks. *IFAC-PapersOnLine*, **2018,** *51* (6), 473–478.

Chakraborty, T.; Ganguly, M. Crafting Engaged Employees through Positive Work Environment: Perspectives of Employee Engagement. In *Management Techniques for Employee Engagement in Contemporary Organizations*; IGI Global, 2019; pp. 180–198.

Chakraborty, T.; Mishra, N. Appreciative Inquiry: Unleashing a Positive Revolution of Organizational Change and Development. *Int. J. Econ. Comm. Bus. Manag.* **2019,** *6* (2), 32–37.

Chakraborty, T.; Saha, R. Proactive and Reactive Role of Human Resource Professionals in an Aftershock Management. *Train. Devel. J.* **2017,** *8* (1), 1–11.

Chakraborty, T.; Saha, R. Digital Innovations: Breakthrough Opportunities to Build Novel Business Models. *Bus. Perspec. Aims and Scope* **2018,** *1*, 46.

Chakraborty, T.; Ganguly, M.; Natarajan, A. Predicting Entrepreneurial Satisfaction: The Role of Non-Financial Incentive Factors and Quality of Life among Women Digital Entrepreneurs. *J. Glob. Bus. Adv.* **2019,** *12* (3), 328–355.

Umachandran, K.; Jurčić, I.; Della Corte, V.; Ferdinand-James, D. S. Industry 4.0: The New Industrial Revolution. In *Big Data Analytics for Smart and Connected Cities*; Dey, N., Tamane, S., Ed.; IGI Global, 2019, pp. 138–156).

Črešnar, R.; Nedelko, Z. Understanding Future Leaders: How are Personal Values of Generations Y and Z Tailored to Leadership in Industry 4.0? *Sustainability* **2020,** *12* (11), 4417.

Crnjac, M.; Veža, I.; Banduka, N. .From Concept to the Introduction of Industry 4.0. *Int. J. Indus. Eng. Manag.* **2017,** *8* (1), 21–30.

Erboz, G. In *How to Define Industry 40: The Main Pillars of Industry 4.0*, Managerial Trends in the Development of Enterprises in Globalization Era, November; 2017, pp. 761–767.

Fonseca, L. M. In *Industry 4.0 and the Digital Society: Concepts, Dimensions and Envisioned Benefits*, Proceedings of the International Conference on Business Excellence, 2018, vol. 12, no. 1; pp. 386–397. https://doi.org/10.2478/picbe-2018-0034

Ghadge, A.; Er Kara, M.; Moradlou, H.; Goswami, M. The impact of Industry 4.0 Implementation on Supply Chains. *J. Manuf. Technol. Manag.* **2020,** *31* (4), 669–686.

Götz, M. The Industry 4.0 Induced Agility and New Skills in Clusters. *Foresight STI Governance* **2019,** *13*(2), 72–83.

Hitpass, B.; Astudillo, H. Editorial: Industry 4.0 Challenges for Business Process Management and Electronic-Commerce. *J. Theor. Appl. Electron. Commer. Res.* **2019,** *14* (1), 1–3.

Jindal, F.; Jamar, R.; Churi, P. Future and Challenges of Internet of Things. *Int. J. Comp. Sci. Inform. Technol.* **2018,** *10* (2), 13–25.

Kainer, L. *University of Nottingham The Evolution of Business Models*. 62.

Kapás, J. Industrial Revolutions and the Evolution of the Firm's Organization: An Historical Perspective. *J. Innovat. Econ. Manag.* **2008,** *2* (2), 15.

Koleva, N. . Industry 4.0's Opportunities and Challenges for Production Engineering and Managemnet. *Int. Sci. J.* **2008,** *6* (1), 17–18.

Leiner, B. M.; Cerf, V. G.; Clark, D. D.; Kahn, R. E.; Kleinrock, L.; Lynch, D. C.; Postel, J.; Roberts, L. G.; Wolff, S. A Brief History of the Internet. *ACM SIGCOMM Comp. Commun. Rev.* **2009,** *39* (5), 22–31.

Mihardjo, L. W. W.; Sasmoko, S.; Alamsjah, F.; Elidjen, E. Digital Leadership Role in Developing Business Model Innovation and Customer Experience Orientation in Industry 4.0. *Manag. Sci. Lett.* **2019,** *9* (11), 1749–1762.

Nagy, J.; Oláh, J.; Erdei, E.; Máté, D.; Popp, J. The Role and Impact of Industry 4.0 and the Internet of Things on the Business Strategy of the Value Chain-the Case of Hungary. *Sustainability* **2018,** *10* (10). 3491.

Ooi, T. Purpose Benefit and Promise. *Intel-Internet of Things Group.*

Radanliev, P.; De Roure, D.; Van Kleek, M.; Santos, O.; Ani, U. Artificial Intelligence in Cyber Physical Systems. *AI Soc.* **2021,** *36*, 783–796.

Rojko, A. Industry 4.0 Concept: Background and Overview. *Int. J. Interact. Mobile Technol.* **2017,** *11* (5), 77–90.

Schuh, G.; Rudolf, S.; Riesener, M. In *Design for Industrie 4.0*, Proceedings of International Design Conference, DESIGN, DS; 2016, Vol. 84; pp. 1387–1396.

Şen, E.; İrge, N. T. Industry 4.0 and Agile Firms. In *Agile Business Leadership Methods for Industry 4.0, January 2021*; 2020; pp. 209–231.

Vaidya, S.; Ambad, P.; Bhosle, S. . Industry 4.0—A Glimpse. *Procedia Manuf.* **2018,** *20*, 233–238.

Werthmann, H.; Blunck, E. . In *Industry 4.0—An Opportunity to Realize Sustainable Manufacturing and Its Potential for a Circular Economy*, DIEM: Dubrovnik International Economic Meeting, 2017, Vol. 3, no. 1, pp. 644–666. http://hrcak.srce.hr/187419

Zezulka, F.; Marcon, P.; Bradac, Z.; Arm, J.; Benesl, T.; Vesely, I. Communication Systems for Industry 4.0 and the IIoT. *IFAC-PapersOnLine*, **2018,** *51* (6), 150–155.

CHAPTER 2

Dynamics of Technological Unemployment, Leadership, and Entrepreneurship during the Industry 4.0 Revolution

ALI EREN ALPER*, FINDIK ÖZLEM ALPER, GURCEM OZAYTURK

Nigde Omer Halisdemir University, Faculty of Economics and Administrative Sciences, Nigde, Turkey

**Corresponding author. E-mail: alierenalper@gmail.com*

ABSTRACT

Throughout the process from the First Industrial Revolution that began with the invention of the steam engine to Industry 4.0, each of the industrial revolutions has led to a technological and economic paradigm shift. The First Industrial Revolution, which included water and steam power for mechanization and began in the 18th century, was followed by the Second Industrial Revolution, which used electrical power for mass production, and the Third Industrial Revolution was experienced in the 1970s along with the introduction of electronics and information technologies. By the dawn of the 21st century, the Industry 4.0 revolution, which can be described as multidimensional, has come into existence covering various modern automation systems and digital software domains. New technological developments that have been accelerated by Industry 4.0, which first emerged in Hannover, Germany in 2011, have caused changes both at the micro-level (productivity in production, cost advantage, profitability, etc.) and at the macro-level (growth, employment, education, investment environment, human resources,

Agile Leadership for Industry 4.0: An Indispensable Approach for the Digital Era, Tanusree Chakraborty, Tahir Mumtaz Awan, Ashok Natarajan, and Muhammad Kamran (Eds.)

entrepreneurship, etc.). The most important discussion of the Industry 4.0 revolution, in which leaders with innovative ideas and entrepreneurial characteristics were quite influential on its occurrence, involves the extent to which it will affect working life and unemployment. There are various views on this issue that can be classified as optimistic and pessimistic. The optimistic view asserts that a highly qualified labor force can persist in working life through human–machine interaction, whereas the pessimistic approach argues that with the inclusion of machines in production, the need for labor force will decline, and thus, unemployment will increase. As a result, while each new technology causes unemployment by adversely affecting certain sectors and business lines, the culture of entrepreneurship and the changing perception of leadership contribute to the emergence of new business lines and new job opportunities. In this study, the development and characteristics of industrial revolutions in the historical process are examined and some detections are made by discussing the impacts of Industry 4.0 in working life within the framework of unemployment, leadership, and entrepreneurship culture.

2.1 INTRODUCTION

Industrial Revolutions have been the most crucial improvements in human history throughout the last 300 years as they continue to affect the political, economic, and military order of the world. The phenomenon of the industrial revolution affected the extent to which humanity would have produced goods and services, generated new methods, and formed new organizations. Industrial revolutions have been a global phenomenon since the very beginning, and each has brought about radical changes. Each of these revolutions had lasted for several decades. During their early stages, while still not having a very pronounced impact on the overall production rates determined by the more traditional methods, they revealed new definitions on the use of machinery and the reorganization of labor and even provided the groups that were not directly involved in such revolutions with the sense of great changes (Stearns, 2013).

After the First Industrial Revolution, the world has witnessed three different industrial revolutions. During the First Industrial Revolution, the main energy source was coal and steam machine, the dominant sectors were the textile and iron-processing industries, whereas the widespread use of electrical energy and the Fordist production-based line system came to the fore as the main change during the Second Industrial Revolution. The Third Industrial Revolution enabled the developments of semiconductor technology, the first personal computers depending on such technology, and the Internet toward the end of the century. Contemporary Industry 4.0 Revolution has begun along with radical changes such as quantum computing technologies, automation, and the Internet of things (Freyer, 2018).

In accordance with the scope of the book, this study summarizes the four industrial revolutions that reflect the development of humanity for three centuries in a chronological order and explains the changes that will be made on the concepts of entrepreneurship and leadership, especially by emphasizing the technological unemployment that emerged by courtesy of the Industry 4.0 Revolution. The first part of the study, which is comprised of seven parts and a series of evaluations and suggestions, describes the First Industrial Revolution, which led to the birth of the working class. In the second part, the Second Industrial Revolution, which was experienced with the replacement of steam power by electricity, and in the third part, the Third Industrial Revolution, which was caused by digitalization and increased production, are presented. In the fourth part, the Industrial 4.0 Revolution and the transformations of the labor force with high-tech and accelerated smart technology systems are emphasized. In other words, the first four parts provide an overview of industrial revolutions. In the fifth part, the concept of technological unemployment, which can be expressed as the focal point of the study, is explained, especially because of the replacement of people by machines due to Industry 4.0. In the sixth part, discussions on how technological unemployment would be prevented, whose nature has changed as a result of Industry 4.0, the risks incurred by entrepreneurs, and the characteristics that entrepreneurs should have as of today are made. In the seventh part, similar to the scope of the sixth part, the fight against technological unemployment is considered from the strategic leadership perspective. The basic parameters of strategic leadership and what needs to be done are emphasized upon establishing the framework of the study. In the last part, suggestions are made on what would be done in the fight against technological unemployment in the new world order created by the Industrial 4.0 Revolution and what should be focused on in the fields of entrepreneurship and strategic leadership for becoming a part of the new order.

2.1.1 FIRST INDUSTRIAL REVOLUTION

The first industrial period started in the 18th century as a result of the introduction of water power and steam power obtained dependent on water to the output. The existence of the steam engine in the same period transformed the economic units producing for their own consumption into the units producing for the markets. The most essential impact of the invention of the steam engine involves the huge rise in the productivity of labor used in the manufacturing process (Sharma and Singh, 2020).

Before James Watt, who went down in history as the inventor of the steam engine, the efficiency of steam-powered engines was quite low. Merely 1% of the energy generated by burning coal could be utilized. By courtesy of Watt's studies conducted between 1765–1776, productivity could be increased threefold (Brynjolfsson and McAfee, 2014).

According to Morris , a single year can entirely change the earth beneath man's feet. The year 1776 is such a year for today's modern economy. A tax revolt evolved into a revolution in the USA, Adam Smith completed his seminal work entitled *the Wealth of Nations*, the first and greatest work of political economy, and Watt's steam engine began to be used. The world had energy even before Watt's steam engine. However, James Watt's energy was more efficient and better, with Matthew Boulton's support. For millions of years, mankind's main energy source was muscle force. Nonetheless, muscles were parts of human beings and animals, and both groups of creatures needed food, shelter, and clothing to survive. Those necessities were fulfilled by obtaining from other animals, and that chain ultimately needed land. Notwithstanding, as the population increased and the land became scarce, muscle force became more expensive. Wind and water energy were used as alternative energy sources, although in limited amounts. In such an environment where muscle force was getting more and more expensive, the idea of draining the water accumulated in coal mines using steam power instead of using horses and people had emerged. Although the first steam engines (for instance, Miner's Friend) began to operate in England in 1698, they were operating at a quite low efficiency, and a significant portion of the coal extracted from the mines was also used as an energy source for that machine. Hoping to increase the efficiency of the steam engine, James Watt introduced the new steam engine, which consumed just one-third of the coal needed by the old steam engine models, to the public in 1776 with the support of Matthew Boulton. In Great Britain, that new technology spread from sector to sector, prompting more technological breakthroughs. One of the most important leaps (along with the textile sector) had been in the iron-processing industry.

The essence of the First Industrial Revolution was quite simple; it included the adaptation of new energy generated by steam power to production processes and specialization and coordination at levels that preindustrial groups could not have predicted. The First Industrial Revolution replaced humans and animals, which were the main sources of power in production processes, with engines powered by fossil fuels (mainly coal). A key point in Western Europe's industrial revolution was the steam engine, which exploited the energy potential of coal. The organizational aspect of the First Industrial Revolution was symbolized primarily by the factories in which large masses of workers worked (Stearns, 2013).

The first improvement of the First Industrial Revolution took place in the weaving industry. The revolution, which began with the change in the weaving industry in Great Britain, subsequently made a breakthrough in the iron-processing industry. Britain is the first industrialized economy in the world. The most important feature of British industrialization is that it came about instantly as a result of a conscious government policy in favor of industrial progress. By the year 1790, the investment levels doubled compared to the 1760s and reached the levels of 11% of GDP for the sustainability of the industrial revolution in Great Britain (Mathias, 1983).

Two major mistakes are made regarding the First Industrial Revolution. The first is that the revolution is too pinned to a certain period. The First Industrial Revolution took place in three main chronological phases. In the first of these phases, it began and directly spread in Western countries. In the second one, it reached a certain level of maturity and exceeded the boundaries of Western economies, and in the third phase, it became significantly global. The lives of entire humanity are being influenced by this ongoing third phase. The second common mistake is made with geographic identification. Although the industrial revolution was initiated in western geography, it is not limited to the history of western countries, it is rather a global phenomenon (Stearns, 2013).

Technological innovations continued following the First Industrial Revolution which was, in essence, cumulative technological innovations. As a result of these continuous technological innovations, a new energy source replaced coal as the main energy source. The name of the new energy source was electrical energy and it heralded the dawn of the Second Industrial Revolution.

2.1.2 SECOND INDUSTRIAL REVOLUTION

The period between 1860 and 1900 is called the Second Industrial Revolution since it coincided with a period when new technologies were invented intensively. These inventions brought about 70 years of rapid technological change. As a result of such technical change, huge increases in productivity and output per hour had been experienced. Historians state that the driving force of the Second Industrial Revolution was electrical energy. The impact of the development of electrical energy on efficiency increases had been experienced in a slow, yet continuous process. There are two main reasons for this. The first one involves the slow spread of new electricity-based technologies among manufacturing facilities in the USA. The second one is that it takes time to learn how to best utilize technology after a facility with new technology is built (Atkeson and Kehoe, 2001).

The innovations that emerged with the Second Industrial Revolution, unlike the First Industrial Revolution, were experienced not only in Britain but also in the USA and Germany. According to Mokyr and Strotz (1998), technological developments in the Second Industrial Revolution can be listed as follows;

The invention of steel, in the 1850s, although the raw material of iron was now widely used, wrought iron was inferior and nondurable compared to steel. Furthermore, wrought iron is less durable and less flexible than steel in many areas where it is used. The main problem was to produce the large quantities of needed cheap steel rather than steel production. This problem was solved by Henry Bessemer in 1856. The main invention of Bessemer was the discovery that the deficiency of iron was due to its impurity and excessive carbon.

In the *chemical industry,* the Germans were pioneers in the development of the chemical industry. The discovery of Justus von Liebig's artificial fertilization method, which led to the birth of modern agriculture, as well as the discoveries that formed the basis of the food industry was followed by the discovery of aluminum (Fereyer, 2018). With the development of the chemical industry, many elements such as rubber and synthetic plastic have been used in modern industry.

Electrical energy involved an area in which a completely new understanding of electricity was applied, the same as chemistry. Electricity was invented by Michael Faraday. The widespread use of electricity had changed human life forever. The widespread use of electric lamps, especially in factories, has greatly reduced the fire hazard caused by gas lightings and has brought a cleaner and safer environment by reducing the heat and pollution of the working environment (Agarwal and Agarwal, 2017). Although based on previous technological innovations, the long-distance telegraph with electrical energy was developed in this period. A series of micro-inventions (such as the use of electrical circuits for magnetization, laying of underwater cables between oceans) has finally been able to establish a long-distance, swift, and reliable telegraph system (Mokyr and Strotz, 1998).

Transportation, the transportation sector, which developed due to the expansion of the railway network along with the First Industrial Revolution, became faster and more reliable with the invention of the diesel-fueled engine invented by Rudolf Diesel in 1897, and the integration of electrical engines into railway transportation. Moreover, the widespread use of steel and the invention of diesel engines during that period enabled the construction of ships with wider hulls and higher load capacities. Henry Ford Motor Company was established in 1903 as a result of the discovery of internal combustion engines. The USA has taken over the leadership in automobile production with a mass-production style called the Fordist production method, which is identified with a continuous flow process by courtesy of the conveyor belt system (Mokyr and Strotz, 1998). In this regard, along with the decline in unit costs as a result of the increase in productivity, the automobile has become a technology that the purchasing power of quite large masses of people could afford.

Another transportation vehicle that radically changed our lives involved the Wright brothers' first flight in the USA on December 17, 1903, as a result of the increase in aerodynamic knowledge. The improvement of airplanes is a macro-scale invention that emerged in the Second Industrial Revolution (Mokyr and Strotz, 1998).

2.1.3 THIRD INDUSTRIAL REVOLUTION

The Third Industrial Revolution has important impacts on the 21st century, just as the First Industrial Revolution did to the 19th century and the Second Industrial Revolution to the 20th century. As in the previous two revolutions, it has influenced our way of working and life in every aspect. It initiated the process of society's economic, social, political, and fossil fuel-based industrial revolutions to evolve into the green industry era (Rifkin, 2019).

The term Third Industrial Revolution has been coined by economic historians as "green industrial revolution," "efficiency revolution," and a fundamental transformation towards "green capitalism." The Third Industrial Revolution is a period through which information and communication technologies, microelectronics, renewable raw materials, environmentally-friendly clean technologies, recycling, renewable energy, energy efficiency, and globalization are discussed (Jänicke and Jacob, 2009).

The Third Industrial Revolution has begun in the 1970s, with programmable devices and computers allowing partial automation. The Third Industrial Revolution is the linkage between the Fordist mass production style and the smart and flexible production systems that would emerge along with the Fourth Industrial Revolution. Technological innovations such as processors, computers, and the Internet that involve the Third Industrial Revolution are made possible by large research and development expenditures of governments and universities (Sharma and Singh, 2020).

As a result of the Third Industrial Revolution, a faster, more precise, and more efficient production style has been possible with less effort, by courtesy of interconnected technological developments such as personal computers, information-communication, and the Internet infrastructure (Sharma and Singh, 2020).

As a result of these changes, the Third Industrial Revolution has provided the spread of the industrialization movement and the development of many different regions (such as India and China). Moreover, it has been a process initiating the transition of societies from the industrial society to the information society. With this process, machine power and information power have been utilized more frequently than human labor in the production process. In parallel with the increase in the use of information, the production of new information has also accelerated. Creativity and accompanying competitiveness have become the main factors that directly affect the increase in welfare (Genc and Cakıroglu, 2019).

Although the First Industrial Revolution began in Britain, the first information technology revolution started in the USA. Innovative companies in the USA not only laid the foundations for the revolution of the 1970s but also pioneered its spread in the 21st century. This change has become synonymous with Silicon Valley, which was established in Northern California with the construction of Stanford Industrial Park in 1951. Two important factors played a role in designating Silicon Valley the center of the Third Industrial Revolution. The first one involves the development of the information technology revolution, through which discoveries and applications are interacted, in a continuous process of trial and error, tested by learning by doing in research centers such as California, higher education institutions, high-tech companies, and venture capital companies that would support new companies should take place in a region. The second one, when such an environment is established, involves the existence of an attractive region that can create its own dynamics and attract knowledge, investment, and talent from all over the world (Castells, 2008).

According to Troxler (2013), the Third Industrial Revolution may be defined as a combination of digital production and personal production. This period has two main features. First, the use of digital facilities for the design and production of products has become widespread, and thus units have begun to find the opportunity to work with each other over the network without distance limitations. Second, with direct digital production, design files are sent directly to the machines that would produce them. The most important advantage of direct digital production is that it is possible to make much larger scale production by courtesy of a pooled machine group with the removal of geographical restrictions on the machines that would produce.

The Third Industrial Revolution has also deeply affected foreign direct investment, which is one of the most important consequences of the globalization phenomenon in the world. The spread of communication infrastructures, the solution of the connectivity problem, and the cost-based competition offered by international companies, combined with low labor costs in developing countries, have caused the shift of production activities to developing countries such as China and India by the end of the 20th century. This inevitable shift in production processes has forced the developed Western economies, especially Germany, to implement the next revolution.

2.1.4 FOURTH INDUSTRIAL REVOLUTION

The term Industry 4.0 was first coined at the Hanover Fair in Germany in 2011 and it refers to the transformation process in global chains of value creation (Schwab, 2016). The main reason for the Industry 4.0 Revolution is to prevent the shift of foreign direct investments from developed economies, especially Germany, to Far East Asia and other developing countries. Since the German economy cannot engage in cost-based competition with developing economies such as India and China, especially in terms of labor costs, this revolution has taken place as a result of the pursuit of a method that would ensure minimum human participation in all stages of production. The Industry 4.0 Revolution has emerged with the development of many technologies that are at the heart of, but not limited to, the Internet of Things concept. As a result of this revolution, fully automated production processes have begun to dominate economic life. In order for these processes to develop and sustain, high-speed internet infrastructure and robotic technologies should be improved. The core of the systematic transformation in the industry has become synonymous with fully automated manufacturing. In addition, the few labor forces that would be involved in production processes will have to be replaced by interventions (such as exoskeleton) that would change their physical properties (Popkova et al., 2019).

The most important feature that distinguishes the Industry 4.0 Revolution from previous industrial revolutions is the elimination of the human factor from the production process. Even though the contribution of the human factor was reduced by the innovations made in the previous industrial revolutions, its central role prevailed. Nevertheless, with the new revolution, the production process was transformed from a socio-technical structure into a full technical system (Popkova et al., 2019). According to Leonhard (2018), following the Industry 4.0 Revolution, the competition between man and machine would intensify through ten major mega transformations. These transformations are digitization, mobilization, screenification, disintermediation, transformation, intelligization, automation, virtualization, anticipation, and robotization. While the impacts of these transformations are seen gradually at first, they will act suddenly once a certain maturity level is achieved. Among these transformations, automation is undoubtedly the most effective aspect of human lives. The automation processes in society progress in five stages, respectively. In the first stage, automation emerges; in the second stage, some operations are consented to be conducted by automation; whereas some obligations that once belonged to people are waived in the third stage. In the fourth stage, the anger and disappointment of a few remaining human service organizations and system users occur, and in the final stage, there is an anomaly due to considering people as numbers and disembodied data sources. If not intervened during the first two stages, the automation will surely reach the fifth and painful anomaly stage.

Another distinctive feature of the Industry 4.0 Revolution is pertinent to its area of influence. By courtesy of artificial intelligence technologies, this revolution leads to profound changes and modernization not only in a certain area of the economic structure but also in all areas such as production, distribution, management, and marketing (Popkova et al., 2019). Along with the introduction of artificial intelligence technology into our lives, the production of machines that can perform cognitive tasks has been initiated. Machines that can undertake cognitive work are much more crucial than machines that can undertake physical work (Brynjolfsson and McAfee, 2014).

Another fundamental feature that distinguishes this revolution from other industrial revolutions involves an increase in inter-unit integration and interaction (Knyaginina, 2017). As of today, we bear witness to the accelerating impact of such interaction on production and development processes in every field. For instance, the efforts to develop vaccines in seven months against the COVID-19 virus, which was seen in China in December 2019 that rapidly spread to the whole world and eventually turned into a global pandemic, have yielded results by courtesy of the interaction among scientists from all over the world and the constant exchange of information with each other.

As a result of this revolution, computers controlled by artificial intelligence may automatically initiate the process by transferring real-time information obtained via the Internet to production units. In this respect, the negative externalities generated by the production processes do not influence the living standards of people in the industrialized regions. By courtesy of the remote control facilities operated by a limited number of employees, production can be made in uninhabited geographical regions without endangering the health of the employees as well as the people in the region. The Industry 4.0 Revolution also has caused changes in the patent system. In previous revolutions, patents focused on concealing information on a particular technology not to be utilized by competitors. Nonetheless, with this revolution, even though technological information is now completely public on the Internet, patents tend to concentrate on prohibiting their legal use (Popkova et al., 2019).

The fundamental infrastructure of the Industry 4.0 Revolution is the *Internet of things*. This system consists of a series of intelligent and connected sensors that collect, process, and transform data upon request. Such data are then transmitted to other devices or individuals to achieve the goals of the system or the users. It is estimated that the number of Internet-based devices in the world, which was 15.4 billion as of 2015, will increase to approximately 75 billion by the year 2025. This figure indicates that economies will switch from machine to machine after all. The destructive potential of the Internet of Things would lead to change in organizations and industries. When the Internet of Things is combined with artificial intelligence and robotics, it would reduce the demand for routine and manual works as well as increase the control over employees (Schwab and Davis, 2018). Therefore, it is certain that these changes would deeply affect various areas such as working styles, working hours, leadership styles, and entrepreneurship culture, as well as the required qualifications.

2.2 INDUSTRY 4.0 AND TECHNOLOGICAL UNEMPLOYMENT

The pretechnology capacity of nations had been evaluated in terms of a mixture of manpower, equipment, geography, economy, and morale

(Kissinger, 2014). As today's conditions evolve from simple to complicated, two groundbreaking changes occur. The first of these is the increase in both individual and social capabilities. The second, and the scary one, is that more and more authority is being devolved on other powers. Technology is undoubtedly the factor that serves both of these changes. Those who generate, store, and enforce technology are stronger than those who use it, whereas those who use technology are also considered to be stronger than those who fail to do so. In this case, the segment with the opportunity and capacity to keep up with technology would have the chance to generate more surplus value compared to those who do not (Auerswald, 2018).

History is the ultimate evidence of this situation. The First Industrial Revolution, which replaced manpower with machinery following the invention of the steam engine, gave way to the dawn of the working class. Mechanical and chemical factors, which replaced human labor, forced millions of unemployed workers to migrate to urban areas to seek skilled/unskilled job vacancies. The Second Industrial Revolution, which took place with the replacement of steam power with electricity, caused the production to be serialized in factories, and thus millions of blue-collar workers became qualified as a part of the white-collar workforce in the flourishing service industry. The Third Industrial Revolution, which means experiencing technological transformations in the field of production-processing-transfer, refers to a period throughout which more production is conducted utilizing less labor and fewer resources as a result of digitalization. Nevertheless, the world is experiencing a new transformation as of today. This revolution, called Industry 4.0, is a transformation during which salaried labor in agricultural, industrial, and service sectors is replaced with boutique, high-tech workforces, and smart technology systems that are gaining depth and speed. This brings the following question to mind: What will happen to millions of salaried labor of the industrial age with this process? (Rifkin, 2019).

In the transformation from handcraft to manufacture, from manufacture to mechanization, from mechanization to automation (Perrin, 1992), it should be admitted that automation technology is booming. The reason is that the manpower is expensive, slow, and usually insufficient, whereas machines are inexpensive, swift, and very effective (Leonhard, 2018). On the one hand, the fright of machines fully substituting manpower and excluding it from the production process is increasing day by day; on the other hand, it is said that technology would provide the labor market with new opportunities. In this context, it can be claimed that the views on the possible impacts of technology on the labor market constitute two opposite poles: optimistic and pessimistic views.

The optimists argue that technology can lead to a certain level of unemployment, but is still a helpful factor in the workforce, or it can increase employment by providing new job opportunities. For instance, Ricardo stated that the use of machinery cannot completely eliminate workers in production in accordance with the technological developments experienced in the production process. According to him, the workforce would always exist to fulfill the needs such as operation and maintenance of the machines in use (Ardor and Varlık, 2009). Because it is labor that produces capital. Furthermore, Ricardo, who perceived capital consisting of tools and machinery as indirect labor, considered the capital inflows to the production process along with labor as equipping labor with tools and machines (Bocutoglu, 2012).

According to Marx (1997), labor instruments that become the form of machines require the substitution of natural forces instead of human power and conscious scientific practices. While manufacturing is a combination of piece-workers, there is a purely objective productive organism in a large industry based on the machine system, and the workers have become something that is included in the already existing material conditions of production. Machines, on the other hand, can only be operated by combined labor or joint labor.

Jean-Baptiste Say has stated that when the production rate increases, the share of the national market increases at the same rate (Oser and Brue, 1988). Accordingly, with the increase in efficiency by courtesy of technology, the suppliers' production of more goods at a lower cost per unit generated their own demand and the increased production demand created additional production. The increasing value of the goods sold was compensating for the initial employment loss of technological developments, with new jobs to fulfill the increased production level (Rifkin, 2019).

According to Clark (2019), it was inevitable that there was some unemployment in the dynamic structure of the current economic system because it was impossible to make progress without enduring temporary unemployment. In other words, the creative forces of capitalism required a ready stock of labor, and that stock was none other than the unemployed workers due to technology (Brynjolfsson and McAfee, 2014).

Accordingly, optimistic views are based on the idea that markets are more efficient, costs are reduced, industries and economies develop, and thus new sectors flourish. In fact, it is stated that each wave of industrialization enables new sectors to occur and thus, a sufficient number of new job opportunities are generated instead of old jobs that have become obsolete. The relationship between the automation system, which is beyond technological progress, and job creation have deviated to a completely different path. Indeed, this has led to the creation of new jobs such as human–machine interface designer, cloud biologist, artificial intelligence supervisor, human genome analyst, or privacy manager that did not exist in the past (Leonhard, 2018).

Nonetheless, the unemployment of many middle- and lower-class wage earners was not in line with the welfare promises of the automation system. It was impossible for the jobs created to keep up with the decreases in manufacturing employment, white-collar, and service sector workers. Over the years, the employment growth rate decreased from 37.7% in the 1940s to 20.2% in the 1980s, to 19.8% in the 1990s, and 1% in the 2000s (Kim et al., 2017). Although these figures reveal the overall situation, it cannot be claimed that all jobs and workers are out of the competition. An expectation may arise that the employability of low-skilled workforce would decrease, whereas high-skilled workforce happens to be the segment that can evaluate employment opportunities of characteristics such as dynamic routing, self-organization, comprehensive connections, big data, and deep convergence through the advanced automation process in a smart factory (Wang et al., 2016). For instance, high-wage occupations, such as managers and entrepreneurs, that benefit from computing to increase their creativity and improve their problem-solving abilities have begun to earn even higher wages. The lowest-paid service jobs, including doormanship and handcraft jobs, also remained, because such occupations did not require special skills and training, and the use of machines only increased costs. Moreover, many middle-class occupations such as administration and office work were retired by courtesy of the higher efficiency of computers (Frey and Osborne, 2017). Moreover, the use of computers in office work caused the qualifications required for employees to decrease. In the past, an accountant's handwriting had to be nice, and he had to be good at multiplication and division. When that man was replaced by a girl using the accounting machine, it took her only a few weeks to become a skillful accountant. In short, although the introduction of machines increased human capabilities, they also dismissed them for the same reason (Auerswald, 2018).

An example of this situation is that 600 stock exchange operators were working in the New York office of Goldman Sachs back in the year 2000, and only two people working with the support of automated trading programs were left as of 2017 (Schwab and Davis, 2018). Another example is that while the demand for skilled jobs peaked in the USA around the year 2000, it then declined and new college graduates had to find job vacancies that require less and fewer skills. Moreover, Noriko Arai, a mathematician from National Informatics Institute in Japan, stated that he was leading a system development project that could pass the Tokyo University's entrance examination. Accordingly, if a computer displays a combination of natural linguistic and analytical skills enough to enter Japan's highest-rated university, then most of the job vacancies that college graduates can apply for will be jeopardized (Ford, 2018).

Keynes defined such a situation as "*Technological Unemployment*." This term refers to a type of unemployment resulting from the speed of realization of new discoveries that provide savings in labor usage than the speed of opening new domains for the workforce (Keynes, 1933). Uncertainties regarding the extent to which the workforce can be substituted and how long such a situation would last determine the degree of concern about the persistence of technological unemployment (Schwab, 2016).

Keynes stated that unlike those who argue that technology will create temporary unemployment, based on economic theories and 200 years of historical evidence, it would generate permanent and structural unemployment. The reason for this is attributed to inelastic demand, rapid change, and severe inequality. According to Keynes, the inference that more goods will be consumed since the price of the goods decrease is not correct, although the technology decreases the costs by increasing the productivity of the labor force and decreases the product prices.

On the contrary, there will be less consumption with satiation. As a result, there is no strict relationship between purchasing power and output, or between output and employment (Neisser, 1942). In that case, there would be a drastic decline in weekly working hours due to technological developments. Even more dangerous than this is the insufficient adaptation skills of organizations and institutions toward rapidly changing technology. The third argument for technological unemployment involves the fact that technological development draws a fine line between winner and loser. Along with skill and capital-oriented technological changes, markets with a "winner takes all" approach emerge in all areas (Brynjolfsson and McAfee, 2014). For instance, in search engine advertising, almost 90% of the global market share is owned by Google, 77% of mobile social traffic is owned by Facebook, and almost 75% of the e-book market is held by Amazon. In fact, if it turns out that it is possible to create artificial general intelligence that would improve itself into super-intelligence, the first-move advantage may lead to domination in many markets (Schwab and Davis, 2018). Jobs created in new sectors require technical expertise and noncognitive skills, whereas reasonable and rational people would try to earn money in other fields if the equilibrium wage for a less skilled workforce drops to near-zero wage level. In short, technology can generate unemployment as well as inequality (Brynjolfsson and McAfee, 2014).

Those who reject the fact that machines substitute people and create technological unemployment perceive the comparative advantage theory as a haven. However, Ford (2018) explained the situation with an example as follows: As long as two people, one of whom is a cook and the other one is a better neurosurgeon than a cook, specialize in the area where they are "least bad," they will always find a job. Moreover, the neurosurgeon's work as a doctor will allow the other cook to specialize and earn more money. The theory of comparative advantage also explains the same thing. But as of today, this story has slightly changed. If the neurosurgeon clones himself, what would happen to the theory of comparative advantage based on opportunity cost resulting from time and space constraints? The ability to easily reproduce machines, especially computer software, that is, cloning and reproducing intelligence would lead to the collapse of opportunity cost. If the neurosurgeon is both a very good doctor and a better cook than his peers and can keep up with both occupations, why should the other cook be needed? Or when information technology clones intelligence from machine to machine and reproduces it within the company, such as a large company training a single employee and then turning him/her into an army of employees, won't machines become alternatives to humans?

According to Vivek Wadhwa, an engineer at Carnegie Mellon University, the truth is that we encounter a future without jobs. A future awaits us where most of the work done by people will be done by machines. Robots will drive our cars, manufacture our products, and do our tasks, but there will not be much work for us (Kapeliushnikov, 2019). Harari (2017) predicted that the creation of artificial intelligence would lead to the emergence of a large nonworking class/useless class/a class that is not only unemployed but also unemployable by the year 2050. This justifies the idea that those who exhibit a pessimistic view of the impact of technology on the labor market, the disruptions that reduce job opportunities with the transformation, and the increase in the number of unemployed would become a norm, not an exception.

2.3 INDUSTRY 4.0 AND ENTREPRENEURSHIP

The industrial revolution is an inevitable phenomenon caused by rapid changes in the design of solutions. The industrial revolution may be described as an innovation process in terms of industrial, social, economic, and technological systems driven by rapid growth in advanced smart scientific solutions (Dombrowski and Wagner, 2014).

At the core of the Industry 4.0 Revolution is the access of all components in production units to integrated computing power and communication capacity. It is a process that is not only about communication among machines, but it also has far-reaching implications concerning the interaction of humans and technology. At the beginning of these implications, the most crucial problems for the entrepreneur are the development of both the duties and the demand spectrum of the employees working in the production units, and taking the necessary measures for the employees to keep up with such change would be the main task of the entrepreneur. For instance, as machines become more autonomous, even complex production processes of custom-made parts will be operated without human intervention. In this case, the classic location-based workplace (such as an office) will lose its significance. Decision-making and monitoring processes can be carried out directly from the production site or remotely, and moreover, different production facilities can be operated remotely. For these reasons, it can be assumed that the entrepreneur will undertake a larger range of operating areas and responsibilities in the future (Gorecky et al., 2014 [cited in Joseph-Auguste, 2017]).

The difficulties that the entrepreneur would encounter along with the Industry 4.0 revolution can be examined from macro and micro perspectives. In the macro perspective, the problem of cross-linking of stakeholders, raw materials, production, and usage of products or services occurs. Basically, it is a cycle that commences with the acquisition of raw materials and other resources and culminates in the expiration of the product. Micro perspective, on the other hand, is characterized by the problems experienced due to micro-level activities conducted in a single smart factory such as logistics, cloud, infrastructure, technology development, human resources, marketing, and production (Stock and Seliger, 2016).

The entrepreneur has to adapt to the Industry 4.0 revolution, which is the source of risk, innovation, and change. This obligation leads to responsibility not only for its own survival but also for the results obtained within the Industry 4.0 revolution to become more beneficial for society (Marinescu, 2018).

Entrepreneurship is a critical phenomenon in terms of maintaining economic growth and generating new job opportunities. Because the entrepreneur is an innovator, more precisely, he/she is someone who is constantly in pursuit of opportunities to create new goods and services. In this regard, the entrepreneur acquires new markets and creates new organizational forms (Baumol, 2002). One of the most important characteristics of the entrepreneur to achieve these goals is the capability of keeping up with change and innovation (Devezas and Sarygulov, 2017). Innovation is a more comprehensive concept than invention. Commercialization of designs involves the modification of ideas and their implementation in existing products, systems, and resources (Kruger and Steyn, 2020).

With the Industry 4.0 Revolution, entrepreneurs began to innovate in two ways. The first one is evolutionary innovation. Evolutionary innovation may be explained as continuous and progressive advantages in technology or processes that foster efficiency. The second way of innovation is a revolutionary innovation. Revolutionary innovation involves the inauguration of new domains of revenue through completely new and disruptive changes in technology and production processes (Lu, 2017).

In-depth learning, identifying opportunities, finding resources for change, and planning change are at the core of entrepreneurship (Kuratko, 2005). Nonetheless, the Industry 4.0 revolution has affected the nature of entrepreneurship through various channels.

According to Mkwanazi and Mbohwa (2018), these channels can be listed as follows:

- After all, the entrepreneur must have knowledge of complex socio-technological transformation issues.
- Entrepreneurs should have knowledge of technical-economic opportunities to develop competitive business ideas.
- Entrepreneurs should include "technical capability" features besides various personal features. This type of personality trait means having knowledge of the overall utilization of technical resources in the entrepreneurial arena.

Since the entrepreneur's ability to adapt to this revolution is a necessity for countries to sustain their economic development, existing entrepreneurs should adapt to the changing entrepreneurial culture, and future entrepreneurs should be trained in accordance with these characteristics.

2.4 INDUSTRY 4.0 AND LEADERSHIP

Leadership is the ability to influence others, to encourage, motivate, and manage activities to achieve organizational goals. Leadership is more than the management of ongoing processes. Leadership deals with change, setting a direction, and identifying the strategies needed to provide people with vision. It means directing the organization toward a constructive transformation (Kotter, 2001).

Leadership is quite crucial for organizations to adapt to the developments experienced as a result of the Industry 4.0 Revolution because, depending on the change in production processes, the capacity to make comprehensive strategic decisions at each level of the organization has gained great importance (Akcay Kasapoglu, 2018). Kelly (2018) listed the basic components that characterize the leadership structure of that period in each industrial revolution. Accordingly, "charismatic leadership" came to the fore during the First Industrial Revolution, whereas "directive leadership" during the Second Industrial Revolution, and "relational leadership" during the Third Industrial Revolution.

Organizations need more than transformational leadership that is open to learning and innovation in the Industry 4.0 Revolution. Transformational leadership is limited to providing inspirational motivation, intellectual stimulation, and vision. Industry 4.0, however, is based on learning and innovation. Therefore, it should adopt the knowledge-based leadership style structure by combining the transformational and transactional leadership styles. Knowledge-based leadership is libertarian in learning and trying new things. Such an expanded structure of knowledge-based leadership would boost the innovation capacity and learning speed of the organization in the sector (Durmus, 2019). Organizations in the present context would require a positive organizational culture (Chakraborty and Ganguly, 2019), will need positive approaches to understand employees by appreciative questioning (Chakraborty and Mishra, 2019; Chakraborty et al., 2019). Organizations are more agile after a disaster, so the role of human resource professionals in managing after effects of a crisis is very crucial (Chakraborty and Saha, 2017). Digital innovations in businesses have made remarkable changes in the way businesses are done (Chakraborty and Saha, 2018).

Due to the Industry 4.0 revolution, organizations should have dynamic, more flexible, and agile capabilities and be able to use them effectively. It is imperative for enterprises to restructure and have a more elastic and agile structure to keep up with technological changes and compete with their competitors (Kocyigit and Akkaya, 2020). Therefore, Industry 4.0 leaders must first increase the elasticity of the organization and keep up with continuous change and development. Agile leadership characteristics include learning by experiment, maximizing customer satisfaction, and constantly searching for the most effective ways to achieve them (Cinnioglu and Salha, 2019). The basic parameters of leadership in the Industry 4.0 Revolution can be listed as follows (Durmus, 2019):

- Openness and trust
- Rapid digital connection
- Utilization of the best knowledge
- Development of employees
- Organization as a community
- Transparency of responsibilities and duties
- High technology

2.5 WAY FORWARD

Automated teller machines (ATMs), which emerged as a result of technological developments, were expected to unemploy a large number of people. While the number of ATMs installed increased from 50,000 to 400,000 over the years 1985 to 2010, the number of human cashiers increased from 485,000 to 600,000 over the same period. This was explained as ATMs are doing a lot of routine work, allowing tellers to concentrate on more complex tasks. In summary, ATM technology has not eliminated bank tellers, it has merely changed the way tellers work and the skills they need. Moreover, once ATMs took over the tasks of bank tellers, fewer human tellers were needed to operate the bank branch. Nonetheless, as the cost of operating the bank branch declined, the banks tried to boost the number of branches to dominate a larger market. While fewer tellers were employed per branch, more branches meant more tellers (Auerswald, 2018).

According to McKinsey and Company (2017), by the year 2030, 75 to 375 million employees representing 3% to 14% of the global workforce will need to shift occupational categories. Furthermore, as all workers' professions improve with increasingly skilled machines, they will need more training to adapt to the new situation.

As can be seen, it is a fact that technology increases profits and generates new job opportunities with labor productivity on the one hand, and eliminates various lucrative jobs on the other hand. Therefore, technological development is creative and destructive in a way that no one can foresee (Cahuc and Zylberberg, 2004). Compared to the automation of existing jobs, more creativity of entrepreneurs, managers, and workers is required to reorganize the company in this regard. Upon examining the optimistic and pessimistic approaches, the common opinion involves the fact that the workforce is absolutely rendered qualified and creative. Being complementary to an increasingly common situation is the only means to stay out of the system. Machines have quite different advantages than humans, as well as weaknesses. It would be wise to pay attention to the areas of machine weaknesses rather than vainly challenging machines, which are competitors with qualifications close to manpower (Brynjolfsson and McAfee, 2014). As technology evolves, machines must be assigned tasks in which their attributes are relatively strong, and people must do the same. As a result, an efficient production activity should concentrate on the digital–human workforce, that is, the contribution of both people and machines should be sought.

Entrepreneurship can be viewed as a series of actions of individuals or groups concentrated on creating new economic opportunities besides the established organizations (Carree and Thurik, 2010). In other words, entrepreneurs create dynamic market environments in which resources gain momentum and value. Another task of the entrepreneur in the markets is to become the source of innovations (Bakator et al., 2018). Due to the impacts of the Industry 4.0 Revolution, a large network of expansion and connections has emerged among industries. This network enables quite close communication among consumers, employers, systems, and suppliers, allowing for increased performance, and new opportunities arise by courtesy of enhanced collaboration (Audretsch et al., 2019).

The key concept for countries to benefit from these advantages is entrepreneurs since entrepreneurs link innovation and opportunities for gaining comparative advantages (Mazzei, 2018). Along with the technological advantages provided by the Industry 4.0 Revolution, entrepreneurs have the potential to find new ways to provide goods and services (van der Westhuizen and Goyayi, 2020). Nevertheless, in order for this to be achieved, it is necessary to acquire knowledge on the provision of goods and services using digital capabilities (Zivdar et al., 2017).

Leadership can be defined as seeing the future, creating a vision, providing flexibility, strategic thinking, and working with others to make changes that would create a feasible future for the organization (Ugurluoğlu and Celik, 2009). Care should be taken to remove psychological stress among employees for better functioning (Deb et al., 2008).

Leaders impress people with their visions. As of today, leaders need to combine three basic components to achieve this with the impact of Industry 4.0 Revolutions. These are seeing the road, walking on the road, and being the road. Seeing the road, which is the first of these components, means being able to predict the future and determining policies about the future as of today. Walking on the road means being able to cooperate for implementing these policies. And finally, being the road is to make people follow oneself by removing or reducing the obstacles on the road (Sevinc, 2009). Leaders who incorporate these three components would be able to successfully see their organizations through during the Industry 4.0 Revolution.

2.6 KEY TAKEAWAYS

The solution proposals to prevent technological unemployment that emerged as a result of the Industry 4.0 Revolution would be summarized as follows:

- To adapt to new business lines that would emerge in the new order, training programs should be improved to make the workforce qualified and creative.
- Efforts should be made to train entrepreneurs who can establish a connection between innovations and opportunities as well as acquiring strategic leadership qualifications.
- More focus should be placed on business lines in which mechanization has certain weaknesses.
- Workspaces compatible with machinery should be developed by providing a digital human labor force.

KEYWORDS

- **industrial revolutions**
- **Industry 4.0**
- **labor market**
- **technology**
- **entrepreneurship**
- **leadership**

REFERENCES

Agarwal, H.; Agarwal, R. First Industrial Revolution and Second Industrial Revolution: Technological Differences and the Differences in Banking and Financing of the Firms. *Saudi J. Humanities Soc. Sci.* **2017,** *2* (11), 1062–1066.

Akcay Kasapoglu, O. Leadership and Organization for the Companies in the Process of Industry 4.0 Transformation. *Int. J. Organ. Leader.* **2018,** *7*, 300–308.

Ardor, N. H.; Varlık, S. David Ricardo ile Joseph Alois Schumpeter'in Teknolojik Gelişme Kuramlarının Karşılaştırılması. *Hitit Üniversitesi Sosyal Bilimler Enstitüsü Dergisi,* **2009,** *2* (1), 15–40.

Atkeson, A.; Kehoe, P. J. The Transition to a New Economy after the Second Industrial Revolution. *Natl. Bureau Econ. Res.* **2001,** W8676.

Audretsch, D. B.; Cunningham, J. A.; Kuratko, D. F.; Lehmann, E. E.; Menter, M. Entrepreneurial Ecosystems: Economic, Technological, and Societal Impacts. *J. Technol. Transf.* **2019,** *44* (2), 313–325.

Auerswald, P. E. *Kod Ekonomisi*; Dogan, M., Trans.; Bogazici University Publication: Istanbul (Original Work Published, 2017), 2018.

Bakator, M.; Đorđević, D.; Ćoćkalo, D.; Nikolić, M.; Vorkapić, M. Lean Startups with Industry 4.0 Technologies: Overcoming the Challenges of Youth Entrepreneurship in Serbia. *J. Eng. Manag. Compet.* **2018,** *8* (2), 89–101.

Baumol, W. J. Entrepreneurship, Innovation and Growth: The David-Goliath Symbiosis. *J. Entrep. Finance.* **2002,** *7* (2), 1–10.

Bocutoglu, E. İktisat Teorisinde Emeğin Öyküsü: Değerin Kaynağı Olan Emekten Marjinal Faydanın Türevi Olan Emeğe Yolculuk. *HAK-İŞ Uluslararası Emek Ve Toplum Dergisi* **2012,** *1* (1), 127–150.

Brynjolfsson, E.; Mcafee, A. *The Second Machine Age: Work, Progress, and Prosperity in a Time of Brilliant Technologies*; WW Norton and Company, 2014.

Cahuc. P.; Zylberberg A. *Labor Economics*; London: The MIT Pres, 2004.

Carree, M. A.; Thurik, A. R. The Impact of Entrepreneurship on Economic Growth. In *Handbook of Entrepreneurship Research*; Springer: New York, NY, 2010; pp. 557–594.

Castells, M. *The Information Age: Economy, Society and Culture*; Kılıc, E., Trans.; İstanbul Bilgi University Publishing: İstanbul (Original Work Published 2003), 2008.

Chakraborty, T.; Ganguly, M. Crafting Engaged Employees through Positive Work Environment: Perspectives of Employee Engagement. In *Management Techniques for Employee Engagement in Contemporary Organizations*, IGI Global; 2019, pp. 180–198.

Chakraborty, T.; Mishra, N. Appreciative Inquiry: Unleashing a Positive Revolution of Organizational Change and Development. *Int. J. Econ. Commerce Bus. Manag.* **2019,** *6* (2), 32–37.

Chakraborty, T.; Saha, R. Proactive and Reactive Role of Human Resource Professionals in an Aftershock Management. *Train. Dev. J.* **2017,** *8* (1), 1–11.

Chakraborty, T.; Saha, R. Digital Innovations: Breakthrough Opportunities to Build Novel Business Models. *Bus. Perspect.–Aims Scope* **2018,** *1*, 46.

Chakraborty, T.; Ganguly, M.; Natarajan, A. Predicting Entrepreneurial Satisfaction: The Role of Non-financial Incentive Factors and Quality of Life among Women Digital Entrepreneurs. *J. Glob. Bus. Adv.* **2019,** *12* (3), 328–355.

Cinnioglu, H.; Salha, H. In *VUCA Dünyasında Turizm Işletmeleri Ve Çevik Liderlik*, The Third International Congress on Future of Tourism, 2019; pp. 26–28.

Clark, J. B. *Essentials of Economic Theory: As Applied to Modern Problems of Industry and Public Policy*; Good Press. 2019.

Deb, S.; Chakraborty, T.; Chatterjee, P.; Srivastava, N. Psychological Stress of Traffic Police Officers Causal Factors Behind the Same and Their Coping Strategies. *J. Indian Acad. Appl. Psychol.* **2008,** *34* (1), 19–28.

Devezas, T.; Sarygulov, A. *Industry 4.0*; Springer: Basel, 2017.

Dombrowski, U.; Wagner, T. Mental Strain as Field of Action in the 4th Industrial Revolution. *Procedia Cirp* **2014,** *17*, 100–105.

Durmus, A. *Endüstri 4.0 Eğitim 4.0 Liderlik 4.0 Toplum 5.0*; Efe Akademi Yayınları: İstanbul, 2019.

Ford, M. *Rise of the Robots*; Duran, C., Trans.; Kronik Publishing: İstanbul (Original Work Published 2015), 2018.

Frey, C. B.; Osborne, M. A. The Future of Employment: How Susceptible are Jobs to Computerisation? *Technol. Forecast. Soc. Change* **2017,** *114*, 254–280.

Freyer, H. *Theorie Des Gegenwärtigen Zeitalters*; Akarsu, B., Batuhan, H., Trans.; Dogubatı Publishing: Ankara (Original Work Published 1955), 2018.

Genc, K. Y.; Cakıroglu, M. First Industrial Revolution. In *Industrial Revolutions*; Atik, A., Ed.; Gece Publishing: Ankara, 2019; pp. 151–197.

Gorecky, D.; Schmitt, M.; Loskyll M. Mensch-Maschine-Interaktion Im Industrie 4.0-Zeitalter. In *Industrie 4.0 in Produktion, Automatisierung Und Logistik*; Bauernhansl, T., Ten Hompel. M., Vogel-Heuser, B., Eds.; Springer Vieweg: Wiesbaden, 2014.

Harari, Y. N. *Homo Deus: A Brief History of Tomorrow*; Harper Collins: New York, 2017.

Jänicke, M.; Jacob, K. A Third Industrial Revolution? Solutions to the Crisis of Resource-Intensive Growth. *Environmental Policy Research Centre,* FFU-Report 02-2009, 2009.

Joseph-Auguste, B. Industry 4.0 and Entrepreneurship of People with Disabilities. *12th IWKM*, Trencin, Slovakia, 2017.

Kapeliushnikov, R. The Phantom of Technological Unemployment. *Russ. J. Econ.* **2019,** *5*, 88.

Kelly, R. *Constructing Leadership 4.0: Swarm Leadership and the Fourth Industrial Revolution*; Springer, 2018.

Keynes, J. Economic Possibilities for our Grandchildren. In *Essays in Persuasion*; Macmillan: London, 1933.

Kim, Y. J.; Kim, K.; Lee, S. The Rise of Technological Unemployment and its Implications on the Future Macroeconomic Landscape. *Futures*, **2017,** *87*, 1–9.

Kissinger, H. *Dünya Düzeni*. 5. Baskı, Boyner Yayınları, 2014.

Knyaginina, V. N. *New Technological Revolution: Challenges and Possibilities for Russia: An Expert Report*. The Center for Strategic Developments: Moscow, 2017.

Kocyigit, Y.; Akkaya, B. The Role of Organizational Flexibility in Organizational Agility: A Research on Smes. *Bus. Manag. Strategy* **2020,** *11* (1), 110–123.

Kotter, J. P. What Leaders Really Do. *Harvard Business Review*, **2001,** *79* (11).

Kruger, S.; Steyn, A. A. Enhancing Technology Transfer through Entrepreneurial Development: Practices from Innovation Spaces. *J.Technol. Transf.* **2020,** *45* (6), 1655–1689.

Kuratko, D. F. The Emergence of Entrepreneurship Education: Development, Trends, and Challenges. *Entrep. Theory Pract.* **2005,** *29* (5), 577–597.

Leonhard, G. *Technology vs. Humanity: the Coming Clash between Man and Machine*; Akkartal, C., Akkartal, İ., Trans.; Fast Future Publishing: New York. (Original Work Published 2016). 2018.

Lu, Y. Industry 4.0: A Survey on Technologies, Applications and Open Research Issues. *J. Ind. Inf. Integr.* **2017,** *6*, 1–10.

Marinescu, P. Entrepreneurship in the Context of the Fourth Industrial Revolution. *Manager* **2018,** (28), 5–5.

Marx, K. . *Kapital;* Bilgi, A., Trans.; Cilt 1, Sol Yayınları: Ankara. (Original Work Published 1867). 1997.

Mathias, P. *The First Industrial Nation: The Economic History of Britain*; Routledge: London, 1983; pp. 1700–1914.

Mazzei, M. J. Strategic Entrepreneurship: Content, Process, Context, and Outcomes. *Int. Entrep. Manag. J.* **2018,** *14* (3), 657–670.

Mckinsey & Company . Jobs Lost, Jobs Gained: Workforce Transitions in a Time of Automation [Online]. 2017, https://www.mckinsey.com.

Mkwanazi, S.; Mbohwa, C. In *Implication of the 4th Industrial Revolution on Entrepreneurship Education,* iProceedings of the International Conference on Industrial Engineering and Operations Management, Washington DC, USA; 2018.

Mokyr, J.; Strotz, R. H. The Second Industrial Revolution, 1870–1914. *Storia Dell'economia Mondiale* **1998,** *21945.*

Morris, I. *Why the West Rules-For Now: The Patterns of History and What They Reveal about the Future*; Profile Books: London. 2010.

Neisser, H. P. Permanent "Technological Unemployment:" Demand for Commodities is Not Demand for Labor. *Am. Econ. Rev.* **1942,** *32* (1), 50–71.

Oser, J.; Brue, S. L. *The Evolution of Economic Thought*; Harcourt Brace Jovanovich: San Diego, 1988.

Perrin, J. *Teknoloji Transferi*; Arnas, T., Trans.; İletişim Yayınları: İstanbul (Original Work Published 1983), 1992.

Popkova, E. G.; Ragulina, Y. V.; Bogoviz, A. V., Eds. . *Industry 4.0: Industrial Revolution of the 21st Century*; Springer, 2019.

Rifkin, J. *The Third Industrial Revolution*; Sıral, P., Başekim, M., Trans.; İletisim Publication: İstanbul (Original Work Published 2011), 2019.

Schwab, K. . *The Fourth Industrial Revolution;* World Economic Forum, 2016.

Schwab, K.; Davis, N. *Shaping the Future of the Fourth Industrial Revolution*; Currency: New York, 2018.

Sevinc, A. A. İşletmelerde Stratejik Liderlik Ve Bir Uygulama. Unpublished Master Thesis, Inonu Universitysocial Sciences Institute, Malatya. 2009.

Sharma, A.; Singh, B. J. .Evolution of Industrial Revolutions: A Review. *Int. J. Innov. Technol. Explor. Eng.* 2020**,** *9* (11), 66–73.

Stearns, P. N. *The Industrial Revolution in World History*; Westview Press: Colorado, 2013.

Stock, T.; Seliger, G. Opportunities of Sustainable Manufacturing in Industry 4.0. *Procedia Cirp* **2016,** *40*, 536–541.

Troxler, P. *Making the 3rd Industrial Revolution. Fablabs: Of Machines, Makers and Inventors*; Transcript Publishers: Bielefeld, 2013.

Ugurluoğlu, O.; Celik, Y. Örgütlerde Stratejik Liderlik Ve Özellikleri. *Hacettepe Sağlık İdaresi Dergisi* **2009,** *12* (2), 121–156.

Van Der Westhuizen, T.; Goyayi, M. J. The Influence of Technology on Entrepreneurial Self-efficacy Development for Online Business Start-Up in Developing Nations. *Int. J. Entrep. Innov.* **2020,** *21* (3), 168–177.

Wang, S.; Wan, J.; Li, D.; Zhang, C. Implementing Smart Factory of Industrie 4.0: An Outlook. *Inter. J. Distrib. Sens. Netw.* **2016,** *12* (1), 3159805.

Zivdar, M.; Imanipour, N.; Talebi, K.; Hosseini, S. R. An Explorative Study of Inputs for Entrepreneurs' Decision-Making to Create New Venture in a High-Tech Context. *Int. J. Entrepreneurship Innov.* **2017,** *18* (4), 243–255.

CHAPTER 3

Role of Big Data in Agile Business Management: A Conceptual Framework

TAYYBA FATIMA and TAHIR MUMTAZ AWAN*

Department of Management Sciences, COMSATS University Islamabad, Islamabad, Pakistan

Corresponding author. E-mail: tahir_mumtaz@comsats.edu.pk

ABSTRACT

This study aims to establish a link between big data and agile management. Agility is a need of the present era to be competitive in the market and big data provide an opportunity for organization to move from traditional methods to agile management. In this chapter, based on literature, a relation between the different characteristics of big data and agile practices was established. The flexibility, timeliness, and quick decision-making which is needed for agile management are derived from big data. Businesses can create value through agile practices; securing this value from data. 5 Vs of big data enabled the flexibility in organization, authentic and quick information, and quick decisions of organizations to be competitive in an organization. The relation between data and agility is not directional, it can go in both ways, both agility and big data are interlinked. Big data help the organization to be agile and agility is needed to entrench data analytics at every level of the firm.

3.1 INTRODUCTION

Industrialization and globalization around the globe have not only brought up better opportunities to grow but also different challenges for organizations. A

Agile Leadership for Industry 4.0: An Indispensable Approach for the Digital Era, Tanusree Chakraborty, Tahir Mumtaz Awan, Ashok Natarajan, and Muhammad Kamran (Eds.)

volatile environment is one of these challenges which need to be addressed for better growth and sustainability (Altay and Ramirez, 2010). This changing world needs a different way of doing business, managing projects, and leadership. Agile management and leadership significance have been mentioned in literature especially in high-generation enterprises (Kurniawan et al., 2020). Agility is one of the most critical components for the success of a business enterprise by immediately responding to surroundings, stripping failed projects, and investing in a successful one. Agile companies are believed to be decreasing product value, multiplying market percentage, and meeting purchaser call for and in the long run company competitiveness (Lin et al., 2006). Agile control is taken into consideration as a winning strategy in modern technology of unpredictability and volatility. The time "agile production" was used in 1991 by way of the Iacocca Institute to describe the critical practices for production methods. It turned into first defined by Goldman et al. (1995) along with its four dimensions: delivering the price to the clients, being prepared for transformation, assessing human know-how and talents, and establishing digital partnerships. It is described as recognizing outside adjustments in the environment, identifying firm sources, reconfiguring resources, and methods to respond to outside exchange (Mathiyakalan et al., 2005). Agility as a system was described by Naylor et al. (1999) as exploiting marketplace opportunities with knowledge and collaboration. Change in the environment is needed to address effectively with the use of modern technology. Advancements in technology not only increased the challenges but also provided the opportunities and capabilities to cope with challenges. Agile practices help to tackle any change in the market from war to the pandemic and other such technological revolutions as well with the help of advanced technology, big data, and the internet of things. Moreover, progression in the field of management and technology still lacks conceptual relation between both ends. Insight from the huge amount of data for management practices need to be extracted. An alignment of the big data and company toward agility is also key to commercial success. New standards for start-ups need a connection between big data and agile practices.

In this chapter, the role of big data is proposed to be discussed for agile management. As Industry 4.0 requires the latest techniques and technologies to shape the market as an "early mover" (Jesse, 2019), strategists need to decide the way of agile management with the use of modern technologies and big data. Considering this important fact, the proposed chapter will cover an extensive review of the published work covering the role of big data in agile management and a conceptual framework will be proposed to understand the mechanism.

3.2 CONCEPT OF INDUSTRY 4.0: AN OVERVIEW

The fourth industrial revolt is known as I4.0. Simple kind of digitalization was also a part of the third industrial revolution but I4.0 is characterized by immense digitalization (Guzmán et al., 2020). Different factors can characterize the fourth industrial revolution: (1) *velocity,* the emergence of new technology, (2) *breadth and depth,* the combination of numbers of different technologies, (3) *systemic impact*, the transformation of the entire system (Hermann et al., 2015). I4.0 is a combined name for technologies and corporations for creating values. Digitalization and change have brought variations in manufacturing, design, and operation system (Schwab, 2019). Extensive digitalization has brought many challenges for an organization such as customers' expectations, shorter product lifestyles, and constant innovation (Kane et al., 2018). Organizations are confronted with technological integration, organizational transformation, decision-making issues (World Economic Forum, 2019). To cope with the challenges strong and skillful leadership is needed to guide employees for cooperation, communication, confidence, and agility (World Economic Forum, 2019). Agility and an agile environment should be supported through this leadership for better business performance and customer satisfaction (Bolte et al., 2018). Innovative approaches through data utilization need better management, collaboration, and communication. I4.0 requires efficient management. Along with the administrative task, the leader has to deal with change and manage it effectively. I4.0 is not possible without IT infrastructure; existing technologies are combined with new infrastructure and production challenges are met. Major components of I4.0 are IoT and CPS supporting systems (Xu et al., 2018). Other components include big data, robotics, cloud computing, straight/upright integration, and augmented (Kurniawan et al., 2020). All businesses are facing fast transformation and digitalization. It is I4.0 that has enabled all the systems and processes to collect data, transform much faster, and cost-effective manner. It has increased the efficiency of the business through an effective management system (Akkaya et al., 2019). In such a volatile and confronted environment, agile leadership and management are being focused.

3.3 AGILE MANAGEMENT IN AGILE BUSINESS ENVIRONMENT

Innovation and digitalization have led to change in doing and managing business including leadership and management (Akkaya et al., 2019). Volatility demand for dynamic leadership and management to be successful all around the globe in all kinds of businesses (Ozdiemir and Cetin, 2019). In 1991, the concept of agility was first introduced by the "Lighco Institute of Lehigh University." This report explains how to become a market leader again (Nagel and Dove, 1991). But this report lacks the explanation of agility in terms of organizational structure and project planning. Later on, agility and agile practices were used in software and computer technology (Akkaya, 2019). Agility is defined as the flexibility to adapt to change in a volatile environment. The elasticity of the whole supply chain drives dexterity. From information flow to recognition, penetration, and manufacturing process strong IT infrastructure helps a lot. A multilayered structure is needed to be agile in the market (Zhang and Sharifi, 2000). Two main characteristics of agile companies are fact and flexibility (Sekman and Utku, 2009). Quickness is associated with the speed of product completion and flexibility is associated with the capability to adjust with market shifts. It refers to working quickly and adapt changes in the environment. So, being agile refer to speed up and being adaptable at the same time (Sekman and Utku, 2009). Organizations to become agile need to adapt all human resources into agile organizational structure starting from the internal management structure. Processes should be designed to make processes easy, simple, and suitable (Denning and Hamel, 2018). The main characteristics of agility include personnel and interaction instead of processes and tools, working product instead of lengthy documentation only, customer collaboration, and responding to change.

Self-organization, simplicity, and decision-making are essential elements of agility behavior. The roots of agile management can be traced back to the software industry (Beck et al., 2001). Agile management is a way of providing a quick response in a volatile environment. In a business environment with changing customer needs and technology, requirements are met with agility. It is defined as an "organizational culture and collaboration method to achieve a higher level of adaptiveness" (Mergel et al., 2018). Agile governance principles are first adopted in the software industry and later in all organizational levels (Overby et al., 2006). This idea steadily moved to other services in the software industry. Scholars have developed different theories and frameworks (i.e., scaled agile framework [SAFe]) to understand agility governance (Merhout and Kovach, 2017). Networking theory is one of the bases of agility (Overby et al., 2006). Interactivity, flexibility, and stakeholder involvement are the key characteristic of agile methods. The productivity of the concept will be higher if all the developments and advancements are included at the process level (Merhout and Kovach, 2017). Companies need to create a balance between rapidness and effectiveness. In the present era, more focus is needed on innovation to adapt according to the current market condition (Rigby et al., 2020). Agile management is important from a customer and innovation point of view. Agile practices are in the routine process of organization, but academic definition and measurement still need more attention (Hummel, 2014). Agile practices need market information for the quick response and information technology has helped a lot in this regard (Wysocki, 2014).

3.4 ROLE OF BIG DATA IN BUSINESS

Data are a very valuable organizational asset as with other assets. It is deemed extremely crucial for the company. Companies need to understand the value of their data to gain full advantage of it. Big data is a large amount of data characterized by its five characteristics (Sagiroglu and Sinanc, 2013). The five characteristics of big data are (volume, veracity, velocity, and variety); however, value is considered the fifth characteristic of the key of big data (Chiang and Yang, 2018). These are called the five Vs of big data. Data without delivering value to the organization have no use. It is generated through a popular mediums such as user content, mobile, internet, and social media. Powerful computations and analytics are needed to find out the trends and information from the data to gain value (Grover et al., 2018). Scholars have discussed the term "big" and how big the term big is. McKinsey defined big data as "large pools of data that can be captured, communicated, aggregated, stored, and analyzed." It is true for historical static data, which is anything static. Business is gaining value from the dynamically changing data (Günther et al., 2017)

In the present era, big data trends and relationships are developed from the data. The volume of big data is a million times greater than normal data. "The speed at which new data is generated is referred to as the velocity of big data." The forms of data are different such as audio, video, graphical, and text data (Faroukhi et al., 2020). This different kind of data provides valuable insight from data through data analytics. Opportunities related to big data are innovation, personalization, customization of service, management, and better decision-making. Information and knowledge from the data increase the efficiency and effectiveness of data (Line et al., 2020). Data deliver the value in form of analytics, information about the different dynamics of business and its context. Data also provide countless opportunities, that is, innovation, customization, and service modernization (Qi, 2020). Data are very disruptive that it can change the whole business. In the business environment change is the only static thing, and business is revolving around it (Ghasemaghaei and Calic, 2020). Different factors such as social, political, environmental, and physical are always changing in the business context and have an impact on it. A business needs to have information on all these changing factors and be able to extract knowledge from them (Grover et al., 2018). In this scenario, decisions need to be taken immediately and accurately to respond to the unstable environment. Big data properties and characteristics need to be utilized in a more realistic manners for business development and decision making. Practical research on big data started in the 21st century (Vera-Baquero et al., 2015). Among other characteristics of big data, volume was first identified and understood that big data is simply of huge volume. Initially, three Vs of big data are specified (volume, velocity, and variety). Volume referred to the huge amount of data; velocity is associated with the speed at which data are increasing, and variety is the different types of data available (Laney, 2001). The other two characteristics of big data are veracity (Schroeck et al., 2012) and economic value of the data (Idc-Vesset et al., 2012). The 5 Vs model of big data is presented (Wamba et al., 2015). Data present a real picture of the company that how it is working and help to make future decisions. Recent technological developments have provided a vast variety and a large amount of data. Traditionally, the company has access to its internal data, but now with the use of social media, a large amount of user content is generated which helps a lot to understand the market situation and competitors (Jha and Subramanian, 2016). Data play important role in business operations. It is not difficult to get data in recent times and make use of it for business and its operations are actual tasks. Getting insights from data by applying different techniques, such as data mining, statistics, machine learning, and pattern recognition, etc., are used to extract value from data. It is very important to get the right data, at right time in the right format to get accurate information from it (Faroukhi et al., 2020). These characteristics of big data make it a valuable and effective use of it can increase business efficiency. In the present volatile environment companies need to make quick decisions; for making rapid decisions right information at right time is necessary (Faroukhi et al., 2020). To be quick and responsive in the present era of impulsiveness, agile management practices are being followed. The relation between agile management and big data is needed to be developed.

3.5 DIGITAL BUSINESS SOLUTIONS FOR SUSTAINABILITY: ROLE OF BIG DATA

In a dynamic and volatile business environment, the need for high-quality decision-making has to sustain in the market. Large scale decision making under ambiguous environment draw strategists and manufacturers attention toward data-driven strategies and decision (Jabbour et al., 2020). External factors, that is, government agencies, trade departments act as critical factors in the digital economy (Dubey et al., 2019). Institutional pressure requires a firm to operate within certain social boundaries and countries have developed specific data-driven strategies to act according to that and measure performance on data-driven methods in this digital era. According to the resource-based view of the organization, the focus should be on developing complex resources that are difficult or would be difficult for competitors to copy. Workforce skills are considered to be the most difficult skills as they are not visible and are achieved through learning and practice (Hart, 1995). In the digital era to achieve stainability digital skills are the ones to be learned. Data analytics and artificial intelligence skills are considered tacit resources. A source-full and skilled person having strong and in-depth knowledge can create value for business through data to sustain (Bag et al., 2021). The big tech industry is recognized as providing common good as the UN puts that data foster, competitiveness, sustainability, innovation, creativity, jobs, and societal good as well (Lucivero, 2020). Big data has changed the long-standing ideas of decision making, strategies formulation, and value creation. Digital infrastructure is a part of numerous fields, that is, business, health, transportation, and finance present the ideas for sustainability. In the 21st century, any actor of the system cannot be visualized independently, they need to be studied in collaboration to create knowledge and innovation. The cross-disciplinary nature of information system drive forces to give meaning to big data and improve relations among data and business model. Contemporary business models are needed for a sustainable business environment (Pappas et al., 2018). New digital business models will not be accurate only but also foster the societal benefit at large affecting the companies, organizations, consumers, and public at large (Porter and

Kramer 2019). Big data plays a significant role in the transformation and combining multiple sources and sharing them variously with various stakeholders and analyzing them in different ways to allow the achievement in digital transformation to create a sustainable market.

3.6 BIG DATA: THE FACE OF INDUSTRY 4.0 AND AGILE MANAGEMENT

The present era of instability needs quicker and effective methods, strategies, and approaches to deal with the challenges of instability. Agility is a solution to unpredictability but to apply agile management timely information and future predictions are really important (Günther et al., 2017). Technological advancements and big data analytics are opportunities for agile practices. The different characteristics of big data play a vital role in the implementation of agile management and, on the other hand, agile management supports the adoption of big data (Bainbridge, 2019). The software industry is the pioneer in the adoption of agile practices while other industries can implement the agile practices into their organizations for better performance. Industry 4.0 is the massive era of digitalization that produces data. It is the most critical resource of the current organization to deal with for value creation (Şen and İrge, 2020). Big data and cloud computing are used in relative processes of the firm to increase productivity, performance, and flexibility. System or organizations need quicker and authentic information to adopt agility and big data provides a whole database to increase productivity analyzing the patterns in data (Şen and İrge, 2020). System integration is vital not only at the horizontal level (within the organization) but also at the vertical level of the organization (from supplier to customer). Data in the organization are coming from all levels from suppliers to customers. Big data can not only stimulate the products but also industrial manufacturing processing as well (Pamuk and Soysal, 2018). Data provide the information to all individuals inside the organizations to make their own decisions based on information available to create value (Faroukhi et al., 2020). This is decentralization which requires agile management and it comes with the data (Jesse, 2019). To incorporate the data into the organizational system, it is needed to work for upthrust of organizational culture toward agility for better management.

The second thing that is the most crucial part of agile management is time; as urgency needs quicker decisions for which quicker information is available through big data. The velocity characteristic of big data enables the implementation of agile management (Şen and İrge, 2020). Visions derived from diverse sources can help to make balanced decisions and support all, that is, customers to make purchases and organization about production. Agility is a cultural trait of an organization that enables the organizations and its individuals to handle the uncertainty and decision-making based on technologies and data (Gunasekaran et al., 2018). Data are an organizational source that enriches its employees. Big data provides an opportunity to work remotely 24/7 which requires new skills and techniques. An agile organization needs to be learned constantly as learning is the party of agility. "Learnability" makes an organization agile through big data as agile values are continuously upgraded in a learning organization. The relation between big data and agile management can be illustrated in Figure 3.1.

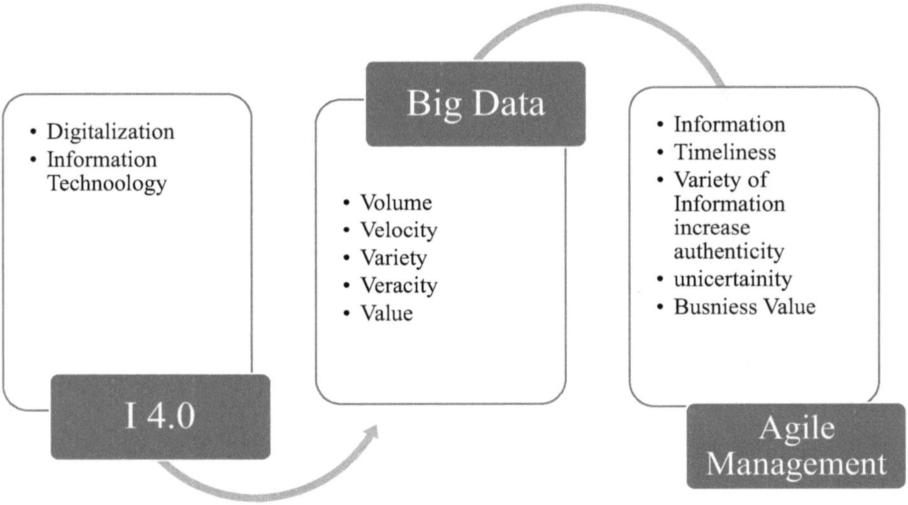

FIGURE 3.1 Big data and agile management in industry 4.0 (*Source: Author Compiled*).

Literature and previous studies proved that big data and agile management practices are interlinked. Few studies have discussed that agile is the mechanism to get value from big data (Gunasekaran et al., 2018). The present chapter based on literature attempts to develop a relation between big data and agile management. Figure 3.2 presents the relation between the characteristics of big data and agile practices.

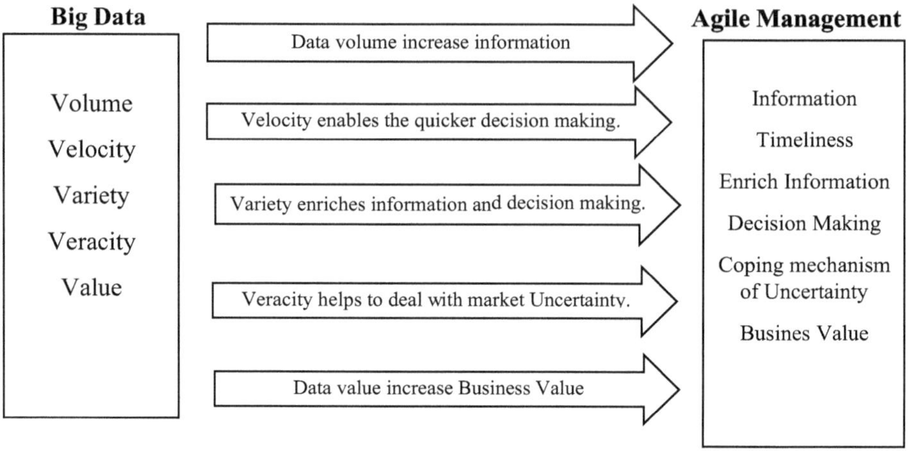

FIGURE 3.2 Agile management through big data (*Source: Author Compiled*).

This figure explains the relation between the different characteristics of big data and agile management characteristics. The practices followed for agility mechanisms are enforced by big data analytics and I4.0 has enabled it. This figure explains that huge data volume comprises a lot of information which is important to deal with uncertain markets situation. Agile management first needs information and a variety of information to implement agility. Big data is generated at a very high pace; which increases the timeliness of decision-making in the agile process. Erich and decentralized decision making are the core of agile practices all the characteristics, that is, information from data, its relevance enabled the organizational employees to make them quicker and authentic decisions based on facts available. As veracity characteristic of data ensures its authenticity which is important in the highly volatile environment to be competitive and sustainable. In the business, value is generated in form of innovation, customization, dealing with challenges with the help of value generated from data.

3.7 VALUE CREATION THROUGH BIG DATA IN INDUSTRY 4.0

The value of big data comes from its characteristics, data are considered a key agent of the third industrial revolution and are now considered a promising area for value creation in I4.0. Despite different challenges, big data provides a boom to science and increase organizational performance (Wamba et al., 2015). IT and technology have assured the accumulation of big data that needs to be analyzed and gain actionable insight to increase business value. Big data bring benefits to the organization, reduce communication cost, improve customer relations, and foster marketing activities on-monetary outcomes (Hallikainen et al., 2019). The connection between value creation and big data is discussed in the literature. It is proved that big data invites value creation opportunities, that is, business strategies, digital business value, informed decision making, strategic marketing, supply chain planning, increasing market share, and improving performance (Wamba et al., 2017). Marketing activities in big data paradigms create more value in the online and offline marketplace. According to Gregor's model of value creation, big data and information technology lead to different kinds of value, that is, transformational value, informational value, transactional value, and strategic value. Big data value creation directions are marketing planning, decision making, cost-saving, skill development, learning, knowledge sharing, IT infrastructure, and organizational benefits (Elia et al., 2020).

3.8 CONCLUSION

This study aims to create a connection between big data and agile management as the uncertainty of the environment needs more innovative and fast pace decisions to be made. Rapid digitalization has challenged the status quo of the organization. Even though technology has provided different opportunities, knowledge, and potential; companies need to work to create and capture value. Big data is one of the matchless opportunities that can deliver enormous value to customers and businesses as well. Big data also needs proper management, flexibility, and organizational restructuring to adapt big data at every level of the organization. At the same time initially, it was considered that agile is a way to embed big data at every level of any firm. But later on, data-based decision-making is practiced. Agility is experienced when employees are empowered to make decisions and have fast and authentic information. The flexibility of organization through agile management and knowledge gained through data analytics enabled a firm to make better and timely decisions. The uncertainty of the environment is undertaken through knowledge and information gained through data. This provided a framework relation between the different characteristics of big data and agility practices. The relation between data and agility is not directional, it can go in both ways, both agility and big data are interlinked. Big data helps the organization to be agile and agility is needed to entrench data analytics at every firm.

KEYWORDS

- **big data**
- **agile management**
- **Industry 4.0**
- **digitalization**
- **data analytics**

REFERENCES

Akkaya, B. Leadership 5.0 in Industry 4.0: LEADERSHIP in Perspective of Organizational Agility. In *Managing Operations Throughout Global Supply Chains*; IGI Global, 2019; pp. 136–158.

Akkaya, B.; Kayalidere, U. A. K.; Tabak, A. Endüstriyel Alanda Üretim Yapan Firmaların Örgütsel Çevikliği ile Firma Yöneticilerinin Sahip Olduğu Dinamik Yetenekler Arasındaki İlişki: Manisa Organize Sanayi Bölgesinde (MOSB) Faaliyet Gösteren Firmalar Üzerine Bir Araştırma. In *Yeni Nesil Girişimcilik ve Ekonomi*; EKİN Basım Yayın Dağıtım, 2019, pp. 19–54.

Altay, N.; Ramirez, A. Impact of Disasters on Firms in Different Sectors: Implications for Supply Chains. *J. Suppl. Chain Manag.* **2010,** *46* (4), 59–80.

Bainbridge, M. Big Data Challenges for Clinical and Precision Medicine. In *Big Data, Big Challenges: A Healthcare Perspective*; Springer, 2019; pp. 17–31.

Beck, K.; Beedle, M.; Van Bennekum, A.; Cockburn, A.; Cunningham, W.; Fowler, M.; .... Jeffries, R. The agile manifesto. In: Feb, 2001.

Bolte, S.; Dehmer, J.; Niemann, J. Digital Leadership 4.0. In *Acta Technica Napocensis-Series: Applied Mathematics, Mechanics, and Engineering,* **2018,** *61* (4).

Bag, S.; Pretorius, J. H. C.; Gupta, S.; Dwivedi, Y. K. Role of Institutional Pressures and Resources in the Adoption of Big Data Analytics Powered Artificial Intelligence, Sustainable Manufacturing Practices and Circular Economy Capabilities. *Technol. Forecast. Soc. Change* **2021,** *163*, 120420.

Chiang, L.-L. L.; Yang, C.-S. Does Country-of-Origin Brand Personality Generate Retail Customer Lifetime Value? A Big Data Analytics Approach. *Technol. Forecast. Soc. Change* **2018,** *130*, 177–187.

Denning, S.; Hamel, G. *The Age of Agile*; Amacom, 2018.

Dubey, R.; Gunasekaran, A.; Childe, S. J.; Fosso Wamba, S.; Roubaud, D.; Foropon, C. Empirical Investigation of Data Analytics Capability and Organizational Flexibility as Complements to Supply Chain Resilience. *Int. J. Prod. Res.* **2021,** *59*, 110–128.

Faroukhi, A. Z.; El Alaoui, I.; Gahi, Y.; Amine, A. Big Data Monetization Throughout Big Data Value Chain: A Comprehensive Review. *J. Big Data* **2020,** *7* (1), 1–22.

Ghasemaghaei, M.; Calic, G. Assessing the Impact of Big Data on Firm Innovation Performance: Big Data is Not Always Better Data. *J. Bus. Res.* **2020,** *108*, 147–162.

Goldman, S. L.; Nagel, R. N.; Preiss, K. *Agile Competitors and Virtual Organizations: Strategies for Enriching the Customer,* Vol. 8; Van Nostrand Reinhold: New York, 1995.

Grover, V.; Chiang, R. H.; Liang, T.-P.; Zhang, D. Creating Strategic Business Value from Big Data Analytics: A Research Framework. *J. Manag. Inform. Syst.* **2018,** *35* (2), 388–423.

Gunasekaran, A.; Yusuf, Y. Y.; Adeleye, E. O.; Papadopoulos, T. Agile Manufacturing Practices: The Role of Big Data and Business Analytics with Multiple Case Studies. *Int. J. Product. Res.* **2018,** *56* (1–2), 385–397.

Günther, W. A.; Mehrizi, M. H. R.; Huysman, M.; Feldberg, F. Debating Big Data: A Literature Review on Realizing Value from Big Data. *J. Strat. Inform. Syst.* **2017,** *26* (3), 191–209.

Guzmán, V. E.; Muschard, B.; Gerolamo, M.; Kohl, H.; Rozenfeld, H. Characteristics and Skills of Leadership in the Context of Industry 4.0. *Procedia Manuf.* **2020,** *43*, 543–550.

Hermann, M.; Pentek, T.; Otto, B. In *Design Principles for Industrie 4.0 Scenarios: A Literature Review (Working Paper 01). Dortmund: Technische Universität Dortmund. Режим доступа: www.snom.mb.tu-dortmund. dehttp://www.snom.mb.tu-dortmund.de/cms/de/forschung/Arbeitsberichte/Design-Principles-for-Industrie-4_0-Scenarios. pdf, дата обращения, 15*, 2016.

Hart, S. L. A Natural-Resource-based View of the Firm. *Acad. Manag Rev.* **1995**, *20* (4), 986–1014.

Hummel, M. (2014). In *State-of-the-Art: A Systematic Literature Review on Agile Information Systems Development.* Paper Presented at the 2014 47th Hawaii International Conference on System Sciences.

Idc-Vesset, D.; Woo, B.; Morris, H.; Villars, R.; Little, G.; Bozman, J.; .... Conway, S. Market Analysis-Worldwide Big Data Technology and Services 2012–2015 Forecast. *IDC Analyze Future* **2012,** *1*, 1–34.

Jesse, N. Agility eats Legacy—the Long Good-Bye. *IFAC-PapersOnLine* **2019,** *52* (25), 154–158.

Jha, Y.; Subramanian, R. Rhizobacteria Enhance Oil Content and Physiological Status of Hyptis Suaveolens Under Salinity Stress. *Rhizosphere* **2016,** *1*, 33–35.

Jabbour, C. J. C.; Fiorini, P. D. C.; Wong, C. W.; Jugend, D.; Jabbour, A. B. L. D. S.; Seles, B. M. R. P.; ... da Silva, H. M. R. First-Mover Firms in the Transition Towards the Sharing Economy in Metallic Natural Resource-Intensive Industries: Implications for the Circular Economy and Emerging Industry 4.0 Technologies. *Resour. Pol.* **2020,** *66*, 101596.

Kane, G. C.; Palmer, D.; Phillips, A.-N.; Kiron, D.; Buckley, N. Coming of Age Digitally. In *MIT Sloan Management Review and Deloitte Insights*; Vol. 5, 2018.

Kurniawan, R.; Budiastuti, D.; Hamsal, M.; Kosasih, W. The Impact of Balanced Agile Project Management on Firm Performance: The Mediating Role of Market Orientation and Strategic Agility. *Rev. Int. Bus. Strategy*, **2020**.

Lin, C.-T.; Chiu, H.; Tseng, Y.-H. Agility Evaluation Using Fuzzy Logic. *Int. J. Prod. Econ.* **2006,** *101* (2), 353–368.

Laney, D. Big 3D Data Management: Controlling Data Volume, Velocity and Variety. *META Group Res. Note* **2001,** *6* (71), 1.

Line, N. D.; Dogru, T.; El-Manstrly, D.; Buoye, A.; Malthouse, E.; Kandampully, J. Control, Use and Ownership of Big Data: A Reciprocal View of Customer Big Data Value in the Hospitality and Tourism Industry. *Tour. Manag.* **2020,** *80*, 104106.

Mathiyakalan, S.; Ashrafi, N.; Zhang, W.; Waage, F.; Kuilboer, J.-P.; Heimann, D. In *Defining Business Agility: An Exploratory Study*, Paper presented at the Proceedings of the 16th Information Resources Management Conference, San Diego, CA, 2005.

Mergel, I.; Gong, Y.; Bertot, J. *Agile Government: Systematic Literature Review and Future Research*; Elsevier, 2018.

Merhout, J.; Kovach, M. In *Governance Practices Over Agile Systems Development Projects: A Research Agenda*, Paper Presented at the Proceedings of the. 12th Midwest Assoc. Inf. Syst. Conf, 2017.

Nagel, R. N.; Dove, R. In *21st Century Manufacturing Enterprise Strategy: An Industry-Led View*; Diane Publishing, 1991.

Naylor, J. B.; Naim, M. M.; Berry, D. Leagility: Integrating the Lean and Agile Manufacturing Paradigms in the Total Supply Chain. *Int. J. Prod. Econ.* **1999,** *62* (1–2), 107–118.

Pappas, I. O.; Mikalef, P.; Giannakos, M. N.; Krogstie, J.; Lekakos, G. Big Data and Business Analytics Ecosystems: Paving the Way Towards Digital Transformation and Sustainable Societies. *Inform. Syst. e-Business Manag.* **2018,** *16* (3), 479–491.

Porter, M. E.; Kramer, M. R. Creating Shared Value. In *Managing Sustainable Business*; Lenssen, G., Smith, N., Eds.; Vol. 89; Springer: Dordrecht, 2019; pp 327–350.

Overby, E.; Bharadwaj, A.; Sambamurthy, V. Enterprise Agility and the Enabling Role of Information Technology. *Eur. J. Inform. Syst.* **2006,** *15* (2), 120–131.

Pamuk, N. S.; Soysal, M. Yeni Sanayi Devrimi Endüstri 4.0 Üzerine bir Inceleme. *Verimlilik Dergisi* 2018, (1), 41–66.

Qi, C. C. Big Data Management in the Mining Industry. *Int. J. Miner. Metall. Mater.* **2020,** *27* (2), 131–139.

Rigby, D. K.; Elk, S.; Berez, S. Develop agility that outlasts the pandemic. In *Harvard Business Review,* 2020.

Sagiroglu, S.; Sinanc, D. In *Big Data: A Review.* Paper Presented at the 2013 International Conference on Collaboration Technologies and Systems (CTS), 2013.

Schroeck, M.; Shockley, R.; Smart, J.; Romero-Morales, D.; Tufano, P. Analytics: The Real-World use of Big Data. *IBM Global Bus. Serv.* **2012,** *12* (2012), 1–20.

Schwab, K. *A Quarta Revolução Industrial*; Edipro, 2019.

Şen, E.; İrge, N. T. Industry 4.0 and Agile Firms. In *Agile Business Leadership Methods for Industry 4.0*; Emerald Publishing Limited, 2020.

Sekman, M.; Utku, A. *Çevik şirketler: Kurumsal ataletiYenmek (6. Baskı)*. İstanbul: Alfa yayınları, 2009.

Vera-Baquero, A.; Colomo-Palacios, R.; Molloy, O. Measuring and Querying Process Performance in Supply Chains: An Approach for Mining Big-Data Cloud Storages. *Procedia Comp. Sci.* **2015,** *64*, 1026–1034.

Wamba, S. F.; Akter, S.; Edwards, A.; Chopin, G.; Gnanzou, D. How 'Big Data' can Make Big Impact: Findings from a Systematic Review and a Longitudinal Case Study. *Int. J. Prod. Econ.* **2015,** *165*, 234–246.

Wamba, S. F.; Gunasekaran, A.; Akter, S.; Ren, S. J. F.; Dubey, R.; Childe, S. J. Big Data Analytics and Firm Performance: Effects of Dynamic Capabilities. *J. Bus. Res.* **2017,** *70*, 356–365.

World Economic Forum [Online] 2019. https://www.weforum.org/

Wysocki, R. *Effective Complex Project Management: An Adaptive Agile Framework for Delivering Business Value*: J. Ross Publishing, 2014.

Xu, L. D.; Xu, E. L.; Li, L. Industry 4.0: State of the Art and Future Trends. *Int. J. Prod. Res.* **2018,** *56* (8), 2941–2962.

Zhang, Z.; Sharifi, H. A Methodology for Achieving Agility in Manufacturing Organisations. *Int. J. Oper. Prod. Manag.* **2000,** *20* (4), 496–513.

CHAPTER 4

HR 4.0 in Industry 4.0: Dynamics of Human Resource Management in Agile Organizations

RAMANDEEP SAINI[1*] and HAMAYUN KHAN[2]

[1]*Director Principal, Chandigarh Business School of Administration, Landran, India*

[2]*Finance Specialist, The Siena School, Silver Spring, MD, USA*

**Corresponding author. E-mail: ramandeepsaini2000@gmail.com*

ABSTRACT

The inevitable acceptance and the fast-track developments in technology, computers, and the internet, in particular, have revolutionized human life to a great extent. Industry 4.0 or the Factory of the Future is the fourth and the final phase in industrial revolution. It largely focuses on manufacturing by bringing together the manufacturing and digital technologies through the utilization of automation and data analytics. This has further opened doors for several other high-tech folds such as artificial intelligence, automation, data analytics, mass customization, etc., resulting into a reduction of human intervention. The amalgamation of physical, digital, and biological worlds is leading to significant changes in the organizations and HR professionals are struggling to cope up with the same. Traditional HR, hence, has a new version HR 4.0 which is revolutionary and has made HR more automated with activities focused on strategic issues and no longer on manual, bureaucratic, and repetitive actions. The chapter explores the dynamics of HR 4.0 in Industry 4.0 while examining the reasons for redefining the HR practices

Agile Leadership for Industry 4.0: An Indispensable Approach for the Digital Era, Tanusree Chakraborty, Tahir Mumtaz Awan, Ashok Natarajan, and Muhammad Kamran (Eds.)

in the beginning. Further, the chapter dwells on the trends in people management before 4.0 where it gained significance and multiple titles such as "personnel relations" which gradually got changed to "industrial relations" then "employee relations" and finally to "human resources." This is followed by a discussion of current trends in HR 4.0 which have conventionalized digitalized tools in people management like people involvement, artificial intelligence, employee training, and cyber security. The chapter culminates with insights into the challenges faced by HR professionals in the implementation of HR 4.0

Highlights of the Chapter:

- Introduction of Industry 4.0 and HR 4.0
- Reasons for HR 4.0
- Trends in HRM before Industry 4.0
- Trends in HR 4.0
- Challenges in Implementation of HR 4.0

4.1 INTRODUCTION

The inevitable acceptance and the fast-track developments in technology, computers, and the internet, in particular, have revolutionized the human life to a great extent. Industry 4.0 or the Factory of the Future is the fourth and the final phase in the industrial revolution. The vision of Industry 4.0 has originated from a vision to make the manufacturing industry in Germany more competitive (Industrie 4.0) and global. It largely focuses on manufacturing by bringing together the manufacturing and digital technologies through the utilization of automation and data analytics. This has further opened doors for several other high-tech folds such as artificial intelligence (AI), automation, data analytics, mass customization, etc., resulting in reduction of human intervention. In nutshell, Industry 4.0 is dependent on the already existing data and other data originating from ongoing processes in order to achieve increased efficiencies, redesign current manufacturing processes, develop new information channels along with the supply chain, and hence leading to new services and business models. Mankind has evolved with these industrial revolutions since the 18th century.

The 18th century witnessed the First Industrial Revolution with the extensive usage of steam power and mechanization of production processes.

The mechanization led to bigger outputs at the same time as compared to what was achieved through simple spinning wheels. Further, another breakthrough was the usage of steam power in industries aimed at increasing the productivity of labor. The same tenure saw major changes where humans and commodities could travel long distances in lesser time with the development of steamship, steam-powered locomotive. The discovery of electricity and production in assembly line were the highlights of the Second Industrial Revolution in the 19th century. Henry Ford (1863–1947) proved to be a change agent undertaking and achieving the mass production which took place in a slaughterhouse located in Chicago. Conveyor belts had pigs hung on them and every butcher was made to perform only the task of butchering the animal. Convinced with the benefits, Ford implemented the methodology in automobile production after introducing changes. The shop floor got transformed as the conveyor belt now moved across the workstations and the automobile was assembled in partial steps at every workstation many times faster at a lower cost. The 20th century had the Third Industrial Revolution which began in the 1970s. This phase was characterized by partial automation which became possible with the usage of programmable controls and efficient computers. These technologies since then have made the automatization of manufacturing process possible without human assistance. Robots are an excellent example of this advancement as they perform programmed sequences without human intervention.

The current scenario is the Fourth Industrial Revolution which is also known as Industry 4.0. This era has a key feature where the industrial process is driven by information and communication technologies. It is an extension and improvement over the achievements of the Third Industrial Revolution. The Fourth Industrial Revolution was first mentioned at the Hannover Fair in 2011. The Fourth Industrial Revolution propostal was submitted to the German Federal Government in October 2012 by a team from The Bosch group. The German government started working in this direction in 2013 with an aim to prepare its own Industry 4.0 transformation roadmap. A network connection is being used to expand production systems working with computer technology so as to have their digital twin on the Internet. This advancement leads to "cyber-physical production systems" and smart factories where the basis is the networking of production systems, components , and people. The digitalization of the manufacturing environment makes it possible for machines to not only predict errors but also initiate maintenance process autonomously. In short, Industry 4.0 is a revolution that will change the way people work and result in increased work efficiency. Innovative technologies like Big Data, AI, and Internet of Things are pushing atomization of all organizational processes including HR processes leading to more efficient and lean teams.

Research undertaken by the European Union 3 has concluded that Industry 4.0 will impact the roles of employees in terms of health, working time, demographics, flexibility, and private life (Smit, 2016). Hence advancements in technology are undoubtedly transforming traditional HR into HR 4.0.

4.2 REASONS FOR HR 4.0

It has been acknowledged that the Industrial 4.0 transformation of the industry will have a drastic impact on human resources (HRs) management practices due to its business environment. All organizations irrespective of big or small will have to adopt new processes and technologies which will completely revamp the way operations are carried out. Not only new technologies are affecting every aspect of the organization, but the structure of jobs is also changing. In order to remain relevant during these unprecedented times, organizations need to modify their HR practices hence leading to a new scenario of HR 4.0.

It is foreseen that technology-integrated business processes will lead to loss of job for unskilled employees while the competent skilled employees will be required and hence need to be retained. Business landscape is changing and so is the HR domain. There will be undoubtedly unparalleled opportunities for HR but the same will bring along numerous challenges and deviations. Internally, HR will need to incorporate changes in the culture and the working environment by building up a strong vision and sense of belongingness. Externally, adjustment to the new business environment and regulatory changes will be a challenge.

4.3 TRENDS IN HRM BEFORE INDUSTRY 4.0

In the past, HR practices were more likely the earliest which developed and evolved the people management function while preceded other significant functions (i.e., finance, operation, and marketing) (Stone, 2014). The industrial revolution started back in the 18th century in Europe and gradually spread to the United States and the remaining countries of the world. This revolution caused a change in the way through which the individuals changed their techniques of earning a living as well as led to a manufacturing society. The introduction of machines and the factory system replaced human skills and craftwork (Dulebohn, 1995). At the same time, many organizations started following patronizing practices but the behavior of many employers with their employees was still not up to the mark, which eventually resulted in labor to join groups and societies that later became known as labor unions (Scarpello, 2008). With the advent in manufacturing practices and the increased growth of unions among labors and employees, HR professionals looked for ways to avert the aftermath of unionization and increase efficiency and productivity of the operations of organizations (Stone, 2014). Eventually, this resulted in the development of key HR concepts that are explained as follows:

4.3.1 SCIENTIFIC MANAGEMENT APPROACH

Scientific approach was developed and incubated by Taylor (1947) who worked in the direction of rationalization of organizational work by scientifically studying and analyzing the job as well as classifying it into different constituents, and ultimately choosing the best alternative to effectively pull off the task. The scientific management approach, however, eliminated employees' autonomy and focused on the close supervision of the employees mainly to ensure that they do the task as expected (Stone, 2014). Nonetheless, Max Weber (1927) proposed the use of legitimate framework and rules for organizational efficiency and productivity. The newly designed jobs and the resultant autocratic style of management produced and increased the levels of disagreements between the employees and the employers. In the late 1930s laws like the National Labor Relations Act, the Norris-LaGuardia Act (1932), and the Wagner Act (1935) facilitated the growth of worker unions that led to an increase in the number of unions and the use of scientific management principles. In the same decade, managers believed that the disagreements between employees and employers were not deep-rooted in labor relations, but mainly resulted due to the poor management system. Following that, many researchers carried out a series of experiments to analyze the impact of difference in corporate environment on employee's productivity at work (Roethlisberger, 1939). Finally, they also found that the social factors and employees' needs significantly influenced the employees' output and their wellbeing (Dulebohn, 1995).

4.3.2 HUMAN RELATION APPROACH

The human relation approach (HRA) was introduced as a result of conducting a series of experiments. This approach majorly focused on the needs and well-being of workers. HRA widened the HR practices and then expanded it to the social structure of organizations (Scarpello, 2008). In the 1950s, the assumption that employees were not willing to work was rejected by the human relations movement. Finally, the term "HR management" replaced "personnel management," which also considered HRs as assets for organizations. Thereafter in the era of the 1960s and 1970s, the HRA was replaced by the quality of work life (QWL). QWL focused on prioritizing the benefits of workers while aligning them with the mission of organizations along with focusing on both the efficiency and well-being of employees. For example, HR team worked and developed new policies that focused not only on job enrichment but also on career development (Stone, 2014).

4.3.3 UNION LABORS

Unionized establishments are believed to benefit safety and health inspections, experience greater scrutiny in account of those inspections and compensate higher penalties for those who violate health and safety regulations as compared to nonunion organizations (Weil, 1991). Firms that provide recognition to unions are more likely to miss estimates than those which do not accept unions (Bova, 2013).

4.3.4 COMPLIANCE TO REGULATION

Despite being one of the significant HR practices, compliance to regulations remains cumbersome for the HR professionals to manage the employees, as there are numerous regulations and laws that need to be followed by the employees. HR professionals should first understand and then impose those laws on the employees to help avoid any pay fines or penalties (David and Bright, 2019). The HR function is at the core of many professionals' attempts to identify, hire, and retain the people in an organization mainly to execute its policies and achieve the set-out goals. Not only this, the HR function should be in accordance to the specific laws and can help create policies and procedures in relation to these laws (Carnovali, 2019).

4.3.5 *PERFORMANCE EVALUATION*

Until the late 1980s during the Industry 3.0, the focus has been on the performance evaluation literature based on the performance measurement and the psychometric characteristics of the various techniques used to evaluate the performance of employees (Landy and Farr, 1980). Research was conducted with an objective of improving techniques like constructing rating scales and also evaluating the advantages and disadvantages of these rating scales. This psychometric tradition witnessed drastic changes specifically in the early 1980s, as the cognitive features of the raters became the focus of research (Feldman, 1981). Finally, during the outset of the 1990s (Murphy and Cleveland, 1991) and the 2000s, research concluded that the difference between promotion and demotion based on performance evaluation could help in identifying the conditions in which the given feedback would be effective demanding further for the tasks that need to be performed to satisfy one's objectives and goals of promotion (Higgins, 1998).

4.3.6 *TRAINING AND DEVELOPMENT*

The concept of training and development was introduced during its need for use in World War II. At that point in time, the demand for training was increasingly built mainly for the expansion of the wartime budget and the technological innovations. The emergence of the US labor movement at that time increased the need for training and development (Glass, 2013). However, the prominence and influence of labor unions have significantly shifted to organized people management which has continuously contributed to the extension of the access to training and education, thought as a broadening horizon for the skill set of employees (Swanson and Torraco, 1995). The instructional systems development model was one of the first and widely followed for organizing training (Campbell, 1984). The ISD model was modified to some extent during the 1980s and 1990s, (Dick and Carey, 1990) and significant applications of the model were added. The ubiquitous application of the structured practices to training and education was combined with the demands for their modifications including important suggestions to reduce the time-consuming and pretraining analysis stage (Rossette, 2009).

4.4 TRENDS IN HR 4.0

Since the start of the revolution of Industry 4.0, the HR function has witnessed a massive shift mainly through the use of digitalized tools in HR practices. The combination of manufacturing and digitalized technologies has remolded the human efforts in HR function of organizations into a digitalized sphere. While the adaptation to the newly introduced digitalized tools has led to HR professionals face difficulties like downturn and competition but there are several trends that have eased the HR function to a great extent (Nageshwari, 2020). The following text elaborates the key trends of HR 4.0.

4.4.1 E-LEARNING

E-learning is an important and newly accepted factor in HR 4.0 that focuses on the provision of the necessary training to the employees while using digital learning tools such as Tennessee Electronic Library, computer-based training, web-based training, etc. The significance of E-learning in HR 4.0 lies in its approach toward individuals across the world that cannot easily find an access to education in the schools (Nageshwari, 2020). Moreover, E-learning facilitates innovative training and learning practices for the employees that enable them to efficiently perform the assigned tasks (Chen, 2011).

4.4.2 E-RECRUITMENT

In HR 4.0, hiring is based on a variety of skills and diverse knowledge and experience that needs to be evaluated in the screening process before the final selection of a candidate (Kaur, 2019). Many organizations tend to pay attention to the usage of digital-based recruitment and selection procedures mostly during selection of the fit candidates for different roles (Prieto and Perez-Santana, 2014). The selection criteria of candidates can be analyzed and measured by employing digitalized tools, for instance, openness to experience of a candidate can be evaluated through psychometric testing mainly used in the selection process (Costa, 1992). In addition, HR employers in organizations need to consider identifying the key attributes that are necessary for the innovative behavior of employees during the hiring process. For example, the innovativeness of employees is characterized by traits such as active imagination, intellectual curiosity, inner feeling, variety of preferences, creativity, attentiveness, and flexible thinking (Barrick, 1991).

4.4.3 BAMBOO HR

Horowitz and Sevilla (2019) define Bamboo HR as a key HR 4.0 tool which is basically used to bring together a large set of functions and practices contained in an easy package. BambooHR is a web-based HR software system developed for small- and medium-sized companies. The software presents an easy-to-use platform that can be used by the organizations to transport data from spreadsheets to a more efficient HR information system that can cater to their basic needs (CompareCamp, 2020). Nageshwari (2020) considers BambooHR as one of the most vital HR programming in the contemporary corporate climate.

4.4.4 HR ANALYTICS

The introduction of analytics creates a room for HR employers in order to gain the required spotlight on how to make prompt and effective decisions (Spadafora, 2019). In other words, HR analytics is a data-driven approach used to identify and analyze issues mainly related to people. It also helps employers combine and then analyze the collected data to find new and useful insights while providing the necessary information for better decision-making process (Sympa, 2019).

4.4.5 ARTIFICIAL INTELLIGENCE

Over the last almost 10 years, AI has been a ubiquitous tool used in almost every domain, this is mainly driven by the magnitude in the volume of innovative applications and trends taking place in various functions including HR (Bench-Capon, 2007). A study commissioned by Entelo (2018) indicates that 70% of the employers admitted the effectiveness of automated sourcing in productivity to increase. In addition, AI helps transform conventional HR practices ranging from recruitment to the selection of employees and their experience. Further, AI has proven to be revolutionary into the functions covering HR trends (Taylor, 2021).

4.4.6 AUTOMATION AS A NEW HELPING HAND

The adaptation of automation in daily HR practices has lessened the burden on HR professionals in terms of saving and then allotting more time to people management and direct connection between the HR employers and the employees (Straz, 2015). In a quite number of instances, the inevitable use and rise of automation can be analyzed by a study commissioned by Conspiracy (2015). It shows that 53% of companies with more than 1000 employees are willing to use HR software for people management. Organizations also strive to be more prudent in talent acquisition and that is why these organizations focus on hiring the employees that are well-off with the upgraded technological tools (educba, 2020).

4.4.7 PRIORITIZING SOFT SKILLS

With the introduction and entry of digitalized tools particularly AI in HR4.0, there is an increased demand for the employees to sharpen their soft skills mainly to cope up with the fast-paced corporate climate (Meister, 2020). The adoption of Industry 4.0 technologies will also motivate and lead people to work in a digitized and well-coordinated workplace that could ease the working with robotics and algorithms while operating in a virtual world (Richert, 2016). This could further lead to a shift, resulting in new job requirements for specific skill sets required in an organization (Grzelczak, 2017). The presence of the required skills and abilities in an organization's workforce has a significant impact on the successful adaptation of Industry 4.0 at all levels. The skill set and qualifications of the employees also play a significant role in increasing the competitiveness of organizations (Benesova, 2017).

4.4.8 VIRTUAL REALITY ONBOARDING

Virtual reality (VR) uses smartphones to help employees, particularly the new ones to know the ambience of office. This way the employees could meet their colleagues, listen to their boss, and get the required messages from the top level on smartphones. This technology in turn results in the speed-up of the process of acculturalization. Google cardboard best exemplifies VR on-boarding tool (Bhaskar, 2017). VR applications have also enabled the employees with mental disabilities to improve their efforts and motivation for the rehabilitation of their health to work smoothly (Bisconer, 1998). The VR application plays out a very beneficial role in social skills training for schizophrenia which may prove to be specifically beneficial for improving the conversational skills and affirmation as well as in nurturing the motivation of employees (Park et al., 2011).

4.5 CHALLENGES IN IMPLEMENTATION OF HR 4.0

HR 4.0 remains at the peak of every sphere of industry in today's world (Rubmann, 2015). It, however, stands out at the focal point of many discussions, both in terms of its pros and cons, but the inference that was drawn from studies indicate that still a shift is very much required (Zhou, 2015). Nonetheless, the resistance that is thought to avert the change is widespread and is the key drawback that must be averted (Bonekamp, 2014). Studies show that HR should be flexible and must adapt to the upgraded trends that come the way, those changes include the flexibility in various professions, etc. The role of HRs in an organization may experience some transition and this can only be successful by following a sufficient system (Schuh et al., 2015). Changes are required in terms of management approach and leadership. Organizations need to develop a positive organizational culture (Chakraborty and Ganguly, 2019) and need positive approaches to understand employees by the use of appreciative questioning (Chakraborty and Mishra, 2019; Chakraborty et al., 2019). Organizations become much more delicate as well as agile after a disaster, so the role of HR professionals in managing the after-effects of a crisis is very crucial (Chakraborty and Saha, 2017).

Digital innovations in businesses have made remarkable changes in the way businesses are done (Chakraborty and Saha, 2018). Therefore, there is a need for HR professionals and practitioners to scholastically outline the necessary framework for HR to both intercept the following challenges and be adapted to the flexible industrial climate.

4.5.1 LACK OF CENTRALIZATION AND STANDARDIZATION

Normally the present system of production is not completely controlled at a sole, and centralized place (Trstenjak, 2018). Although many organizations may have somewhat a systematic framework, but still there is a need for an access to the detailed information. The absence of detailed information can cause hurdles in terms of sidelining the important operations, going on in organizations (Pfeiffer, 2016). Since digitalization is a vital part of Industry 4.0, there should be a systematic control over every single step in the outright procedure and subprocesses, such as hardware, software, etc. (Varghese, 2015). Therefore, multiple users can collect the essential information about every step of the process and other operational activities in the organizations (Almada-Lobo, 2015). Overall, the control of the digitalized tools turns the human role into a more complex system because humans should obtain and then process significant information in the real time (Tseng and Lee, 2009).

4.5.2 DECISION-MAKING

Decision making in the conventional corporate environment is concentrated in a few hands and is controlled from a single place, normally done by the top management in general (Power and Sharda, 2009). Many a time, the collection of the required information from different sections of an organization makes a centralized decision-making process more arduous and trivial. Therefore, in such cases, a decentralized decision-making is required to ensure the optimization of the processes and prompt response as demanded (Lee et al., 2014). This demand can be about the customer needs, requirements of the manufacturing technology, or planning demands, decisions should be taken promptly and accurately (Lee et al., 2014). Smaller decisions should be paid attention to especially those which are taken with the specific information aimed to prevent the likelihood of suboptimal output. This is mainly considered as a human effort that focuses on the increment of the complexity of operations of HR professionals and employees. Nonetheless, the HR practices in Industry 4.0 calls for the accurate interpretation of the information and data (Weyer, 2015). Thus, the system that supports the decision-making process should be conventionalized and then used on a daily basis (Schuh et al., 2014). These systems need to be implemented in the operations to have the decisions be made accurately and instantly. Decision support systems need to be made very much user-friendly just to ease the perplexed tasks assigned to the employees (Trstenjak, 2018).

4.5.3 HUMAN–ROBOT, ROBOT–ROBOT, AND HUMAN–HUMAN INTERACTION

Despite having pros in terms of convenience, many employees see robots as a threat that they deem may replace the HRs, due to which they could lose their work (Weiss, 2016). For some people, initially, the human–robot interaction may look unacceptable mainly due to the fact that it is possible to replace both the human and manual work with robots (Gorecky, 2014). On the other hand, the robot–robot interaction also remained an important feature in the Industry 4.0. With the usage of the software system, robots are expected to be able to properly analyze the contemporary situation of the core manufacturing process. In addition, the robot–robot communication can also augment the automatization of the planning that may be framed for prospective operations, as well as the automatization of logistics and other manufacturing activities (Wang, 2016).

4.5.4 SECURITY OR KNOW-HOW APPLICATIONS

In Industry 4.0, every system needs to ensure its part of security and enforce its security policy if there is any. A properly designed security system based on Industry 4.0 guidelines could help HR professionals in meeting the expected needs (Sysgo, 2016). Protection of the sensitive and significant data of an organization and its staff members is must along with ensuring that the security of the information is properly controlled and is in alignment with the company's mission and objectives. It is also important to define and upgrade the outlined responsibilities regarding the access to data and information. Moreover, all the staff members must be made aware of the overall procedure of reporting the breach of data and how to handle any essential data for responding to an expected event (Chavez, 2018).

4.6 TAKEAWAYS

The major takeaways from this chapter are the impact of Industry 4.0 on HR practices, origin of HR 4.0, and dynamics of HR 4.0. At every possible juncture, care should be taken to remove psychological stress among employees for better functioning (Deb et al., 2008).

KEYWORDS

- **human resource management**
- **Industry 4.0**
- **organizational change**
- **HR 4.0**
- **digitalization**
- **soft skills**
- **robotics**

REFERENCES

Act, N. L. SHRM. Better Workplaces Better World [Online]. 2016.

Aitchison, T. J. Wage and Salary Administration in Personnel Administrator; 2003.

Armada-Lobo, F. The Industry 4.0 Revolution and the Future of Manufacturing Execution Systems. *J. Innov. Manag.* **2016,** *3* (4), 16-21.

Barrick, M. The Big Five Personality Dimensions and Job Performance: A Meta-Analysis. *Person. Psychol.* **1991,** *44*, 1–26.

Bench-Capon, T. J. M.; Dune, P. E. Augmentation in Artificial Intelligence. *Artif. Intell.* **2007,** *171*, 619–641.

Benesova, A. Requirements for Education and Qualification of People in Industry 4.0. *Procedia Manuf.* **2017,** *11*, 2195–2202.

Bhaskar, V. Industry 4.0 and Future of HR [Online]. 2017. https://www.linkedin.com/pulse/industry-40-future-hr-varun-bhaskar/

Bonekamp, L. S. *Change Through Digitalization—Value Creation in the Age of Industry 4.0. Management of Permanent Change*; Springer, 2014; pp. 23—45.

Bova, F. Labor Union and Management's Incentive to Signal a Negative Outlook. *Contemp. Account. Res.* **2013,** *30* (1), 14–41.

Bulte, A. T. In *What is Industry 4.0 and what are Its Implications HRM Practices*, 11th IBA Bachelor Thesis Conference, University of Twente: Enschede, 2018.

Campbell, C. P. Procedures for Developing and Evaluating Vocational Training Programs. *J. Indus. Teach. Educ.* **1984,** *21* (4), 31–42.

Chakraborty, T.; Ganguly, M. Crafting Engaged Employees Through Positive Work Environment: Perspectives of Employee Engagement. In *Management Techniques for Employee Engagement in Contemporary Organizations*; IGI Global, 2019; pp. 180–198.

Chakraborty, T.; Mishra, N. Appreciative Inquiry: Unleashing a Positive Revolution of Organizational Change and Development. Int. J. Econ. *Commerce Manag.* **2019,** *6* (2), 32–37.

Chakraborty, T.; Saha, R. Proactive and Reactive Role of Human Resource Professionals in an Aftershock Management. *Train. Devel. J.* **2017,** *8* (1), 1–11.

Chakraborty, T.; Saha, R. Digital Innovations: Breakthrough Opportunities to Build Novel Business Models. *Bus. Perspect.–Aims Scope* **2018,** *1*, 46.

Chakraborty, T., Ganguly, M.; Natarajan, A. Predicting Entrepreneurial Satisfaction: The Role of Non-Financial Incentive Factors and Quality of Life Among Women Digital Entrepreneurs. *J. Glob. Bus. Adv.* **2019,** *12* (3), 328–355.

Chavez, R. SHRM. Retrieved from The Role of HR in Cybersecurity—Communication and Collaboration are Key to Ensuring Your Organization's Information Security [Online] 2018. https://www.shrm.org/ResourcesAndTools/hr-topics/behavioral-competencies/Pages/The-Role-of-HR-in-Cybersecurity.aspx

Chen, C. J. Promoting Innovation in Hospitality Companies Through Human Resources Management Practices. *Int. J. Hosp. Manag.* **2011,** *30* (4), 812–818.

CompareCamp. (BambooHR) 1 2021 [Online]. 2020. https://comparecamp.com/bamboohr-review-pros-cons-pricing-top-hr-management-software/

Costa, P. *Revised NEO Personality Inventory and NEO Five-Factor Inventory Professional Manual*; Psychological Assessment Resources: Odessa, FL, 1992.

Deb, S.; Chakraborty, T.; Chatterjee, P.; Srivastava, N. Psychological Stress of Traffic Police Officers Causal Factors behind the Same and their Coping Strategies. *J. Indian Acad. Appl. Psychol.* **2008,** *34* (1), 19–28.

Desoutter [Online]. 2021. https://www.desouttertools.com/industry-4-0/news/503/industrial-revolution-from-industry-1-0-to-industry-4-0

Diana L., D. Human Resource Management: Past, Present, and Future. *Hum. Resour. Manag. Rev.* **2014,** *24*, 193–195.

Dick, W.; Carey, L. *The Systematic Design of Instruction,* 3rd ed.; HarperCollins: New York, NY, 1990.

Doyle, A. *The Wagner Act of 1935*; National Labor Relations Act, 2020.

Dulebohn, J. F. The History and Evolution of Human Resources Management. In *Handbook of Human Resources Management*; 1995; pp. 18–41.

educba Latest HR Management Trends for the year 2020 [Online], 2020.https://www.educba.com/hr-management-trends/

entelo 2018 recruiting trends report, 2018.

Erol, S.; Jager, A.; Hold, P.; Ott, K.; Sign, W. Tangible Industry 4.0: A Scenario-Based Approach to Learning for the Future of Production. *Procedia CIRP* **2016,** *54*, 13–18.

Feldman, J. M. Beyond Attribution Theory: Cognitive Processes in Performance Appraisal. *J. Appl. Psychol.* **1981,** *66* (2), 127–148.

Glass, F. Work, Money and Power: Unions in the 21st Century, 3rd ed.; The UC Berkeley Center for Labor Research and Education: Berkeley: , 2013.

Gorecky, D. E. In *Human-Machine-Interaction in the Industry 4.0 era*, 12th IEEE International Conference on Industrial Informatics (INDIN). Brazil: Port Alegre, 2014.

Kergroach, S. Industry 4.0: New challenges and opportunities for the labor market. *Foresight STI Governance* **2017,** *11* (4), 6–8.

Higgins, E. T. Promotion and Prevention: Regulatory Focus as a Motivational Principle. In *Advances in Experimental Social Psychology,* Vol. 30; Zanna, M. P., Ed.; Academic Press: New York1998; pp. 1–46.

I-SCOOP. Industry 4.0: the fourth industrial revolution -guide to industries 4.0 [Online] 2021. https://www.i-scoop.eu/industry-4-0/

Kaufman, B. E. *The Origins and Evolution of the Field of Industrial Relations in the United States*; ILR Press, 1993.

Kaur, A. In *Changing Dynamics to Make HR 4.0 in Sync with Industry 4.0*, Conference: National Seminar on Changing Global Dynamics and management practices. IIMT University Meerut, 2019; pp. 6–11.

Landy, F. J.; Farr, J. L. Performance Rating. *Psychol. Bull.* **1980,** *87* (1), 72–107.

Lee C. K. M.; Zhang. S.; Ng, K. H. K. Development of an Industrial Internet of Things Suite for Smart Factors Towards Re-industrialization. *Adv. Manufac.* **2017,** *5* (4), 335–343.

Lee, J.; Kao, H. A.; Yang, S. Service Innovation and Smart Analytics for Industry 4.0 and Big Data Environment. *Proceedia CIRP* **2014,** *16*, 3–8.

Meister, J. C. Power Skills for Jobs of the Future [Online] 2020. Investinwork.org

Murphy, K. R.; Cleveland, J. N. *Performance Appraisal: An Organizational Perspective*; Allyn and Bacon: Boston, 1991.

Nageshwari, D. R. E-HRM and Its Trends in Industry 4.0. *Dogo Ranging Res. J.* **2020,** *10* (6), 241–250.

Park, K. M.; Ku, J.; Choi, S. H.; Jang, H. S.; Park, J. Y.; Kim, S. I.; Kim, J. J. A Virtual Reality Application in Role-Plays of Social Skills Training for Schizophrenia: A Randomized, Controlled Trial. *Psychiatry Res.* **2011,** *189* (2), 166–172.

Pfeiffer, S. Industry 4.0 and Humans, or Why Assembly Work is More than Routine Work. Societies **2016,** *6* (2), 16.

Power, D.; Sharda, R. Decision Support Systems. In *Handbook of Automation*; Springer: Berlin, 2009; pp. 1539–1548.

Prieto, I. M.; Perez-Santana, M. P. Managing Innovative Work Behavior: the role of Human Resource Practices. *Pers. Rev.* **2014,** *43* (2), 184–208.

Richert, A. S. In *Educating Engineers for Industry 4.0: Virtual Worlds and Human-Robot-Teams: Empirical Studies Towards a New Educational Age*, 2016 IEEE Global Engineering Education Conference (EDUCON), Vol. 11, Abu Dhabi, United Arab Emirates, IEEE, 2016; pp. 1501–1509.

Roethlisberger, F. J. Management and the Worker, 1939.

Rosette, A. *First Things Fast: A Handbook for Performance Analysis*, 2nd ed.; Pfeiffer: San Francisco, CA, 2009.

Rubmann, M. E. *Industry 4.0: The Future of Productivity and Growth in Manufacturing Industries*; Consulting Group: Boston, 2015.

Sarah, W.; Bisconer, J. N. Applications in Diverse Populations. *Comprehensive Clin. Psychol.* **1998,** *9*, 1–17).

Scarpello, V. G. Parallel Approaches to Development of the HRM Field and HRM Education. In *The Handbook of Human Resources Management Education: Promoting an Effective and Efficient Curriculum*, 2008, pp. 3–37.

Schuh, G.; Gartzen, T.; Rodenhauser, T.; Marks, A. Promoting Work-based Learning Through Industry 4.0. *Procedia CIRP* **2015,** *32*, 82–87.

Schuh, G.; Potente, T.; Wesch-Potente, C.; Weber, A. R; Prote, J-P. Collaboration Mechanisms to Increase Productivity in the Context of Industries 4.0. *Procedia CIRP* **2014,** *19*, 51–56.

Sevilla, B. T. PC. Retrieved from pcmag [Onmline] August 6, 2019. https://www.pcmag.com/reviews/bamboohr

Spadafora, A. TechRadar Pro [Online] 2019. https://www.techradar.com/sg/news/data-analytics-could-be-an-hr-breakthrough-for-your-business

Stone, D. L. Human Resource Management: Past, Present, and Future. *Human Res. Manag. Rev.* **2014,** *24*, 193–195.

Straz, M. Entrepreneur. HR Trends that are Becoming Best Practices [Online] 2015. https://www.entrepreneur.com/article/251715

Swanson, R. A.; Torraco, R. J. The History of Technical Training. In *The ASTD Technical and Handbook*; Kelly, L., Ed.; McGraw-Hill: New York, NY, 1995; pp. 1–47.

Sympa. Fast Forward to the Future—Your Guide to Future HR Trends in 2020 [Online] 2019. https://www.sympa.com/insights/guides-and-reports/top-hr-trends-in-2020/

Sysgo. Will we have Security in Industry 4.0 applications. Retrieved from Safety and Security for Industry 4.0 [Online], 2016. https://blog.sysgo.com/industry4-0

T. S. The Star Conspiracy [Online]. Retrieved 2020, from Hottest HR technology solutions in 2015, 2015. https://blog.thestarrconspiracy.com/hottest-hr-technology-solutions-in-2015/

Taylor, A. People Managing People [Online] 2021. https://peoplemanagingpeople.com/tools/talent-management-system/

Taylor, F. Principles of Scientific Management, 1947.

Trstenjak, M. *Challenges of Human Resources Management with Implementation of Industry 4.0*; University of Zagreb, Mechanical Engineering and Naval Architecture: Zagreb, Croatia: Industrial Engineering, 2018.

Tseng, Y.-F.; Lee, T.-Z. Comparing Appropriate Decision Support of Human Resources Practices on Organizational Performance with DEA/AHP Model. *Expert Syst. Appl.* **2009,** *36* (3), 6548–6558.

TXM. The Advantages & Challenges: Implementing Industry 4.0 Technologies [Online] 2020. https://txm.com/the-advantages-challenges-implementing-industry-4-0-technologies

Varghese, A. T. In *Wireless Requirements and Challenges in Industry 4.0*, International Conference on Contemporary Computing and Informatics. Mysore: (IC3I), 2015.

Wang, S. E. Towards Smart Factory for Industry 4.0: A Self-organized Multi-agent System with Big Data-based Feedback and Coordination, *Comput. Networks* **2016,** *101*, 158–168.

Weber, M. *General Economic History*; Greenberg: New York, 1927.

Weil, D. Enforcing OSHA: The Role of Labor Unions. *J. Econ. Soc.* **1991,** *30* (1), 20–36.

Weiss, A. E. First Application of Robot Teaching in an Existing Industry 4.0 Environment: Does It Really Work. *Societies* **2016,** *6* (3), 20.

Weyer, S. S. Towards Industry 4.0—Standardization as the Crucial Challenge for Highly Modular, Multi-Vendor Production Systems. *IFAC-PapersOnLine* **2015,** *3*, 48, 579–584.

Zhou, K. L. In *Industry 4.0: Towards Future Industrial Opportunities and Challenges*, 12th International Conference on Fuzzy Systems and Knowledge DISCOVERY. Zhangjiajie, China: (FSKD), 2015.

CHAPTER 5

Agile Leadership in Managing Human Capital in Industry 4.0

MAHESH LUTHIA

Chetana's Institute of Management and Research, Mumbai, Maharashtra 400051, India
E-mail: mahesh.luthia@cimr.in

ABSTRACT

Smart Human Resources 4.0 (SHR 4.0) as part of the industry 4.0 requires leaders to revisit the approach to support adopting of Digital Technologies, manage diverse workforce, and empower employees. There is a need for agility in the leaders to be more open to learning, promote innovation, adapt to changes. Leaders must focus on developing their credibility not based on their status or professional knowledge but their ability to transfer managerial skills. Such agile leaders can succeed only when they are able to create a learning culture, facilitate team collaboration, and adopt emerging technologies. This chapter outlines the characteristics of agile leaders in the current context. It outlines the importance of leadership in an agile environment and presents some contemporary perspectives on leadership frameworks by analyzing leadership theories and behaviors to achieve success in an agile environment. Competencies required in the SHR 4.0 context are examined to help identify different leadership styles. It presents the important of innovative leadership to drive innovation across the organization. A two-dimensional matrix on Servant Leadership Attributes and behaviors in an agile environment was developed based on the analysis. The chapter also presents a brief case let of Tech Mahindra, on Digital leadership can facilitate digital transformation to enhance employee experience.

Agile Leadership for Industry 4.0: An Indispensable Approach for the Digital Era, Tanusree Chakraborty, Tahir Mumtaz Awan, Ashok Natarajan, and Muhammad Kamran (Eds.)

5.1 INTRODUCTION

Smart Human Resources 4.0 (SHR 4.0) is an emerging approach powered by the Industrial revolution 4.0 that is characterized by the adoption of innovative Digital Technologies to transform organizations. While Industry 4.0 is a new era about developing competitiveness and innovation by focussing on interconnections, automation, machine learning (ML), and real-time data, SHR 4.0 is about using these technologies to automate the people management process and be more digital focused to remain competitive. Through these technologies' organizations can attract, develop, and retain the next-generation workforce, make HR operations more efficient, and operate in a lean model. Various Emerging technologies such as 5G with the Internet-of-Things, Artificial Intelligence (AI) and ML, and Business Analytics is generating a large amount of data in real-time, which is creating a new set of challenges in terms of storage, management, and analysis (Verma et al., 2020). Coupled with the changes in the technological landscape, there are changes in the workforce in terms of diversity. The young population joining the workforce, who have grown in the social media age that have diverse expectations from their organizations in terms of instant gratification, feedback, and data-driven decision-making. The combined forces of technological advancement and changing workforce have resulted in new challenges for organizations in SHR 4.0 context mainly around defining the organization culture, aligning HR processes, and managing workforce expectations. However, to manage these technologies and changes, Human Capital is the key resource. To manage Human Capital in the new face-paced technological environment there is a need for agility and a transformational approach in leadership that can support cultural change, managing people and technology. An agile leader develops their credibility not based on their status or professional knowledge but their ability to transfer managerial skills. It is important that leaders are more open to drive learning and promote innovation across the organization (Sivathanu and Pillai, 2018; Guzmán et al., 2020; Ghobakhloo, 2020).

One of the key challenges of Industry 4.0 is not just about implementing the right technology but is related to the skill in organizations and the absence of digital culture. This has resulted in a significant skill gap in organizations between the current competencies and the roles that are evolving due to the introduction of advanced technology. There is a need to have more effective interventions for talent management to equip employees with the required skill sets. Considering the challenges associated with Smart HR 4.0, the identification of agile leadership characteristics to manage Human Capital becomes an important issue. It will help organizations and their leaders to transition and gain deeper insights into the leadership style in the current context (Rana and Sharma, 2019). The chapter aims to discuss leadership frameworks namely, *Innovative, Swarm, Digital, and Servant* Leadership that support managing human capital, while transitioning toward SHR 4.0. It would enable leaders to navigate from traditional leadership style to agile leadership that would, create a learning culture, facilitate team collaboration, and adopting emerging technologies.

5.2 SMART HUMAN RESOURCES 4.0 (SHR 4.0)

Managing human resources is not just an operational but a strategic process to achieve organizational goals by employing and developing a workforce that is committed and dedicated. HR professionals play a key important role in attracting, recruiting, developing, and motivating employees in the organization. Human resource development aims to enhance the effectiveness of individuals and groups; enhance a firm's efficiency and performance; develop competencies comprising of knowledge skills and abilities and enhance the potential of human resources or human capital. Adopting technology and practices associated with industry 4.0 will enhance the ability of firms to innovate and engage competent human resources. It will transform the way jobs are designed and performed and there will be more automation at the workplace. This trend is transforming the role of HR. Industry 4.0 will change the way jobs are executed by human and will involve robots programmed and directed by humans. New ways of doing things will require a new mindset of viewing tasks and would require training of employees. The contemporary concept of SHR 4.0 is the outcome of industry 4.0 process, which involves the use of cutting-edge digital technologies, such as Internet-of-Things, Big Data and Analytics, AI, and ML along with 4G and 5G, data networks to manage the new workforce. Adopting SHR 4.0 practices will enable organizations to attract, develop and retain "new-age talent," optimize HR operations and make them efficient, and operate with lean HR departments due to digitalization. It would also provide an opportunity to for employees use data effectively and make informed data-centric decisions (Rana and Sharma, 2019). Industry 4.0 creates new challenges related to managing human resources. The immediate need is to attract competent young employees at the same time retain the knowledge base created by experienced employees.

There is an increased emphasis in work-life balance and the expectations of remote or virtual working. Organizations need to invest in IT infrastructure to support boundaryless organization structures and virtual networks. To overcome with the challenges associated with Industry 4.0, an effective SHR 4.0 strategy is required to automate HR processes through advanced technology. The next leap in this direction would be the use of smart mobile apps with artificial or virtual reality to attract the new-age talent and provide a platform for them to collaborate and maintain remote team interaction (Hecklaua, 2016; Sivathanu and Pillai, 2018).

SHR 4.0 can support organization success through digital transformation and increased virtualization by adopting robotics, nano-technology, and cloud-based computing on high-speed data networks. The new workforce comprising of gig workers will work remotely and extensively use technology to carry out their job duties. HR processes such as talent acquisition, managing performance and compensation, employee development, and talent management will be supported by data analytics and metrics. In the post-Covid-19 period, the way to develop a sustainable competitive advantage would be through human capital that can take advantage of the benefits of Industry 4.0 (Flores et al., 2020; Mefi and Asoba, 2021).

While there are many benefits of adopting SHR 4.0 through technology, HR professionals would require organizational support and changes in structure and leadership style. Firms will have to carry out techno-structural interventions to established a flatter and agile structure to create a suitable environment for SHR 4.0 implementation. Such a structure would smoothen communication between employees and leaders, decentralize power from leaders to employees and empower them. There is a need for a situational leadership style that would promote innovation and organization learning to support a diverse workforce to feel valued (Sivathanu and Pillai, 2018; Verma et al., 2020).

5.3 IMPORTANCE OF LEADERSHIP IN AGILE BUSINESS

Leadership has been defined in various ways by different authors. While some define it as a trait, some characterize it as a skill or ability. But when we examine the common role of leaders, it can be referred to as goal-directed behavior that creates an influence over others. When we look at leaders around our social, personal, and professional pursuits we see leaders articulate vision, embody values, and create an environment for accomplishing a goal. Leadership in an organization is about the capability to build cultures, and initiate the process of bringing in evolutionary changes that are adaptive in nature. It helps in getting people in a society or organization together and develops a sense of mutual commitment. One of the most important functions of a leader that re-defines a leadership style is the ability of the leader to motivate and inspire others in an organization to achieve common goals and overall effectiveness. Hence, setting the vision for the organization and teams is key to bring about any change and motivate employees to adapt to the new environment (Akkaya, 2020; Guzmán et al., 2020). Dynamic organizations require a positive organizational culture and exemplary leadership (Chakraborty and Ganguly, 2019), will need positive approaches to understand employees (Chakraborty and Mishra, 2019; Chakraborty et al., 2019). Organizations are more agile after a disaster, so the role of human resource professionals in managing after effects of a crisis is very crucial (Chakraborty and Saha, 2017).

SHR 4.0 required a different approach by leaders. To support the technological advancement and changing workforce, an agile leadership style is required to be more focused on driving learning and reward innovation across the organization. Perker et al. (2015) defined agile leadership as a leadership style that involves shared responsibility and an effective way to recognize problems, make decisions, adapt to the changes or systems, and adopt flexible structure.

While initiating a cultural change in the organization, they will need to ensure that there are minimum conflicts in terms of values and beliefs among diverse employees. To attract and retain the best talent, technology upgradation has to be focused on the long-term goals of the organization. The role of the agile leader is not to just apply new technologies for automation but facilitate the digital transformation that would help the organization to serve its customer better. To achieve all this, leaders need to be more agile and adaptive. It requires leaders to possess the agility to understand complex issues and manage them effectively. Agile leadership requires a very different mindset compared to Hubristic and autocratic leaders. While Hubristic leaders demonstrate unrealistic self-evaluations and over-confidence, autocratic leaders do not engage with other members of the organization for decision-making (Akkaya, 2020).

Agile leaders focus on developing credibility not just based on expert power or referent power but through motivation, communication, teamwork, and other transferable skills. Agile leaders invest their time and efforts in continually guiding their team through teamwork and setting a vision. They create a sense of urgency to make the vision a reality. They set clear expectations with their teams in terms of group norms, shared responsibility trust, and mutual accountability. While motivating their teams they maintain a balance between individual and group needs. They focus on cohesiveness among team members and reduces conflicts through effective communication. They can cope with change effectively and can quickly gain control of complex issues. Agile leadership can create an organization climate where employees are motivated and inspired by their leader to be open to change and innovation. In addition to collaboration and nurturing talent they accept diversity, foster innovation, and creativity and put in place a flexible organization structure that would promote delegation, transparency, and trust. It will create an environment where the team will be more prepared to take risks. Based on existing research, agile leadership has emerged as an essential factor that strengthens digital transformation in organizations. The leadership style would have to be forward-thinking to enable firms to respond faster to new opportunities and challenges that occur due to technological advancements and implement SHR 4.0 (Barclay, 2020; Revutska and Maršíková, 2021; Malavika and Mohana, 2021).

5.4 LEADERSHIP FRAMEWORKS IN SHR 4.0 CONTEXT

Traditionally Human Capital was compared with physical assets of an organization, however, recently Goldin (2016) suggested that human capital encompasses skills, health, education, and expertize. They must assess the changes in competencies due to digitization, adapt their training methods to suit the diverse workforce and digital environment, and encourage behaviors required for the digital transformation. Leaders would require to develop a new mindset and develop new skills sets for agile leadership to manage the 21st Century Human Capital (Flores et al., 2020). Organizations such as Google, Microsoft, and IBM provide an example of how leadership can drive innovation and a participative culture in an agile environment where leaders focus on digital transformation for developing unique business models and responding to change with speed and agility. Re-imaging human capital with appropriate leadership will enable firms to create social structures that are sustainable and stable and employees will have to opportunity to realize their potential to the fullest. The following section describes leadership frameworks that support SHR 4.0 practices in an agile environment.

5.4.1 INNOVATIVE LEADERSHIP

Implementation of SHR 4.0 requires leaders to fundamentally think differently about what HR can deliver and built it to deliver results. It requires creating a culture of innovation. To develop this cultural environment the leaders, need to encourage the application of innovation, and creativity across the organization to develop solutions to problems and generate new products and services (Bawany, 2019). Innovative leadership an alternative to contemporary leadership provides fundamental changes through new ideas and techniques to solve future organizational problems. GE's Jack Welch, and Joseph Wilson of Xerox and Jorma Ollila from Nokia are some of the leading innovative leaders. Innovative leadership plays an important role in implementing innovative practices as digital transformation does not just involve the application of new technology but also how it can be used effectively for organizational success. Through their knowledge, which is key for innovative leadership, helps them to develop innovative solutions. Hence, innovative leaders constantly focus and securing new information and knowledge. In such situations, leaders play a pivotal role in creating a learning environment and encourage employees to think beyond traditional realms (Sen and Eren, 2012).

This innovative leadership approach will enable the organization to quite the right talent to develop a competitive advantage through increased productivity. The agile way of working will help leaders to continuously adapt their process of acquiring talent acquisition based on the changing labor market, emphasize innovation, and continuous learning. Such leaders will need to view communication, not as a competence but as a characteristic of their leadership style. It would involve promoting innovation and leading change to develop new products, uncover nontraditional revenue streams, and create pathways for shared experiences and mutual coaching. Agile management requires leaders to give priority to individuals and communicating with them rather than focussing on processes. With the millennials joining the workforce, such an approach of leadership is the need of the hour. In an agile environment, they will bring in new trends of leadership that will involve real-time detailed feedback, agile networks in a less hierarchal structure, use of analytical tools, and learning through personalized and microlearning components. To foster learning and innovation, the leader will have to be more of a coach than a superior of a boss. The agile model of HRM involves supporting organizational agility. It redefines the role of the leader to focus on promoting adaptability, innovation, and collaboration among members with speed (Boneau and Thompson, 2013; Guzmán et al., 2020; Revutska and Maršíková, 2021). An innovative leader in an agile environment should be creative and innovative in communication and focus on transforming complex situations for positive results. Such leaders need to be people-centric and emphasize teamwork by building trust, a delegation of responsibility and authority (Srivastava and Jain, 2017). Within the HR function, leaders must focus on engaging teams, seeking feedback, and adapting to changing customer needs. Adaptability must be given more emphasis over prescriptiveness. Leaders must take greater responsibility for human resources, and reward them for learning, driving innovation, and adapting to support organizational change (Revutska and Maršíková, 2021).

5.4.2 SWARM LEADERSHIP FOR TRUST-BASED RELATIONSHIPS

In the context of SHR 4.0, there is need for leaders to be adaptive and collective in approach. Swarm-based leadership transcends from collaborative competition to competitive collaboration. It involves listening first and building a shared context for team members (Gloor, 2017). Kelly (2018) defined Swarm leadership as an "adaptive, emergent, connected, responsive, and collaborative model" (p. 15). He suggested that in the process of adoption of Industry 4.0, there is a need for leadership that would be self-organizing in a complex environment and lead people through collaborative networked efforts. A responsive way of functioning as leaders have to act in an agile environment, to operate systematically in volatile and uncertain stations. Rather than distributing leadership (distributed leadership), swarming involves collective innovation, organization, and collaboration. Such an approach is required to navigate in a complex environment. Swarming involves a collective behavior, where members move in the same direction through a process of mutual trust, communication, and interaction among them. Swarm leadership involves building enduring relationships with of trust and loyalty instead of power (Nandita, 2013). Leaders in an agile environment may use Swarm Intelligence involving flexibility, robustness, decentralization of power, and being self-organized to solve complex problems. With increasing digitalization, there is a need for leaders to adopt a Swarm behavior as it would decentralize data-driven decision making, delegate work based on functional expertize, develop the ability to sense the internal and external environment, and develop the readiness for managing change. Swarm intelligence in the leader, will enable the leader to empower and motivate teams to assume responsibility for whole task and utilize a variety of skills that each team member possesses. The process by which Swarm intelligence can help leaders to function would involve increasing task interdependence. Through task interdependence team members would require regular interaction among themselves, leading to high group potency, reinforcing the belief that they can be effective, and the outcome is a consequence of teamwork. Leaders will have the trust to empower teams to determine the methods they wish to use to complete the tasks and delegate them to team members based on their competence. It will help the leader to set a common vision for teams and improve overall performance. From an HR perspective, Swarm leadership will help effective team-based performance management, reward teamwork, and provide feedback to teams. Leaders will lay greater emphasis on developing self-managed teams, training, and development for working in a collaborative environment, and ensuring employee engagement and job satisfaction. They will invest more time in productive communication laying operating policies and procedures for clear guidelines (Nandita, 2013; Gloor, 2017; Kelly, 2018; Guzmán et al., 2020). However, there is a need to carefully adopt Swarm leadership as it could involve bias or perceptual errors in the leader. This could impact group cohesion and the leader splitting away from the group (Bernardi et al., 2021). Swarm leadership will help organizations to create a culture based on cooperation and collaborative problem-solving, innovation, and encourage workplace diversity while adopting SHR 4.0 practices. In this context, the leader is also a digital leader who is involved in creating their digital transformation strategy to align business and organizational goals (Bawany, 2019).

5.4.3 DIGITAL LEADERSHIP

Like any other transformation, digital transformation requires leaders who are well prepared to communicate that digitalization will help the firm develop a competitive advantage (Leading Human Capital in the 2020s). Digital innovations in businesses have made remarkable changes in the way businesses are done (Chakraborty and Saha, 2018). A digital leader builds cohesive teams and keeps them engaged and connected to drive innovation, develop tolerance toward risk, and focus on continuous improvement (Bawany, 2019). Technology and Digitalization transform Communication, Production, and Human Resource Management. They influence the way people and processes are integrated to achieve business outcomes. Digital leadership involves, enabling the organization to navigate toward digital transformation for maintaining competitiveness in an agile environment. The leaders achieve this through flatter organization structures, team-based job design, and a supportive environment to foster innovation. Digital leaders also referred to as leadership 4.0 attempt to view business strategy through the integration of technology; supports innovative digital business models and aligning the IT function. These leaders demonstrate openness to change especially with regards to the usage of digital tools to implement digital transformation. Considering the VUCA (Volatile, Uncertain, Complex, and ambiguous) environment, leaders need to demonstrate agility by enabling the organization to drive innovation, move at a faster pace and adapt to and change the environment. To achieve this leaders must observe and be aware of the environment, get themselves aligned, make strategic decisions on how to act, and finally to take action. Digital leadership can create a competitive advantage for an organization by enhancing Human Capital. The leaders can strategically attract and retain employees, develop employees, and managing all HR processes efficiently. Digital leaders set a clear and strategic vision for their organization. They align new digital technology with existing IT infrastructure, create a collaborative team-based environment, focus on organization learning by empowering their employees. However, it must be noted, that empowering employee is one of the biggest challenges leaders face. Empowerment requires a "We" approach rather than "I." Hence to deal with this leadership crisis the premise of Servant leadership has been advocated in an agile context (Sharma and Sharma, 2011; Oberer and Erkollar, 2018; Wasono and Furinto, 2018).

5.4.4 SERVANT LEADERSHIP IN AN AGILE CONTEXT

Servant leadership is an emerging approach relevant to the current time. A concept pioneered by Greenleaf's (1970, 1977, 2002) based on his experiences and on the principle of serving and placing the subordinate first rather than making a conscious choice of aspiring to lead them. He advocated that followers must respond only to competent leaders who should serve their followers with talent, compassion, and courage. Conceptually it is different from Bass's (1985) transformational leadership as servant leadership focuses on serve followers mainly marginalized people compared with empowering and elevating followers. The approach adopted by servant leaders is on empowering and nurturing followers to help them progress in professional and personal pursuits. The emphasis is on motivating, delegating, and serving their teams through skills communication that involves active listening, seeking feedback, building trust, demonstrating empathy, and encourage collaboration rather than competition among team members. Delegation involves, sharing responsibility and authority which helps improved decision-making. This also fosters a sense of control over work among employees. Most importantly they make the ethical use of power to create the influence. Hence, it possesses the individual attributes required to build organizations that support innovation, sustainability, and profitability (Tanno, 2017; Langhof and Güldenberg, 2020).

The process adopted by them involved articulation of the vision and building trust that can empower the team. Developing a strategic vision for teams can help leaders facilitate organization change and transformation. This vision must be clearly articulated through coaching instead of controlling servant leaders can unleash the energy of their followers. Trust is the basis for all leadership. The building block of building interpersonal trust is honesty and integrity, Honesty and integrity also help to build the credibility of the leader. Building confidence, developing credibility, and demonstrating competence are very important to build trust. Empowering involves making individuals responsible and make them accountable in the decision-making process. Integrity is also very important for building teams and effective leadership. Servant leadership follows a more "follower-centric approach," (Huang et al., 2016) rather than leader-centric. Hence in an agile environment, servant leadership helps organizations to respond faster to the emerging business needs. With cross-functional experts in agile teams, servant leadership supports efficient communication and decisions based on diverse perspectives. This approach is most suitable when an organization have a diverse workgroup and there is a need to encourage self-organizing and interchangeability (Russell and Stone, 2002; Tanno, 2017).

The central premise of servant leadership is service. Patterson (2003) suggested, "*empowering people, with the best interest of those served in mind, is at the heart of servant leadership*" (p. 23). He proposed that servant leadership involves serving followers by providing them resources, information, and meaning to their work. By demonstrating stewardship, they empower their teams by building partnerships. Stewardship also involves taking responsibility for common interests and being willing to be accountable through service rather than control and command. In addition, they also act as a role model. They manifest the cultural values of the organization. Servant leadership involves demonstrating ethical behaviors. The visibility of appropriate behavior and interaction is important to create the necessary influence. Followers must be able to view the actions and behavior of their leaders before they imbibe them. In an agile environment, servant leaders must be pioneers of the change management process. They must be prepared to take the risk, innovate, and experiment. Agile leaders must be prepared with an open mind to learn, and gather inputs from within and outside the organization (Langhof and Güldenberg, 2020).

The influence must be aimed at persuading followers to adopt the new direction. Servant leaders use nonmanipulative and noncoercion strategies to influence followers. They use persuasive communication as a means to influence, which is critical to servant leadership (Russell and Stone, 2002).

One of the most evident unique attributes of servant leadership is cherishing the joy of seeing others succeed and appreciating the same. This appreciation encourages followers and exhibits empathy and care. Listening to those who serve and encouraging them through appreciation helps the leader to show concern for others and build trust. Listening is also key to delegation and empowerment. In such agile situations, the leader can coach teams to collaborate, acts as a facilitator, and encourages team members to express opinions (Table 5.1).

TABLE 5.1 Indicates Servant Leadership Attributes and Behaviors in an Agile Environment.

Functional attributes	**Leadership behaviors**
Empowerment	Delegation
Strategic vision	Articulate and communicate vision
Honesty and integrity	Establish credibility
Build trust	Demonstrate competence
Service	Stewardship
Role model	Visibility of behavior
Pioneers of change	Influence others through persuasion
Appreciation others	Listening and encouraging

Source: Based on a review of existing literature (Russell and Stone, 2002; Tanno, 2017; Bavik, 2020).

The VUCA environment required a leadership approach that would foster collaboration, communication, and empowerment without any biases would enable organizations to make appropriate decisions and create a culture that would support innovation and flexibility in a ambiguous environment. Servant leadership would help leaders to develop a vision that is aligned to stakeholder needs and goals (Kornelsen, 2019).

5.5 CONCLUSION

To sum up, this chapter has presented some of the key competencies and traits required by leaders in the SHR 4.0 context. It outlines the importance of an agile and transformational approach in leadership to support cultural change, managing diverse people, and supporting digital transformation. Communication and building trust are key to motivate teams to adopt innovation. To foster an open culture, communication between leaders and followers is very important especially for encouraging innovation across the organization. Working with the millennial workforce requires a more open mindset and the ability to give faster feedback. Possessing explicit and tacit knowledge is very important for leaders to foster innovation. Communication and knowledge provide a means to develop a shared vision. Formulating a shared vision will help leaders to construct a learning culture that creates an environment of collaboration, participation, and decentralized in terms of structure. Swarm-based leadership supports collaboration in an agile environment and focuses on adaptation with collaboration to operate in an uncertain and volatile environment. It provides followers, the opportunity to operate in a flexible, robust, and self-organized manner. Digital leadership creates an environment to support organizations in the digital transformation journey to maintain competitiveness in an agile environment. The main role of the digital leader is to align key stakeholders and communicate the benefits of digital transformation, manage resistance to change, and prepare the organization for change. In an agile environment, there is a need for leaders to empower and develop people through Servant leadership to facilitate quick adaptation and change. Servant leadership is an evolving approach in which leaders focus on active listening, seeking feedback, building trust, demonstrating empathy, and encourage collaboration rather than competition among team members. Future leaders would require these competencies to align HR processes in the Industry 4.0 to leverage the capabilities of advanced technology and manage the diverse workforce.

Case of Tech Mahindra

Tech Mahindra Limited, a leading player in digital transformation, business consulting, and re-engineering services, is part of the Mahindra Group. A fast-growing firm that focuses on leveraging next-generation technology such as blockchain, AI, and 5G are known to empower enterprises globally through technology.

Tech Mahindra used its technology expertise to digitalize HR process and build an agile system where humans and machines can collaborate for higher productivity, foster innovation, and support growth. The key challenges were related to, employee engagement, making workforce more efficient and agile, and equip employees to use these advanced technologies for problem solving. In addition, Tech Mahindra had to look at ways to attract and retain talent during the digital transformation phase. The strategic objective was very clear, develop a competitive advantage by improving productivity, saving costs, and driving innovation in the organization.

The organization developed a marketplace to upskill, reskill, and managing talent within the organization. They set up an AI-based marketplace for talent exchange called Talex. This platform enabled the firm to use AI to match a profile with a particular job opening, on posting of a Job Opening in Talex, the 131,000 employees receive a flash notification and attempts to match the profile and indicate the percentage match. To motivate employees, they offered 100 crypto coins so that the right candidates apply (Bids) for the job of their choice. The employee with the highest bidder is offered an interview. The AI platform also replaced physical ID cards with AI-based face recognition systems with moodometer that indicated the mood of the employees.

The firm also launched an UVO and internal HR bot that automated administrative HR processes such as leave approvals, cash advances, personal calendar for wellness activities, etc. The first HR humanoid K2 platform automated transactional aspects through automated responses for 20,000 HR-related queries. It took less than 8 s to respond to queries. Thus enabled the HR team to focus on core aspects and spare time for what they wanted to do. To build an agile structure within the organization, the digital transformation helped them provide online training, reskill, and upskill employees. The HR Humanoid platform enabled the HR team to deliver a better employee experience.

The above initiatives of the Chief People Officer at Tech Mahindra demonstrate hows Digital leadership enables firms to use advanced technology in an agile environment, to communicate and employer employees. It helps to understand how leaders can drive organizational growth and lead digital transformation.

(Verma et al., 2020; Future Skilling, 2021).

KEYWORDS

- **agile HR**
- **swarm leadership**
- **digital leader**
- **servant leaders**
- **smart HR 4.0**

REFERENCES

Akkaya, B. Review of Leadership Styles in Perspective of Dynamic Capabilities: An Empirical Research on Managers in Manufacturing Firms. *Çanakkale Onsekiz Mart Üniversitesi Yönetim Bilimleri Dergisi* **2020**, *18*(36), 389–389, 438. doi: http://dx.doi.org/10.35408/comuybd.681427

Barclay, R. D. Servant Leadership Behaviors that Positively Influence on-Time Delivery of Committed Work by Agile Teams in a Scaled Agile Framework. Doctoral Dissertation, University of Missouri-Saint Louis, 2020.

Bavik, A. A Systematic Review of the Servant Leadership Literature in Management and Hospitality. *Int J. Contemp. Hosp. Manage.* **2020,** *32*(1), ISSN: 0959–6119.

Bawany, S. Leadership 4.0: How Ready Are you to be A Digital Leader? *Leadership Excellence* **2019**, *36*(2), 28–30.

Bernardi, S.; Eftimie, R.; Painter, K. J. Leadership Through Influence: What Mechanisms Allow Leaders to Steer a Swarm? *Bull. Math. Biol.* **2021,** *83*(6), 1–33.

Boneau, J.; Thompson, G. Leadership 4.0. *Leadership Excellence* **2013,** *30*(1), 6.

Chakraborty, T.; Ganguly, M. Crafting Engaged Employees Through Positive Work Environment: Perspectives of Employee Engagement. In *Management Techniques for Employee Engagement in Contemporary Organizations*; IGI Global: Pennsylvania, USA, 2019; pp 180–198.

Chakraborty, T.; Ganguly, M.; Natarajan, A. Predicting Entrepreneurial Satisfaction: The Role of Non-Financial Incentive Factors and Quality of Life Among Women Digital Entrepreneurs. *J. Global Business Adv.* **2019,** *12*(3), 328–355.

Chakraborty, T.; Mishra, N. Appreciative Inquiry: Unleashing a Positive Revolution of Organizational Change and Development. *Int. J. Econ. Commerce Business Manage.* **2019,** *6*(2), 32–37.

Chakraborty, T.; Saha, R. Proactive and Reactive Role of Human Resource Professionals in an Aftershock Management. *Train. Develop. J.* **2017,** *8*(1), 1–11.

Chakraborty, T.; Saha, R. Digital Innovations: Breakthrough Opportunities to Build Novel Business Models. *Business Persp. Aims Scope* **2018,** *1*, 46.

Chandrasekhar, S. (Ed.). *Leading Human Capital in the 2020s: Emerging Perspectives*; ProQuest Ebook Central, SAGE Publications: USA, 2020.

Della Corte, V.; Del Gaudio, G.; Sepe, F. Leadership in the Digital Realm: What Are the Main Challenges? In *Digital Leadership-A New Leadership Style for the 21st Century*; IntechOpen, 2019.

Fachrunnisa, O.; Adhiatma, A.; Lukman, N.; Majid, M. N. A. Towards SMEs' Digital Transformation: The Role of Agile Leadership and Strategic Flexibility. *J. Small Business Strategy* **2020,** *30*(3), 65–85. https://www.proquest.com/scholarly-journals/towards-smes-digital-transformation-role-agile/docview/2487473134/se-2?accountid=185290

Flores, E.; Xu, X.; Lu, Y. Human Capital 4.0: A Workforce Competence Typology for Industry 4.0: IMS. *J. Manufactur. Technol. Manage.* **2020,** *31*(4), 687–703. DOI: http://dx.doi.org/10.1108/JMTM-08-2019-0309

Future Skilling. August 11, 2021. https://www.techmahindra.com/en-in/people/future-skilling/

Ghobakhloo, M. Industry 4.0, Digitization, and Opportunities for Sustainability. *J. Cleaner Product.* **2020,** *252*, 119869.

Gloor, P. A. *Swarm Leadership and the Collective Mind: Using Collaborative Innovation Networks to Build a Better Business*; Emerald Group Publishing: Bradford, UK, 2017.

Guzmán, V. E.; Muschard, B.; Gerolamo, M.; Kohl, H.; Rozenfeld, H. Characteristics and Skills of Leadership in the Context of Industry 4.0. *Procedia Manuf.* **2020,** *43*, 543–550.

Hecklau, F.; Galeitzke, M.; Flachs, S.; Kohl, H. Holistic Approach for Human Resource Management in Industry 4.0. *Procedia Cirp.* **2016,** *54*, 1–6.

Kelly, R. *Constructing Leadership 4.0: Swarm Leadership and the Fourth Industrial Revolution*; Springer: USA, 2018.

Kornelsen, J. The Quest to Lead (with) Millennials in a VUCA-World: Bridging the Gap Between Generations. In *Leading in a VUCA World*; Springer: Cham, 2019; pp 27–41.

Langhof, J. G.; Güldenberg, S. Servant Leadership: A Systematic Literature Review—Toward a Model of Antecedents and Outcomes. *German J. Hum. Resource Manage.* **2020,** *34*(1), 32–68.

Malavika, M. K.; Mohana, P. A Global Revolution of Work With Smart Hr 4.0. *Psychol. Educ. J.* **2021,** *58*(2), 2829–2835.

McPherson, B. Agile, Adaptive Leaders. *Hum. Resour. Manage. Int. Dig.* **2016,** *24*(2), 1–3. DOI: http://dx.doi.org/10.1108/HRMID-11-2015-0171

Mefi, N. P.; Asoba, S. N. Sustainable Human Resource Practices for Organizational Competitiveness Post The Covid-19 Pandemic. *Acad. Entrepreneurship J.* **2021,** *27*(2), 1–7.

Nandita, S. Swarm Behaviour - an Intelligent Tool in Tackling Challenges. *Adv. Manage.* **2013,** *6*(10), 20–23.

Oberer, B.; Erkollar, A. Leadership 4.0: Digital Leaders in the Age of Industry 4.0. *Int. J. Org. Leadership.* **2018,** *7*(4), 404–412.

Rana, G.; Sharma, R. Emerging Human Resource Management Practices in Industry 4.0. *Strategic HR Rev.* **2019,** *18*(4), 176–181.

Revutska, O.; Maršíková, K. Agile Approach in Human Resource Management: Focus on Generation Y. **2021.** DOI: 10.15240/tul/001/2021-2-005

Russell, R. F.; Stone, A. G. A Review of Servant Leadership Attributes: Developing a Practical Model. *Leadership Org. Develop. J.* **2002,** *23*(3), ISSN: 0143-7739.

Şen, A.; Eren, E. Innovative Leadership for the Twenty-First Century. *Procedia Soc. Behav. Sci.* **2012,** *41*, 1–14.

Sendjaya, S.; Sarros, J. C.; Santora, J. C. Defining and Measuring Servant Leadership Behaviour in Organizations. *J. Manage. Stud.* **2008,** *45*(2), 402–424.

Sharma, A. P.; Sharma, R. K. Agile Leadership and Organisational Development. *ASBM J. Manage.* **2011,** *4*(1), 53.

Sivathanu, B.; Pillai, R. Smart HR 4.0 – How Industry 4.0 is Disrupting HR. *Hum. Resour. Manage. Int. Dig.* **2018,** *26*(4), 7–11. DOI: http://dx.doi.org/10.1108/HRMID-04-2018-0059

Srivastava, P.; Jain, S. A Leadership Framework for Distributed Self-Organized Scrum Teams. *Team Perform. Manage.* **2017,** *23*(5), 293–314. DOI: http://dx.doi.org/10.1108/TPM-06-2016-0033

Tanno, J. Servant Leadership: What Makes it an Effective Leadership Model. Doctoral Dissertation, Walden University, 2017.

Verma, A.; Bansal, M.; Verma, J. Industry 4.0: Reshaping the Future of HR. *Strategic Direction* **2020,** *36*(5), 9–11. DOI: http://dx.doi.org/10.1108/SD-12-2019-0235

Wasono, L. W.; Furinto, A. The Effect of Digital Leadership and Innovation Management for Incumbent Telecommunication Company in the Digital Disruptive Era. *Int. J. Eng. Technol.* **2018,** *7*, 125–130.

Wiraeus, D.; Creelman, J. How to Ensure a Strategy-Aligned Leadership. In *Agile Strategy Management in the Digital Age*; Palgrave Macmillan: Cham, 2019; pp 191–206.

CHAPTER 6

Institutionalization of Leadership Culture (ILC): A Key to Successful Industry 4.0 Transformation

ADITYA DHIMAN*, ASHISH ARYA, AND PANKAJ MADAN

FMS, Gurukula Kangri (Deemed to be University), Haridwar, India

**Corresponding author. E-mail: adityadhiman7@gmail.com*

ABSTRACT

Though change has always been around, but the rate at which the Fourth Industrial Revolution (Industry 4.0) is progressing is exponential. The prevailing volatility, uncertainty, complexity, and ambiguity environment imposed by the current technological advancement has never been witnessed earlier. Being a journey and not a shift, Industry 4.0 transformation is posing numerous challenges to the industries and affecting every element of entire manufacturing, management, and governance structures across the globe.

The fourth revolution is inevitable and economies around the world cannot afford to overlook it. Hence, adopting technology is no longer an option though it's a necessity. Now, the question arises "How to start?" As we know that everything starts with *Intent*, that is, the vision, mission, mindset, and approach. In an organizational setting, intent *(strategic intent)* is determined by leaders at the top level. A leader not just determines the organization's intent but also plans for the actionable steps to address that intent. Therefore, leadership appears to be the most significant factor in driving the change. Evidently, the existing leadership practices are not being helpful to drive the change in the organization and this compels the organizations to rethink on the right leadership competencies.

Agile Leadership for Industry 4.0: An Indispensable Approach for the Digital Era, Tanusree Chakraborty, Tahir Mumtaz Awan, Ashok Natarajan, and Muhammad Kamran (Eds.)

This chapter maps the significant contribution of leadership in managing transformational challenges. The chapter will also theoretically examine the concept of *"Institutionalization of Leadership Culture" and "It's Implementation," "Reverse Mentoring," "Self-Leadership," and "Key leadership competencies"* in relevance to Industry 4.0. The chapter will contribute to the understanding of future leadership, HR personnel may look for outlined leadership competencies while acquiring talent or planning for training and development; and top-level managers can utilize the findings in establishing better workplace culture in line with Industry 4.0.

6.1 INTRODUCTION

Today, the world is witnessing a paradigm change in technological and manufacturing settings, which is never been witnessed earlier and is actually disrupting the way we live, shop, and produce products. These technological advancements clearly indicate that we have entered the fourth phase of the Industrial Revolution. The fourth industrial revolution is inevitable and economies around the world cannot afford to overlook it. Hence, adopting technology is no longer an option though its necessity. Every single entity is in pursuit of implementing the Industry 4.0 technologies to mark the competitive gain in their particular industry. But the question arises "How to start?" because this revolutionary era has made the business environment much more complex and unpredictable.

In this chapter, authors answered this question. Along with this, authors comprehend the concept of Industry 4.0, its current challenges and how it is different from previous Industrial Revolutions (i.e., Industry 1.0, Industry 2.0, and Industry 3.0). Authors postulated the role and importance of leadership, reverse mentoring & self-leadership in managing & successfully executing the Industry 4.0 transformation; and key leadership competencies as well as styles required in today's era. Finally, authors suggest that institutionalization of leadership culture (ILC) will enable organizations to initiate the Industry 4.0 transformation; and outlined a ILC process for the implementation of the same.

6.2 INTRODUCTION TO INDUSTRY 4.0

Before reaching the fourth industrial revolution, civilization has experienced three major industrial revolutions and all these transformations are of historical significance. All these developments brought a paradigm shift in our society and changed the way we produce things. Figure 6.1 shows the emergence of Industry 4.0 and it can be clearly seen that the revolution by revolution the conventional ways of doing business are transforming rapidly.

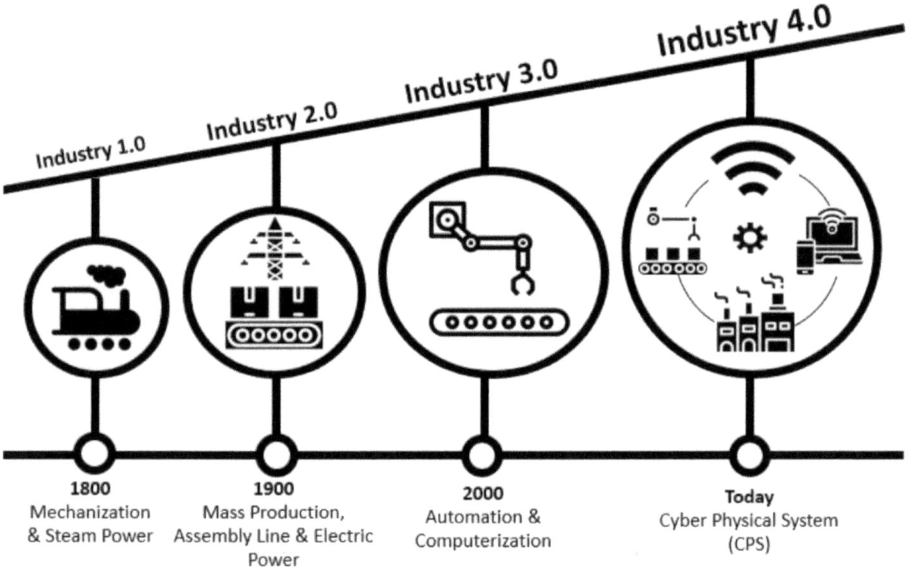

FIGURE 6.1 Industrial revolutions from Industry 1.0 to Industry 4.0.

Source: Authors' own work.

In the late 18th century, the First Industrial Revolution (Industry 4.0) enabled mechanization and the use of steam power that changed—the workshop to factory and manual to mechanical production style, which greatly increased the production amount. The transition in the structure of production has changed both the social and the economic structure to such an extent that society's average life expectancy and quality of life have improved (Kharb, 2018). That is why, this time has been described as a revolution in past.

In the Second Industrial Revolution (Industry 2.0), which started in 1870 and continued until 1989, the world witnessed the use of electrical technology (superior to steam power) in production lines and got familiar with the notion of mass production.

The Third Industrial Revolution, also known as Industry 3.0, introduced in the 1970s, has benefited from the development of electronics and IT. Automation has been accomplished in manufacturing in this revolution.

There has been the growth of digital technology and the advent of information technologies. Arklan and Taşdemir (2008) postulated in their study that availability of information has increased, and information has now been used as the factors of production. In this time, the growth of the internet has remarkably increased communication channels (Kharb, 2018).

The fourth revolution is a technological disruption that has occurred just within the 50 years since the last revolution. The German government launched the concept "Industry 4.0" in 2011 as a national initiative to improve manufacturing industry research and growth. It is therefore obvious that the highest rate of Industry 4.0 implementation is seen in Germany and, in particular, in MNCs. Industry 4.0 technology is now commonly used in most developed countries (Bauernhansl et al., 2014). Industry 4.0 has now become a buzzword and is attracting not even entrepreneurs and industrialists but also academicians (Sae-Lim and Jermsittiparsert, 2019). The fourth revolution is believed to intensify productivity and flexibility to an exponential extent in comparison to the previous three. The concept is to take advantage of recent developments in IT and the Internet to interconnect computers, instruments, equipment, sensors, and people into intelligent, decentralized systems that can sense and adapt to the setting. Bienzeisler et al. (2014) advocated in their study that high-quality goods could be produced more quickly if an individual, system, and processes are organized in intelligent, smart networks, and competitive advantages can be obtained by making production costs more competitive. The revolution of Industry 4.0 reviled the automated mass production of personalized goods and services for modern consumers through interconnected physical and cyber-physical structures (Hofmann and Rüsch, 2017). There is no definite mutually agreed definition, but if we go with the concept, it consists of integrated production facilities, supply chains, and service systems, in order to promote the development of value-added networks. There are studies advocated that simulation, virtual reality, autonomous robots, the industrial Internet of Things, the cloud, cybersecurity, additive manufacturing (3D printing), convergence of horizontal and vertical structures, and Big Data and analytics are the nine technologies that constitute Industry 4.0 (Rüßmann et al., 2015; Kerin and Pham, 2019). The difference between this transition and the previous ones is that technological advances are interconnected, they function in a collaborative manner, and all fields are mutually affected (Schwab, 2016).

So, stakeholders expect huge from prevailing revolutionary era and believe that these technological advancements will make the future even more exciting. Therefore, organizations are striving to implement Industry

4.0 technologies in their organization as early as possible to gain a competitive advantage over their competitors. However, the other side of the coin poses challenges and difficulties stakeholders are facing. In the section, we shall be looking at various challenges and issues, Industry 4.0 has brought along (Soylu, 2018).

6.2.1 CURRENT INDUSTRY 4.0 CHALLENGES

Researchers and consultants talk about the benefits and opportunities, the fourth industrial revolution offers. But, on the other side, Industry 4.0 transformation is posing numerous disruptions and challenges to the industries, and affecting each and every element of entire manufacturing, management, and governance structures across the globe. Industry 4.0 has even disrupted the business environment in which businesses operate. Today's business environment cannot escape from volatility, uncertainty, complexity, and ambiguity (VUCA). This was originated in the early 1990s by the US military to denote the current security climate and has also become a trendy managerial concept in business literature. The prevailing VUCA environment imposed by the current technological advancement has never been witnessed earlier. VUCA identifies a market climate defined by four distinct types of problems requiring new orientations, diverse solutions, and management approaches (Horney et al., 2010; Bennett and Lemoine, 2014).

- *Volatility* relates to the essence, speed, and degree of changes in an environment.
- *Uncertainty* refers to a company's inability to predict major future adjustments that could arise because of insufficient information availability.
- *Complexity* refers to the number of variables, their range, and the relationships between them. The more variables, the more complicated an environment is and the more difficult it is to evaluate.
- *Ambiguity* refers to a lack of understanding of, how something should be perceived.

The market climate in the VUCA world is complex in nature, and the scope and pace of change is difficult to predict. The shift is so sudden that knowing your path as an entity is difficult (Lawrence, 2013). Currently, exactly similar environment is prevailing around us, which is brought by Industry 4.0, and many organizations are clueless and do not even know how and from where to start' Industry 4.0 technology implementation. One the best example can be traced from the statement of Mr. Vinnie Mehta, Director general, Automotive Component Manufacturers Association. According to him, "Automated shopfloors integrated with Industry 4.0, personnel and machines talking to each other, robots and intelligent machines making decisions, AR and VR helping in product design, and understanding their behavior will be the norm of a factory in the not-too-distant future. The moot question we need to ask ourselves is: Can we afford to ignore these changes?" (Molly, 2020).

After reviewing the existing academic as well as grey literature, the authors have listed the following main challenges that are hindering organizations in implementing Industry 4.0 technologies (Table 6.1).

TABLE 6.1 Challenges for Implementing Industry 4.0.

Code	Challenges	References
C1	Lack of awareness of Industry 4.0 and its repercussions	Almada-Lobo, 2015
C2	Lack of qualified and technically skilled talent	Kumar, 2020
C3	Fear of failure of Industry 4.0 technologies (Risk Appetite)	Cezarino et al., 2021
C4	High initial cost of Industry 4.0 technologies	Yadav et al., 2020
C5	Lack of long-term planning on the adoption of Industry 4.0	Kumar, 2020
C6	Lack of clear digital strategic intent (vision, mission and strategy)	Erol et al., 2016
C7	Low support and dedication from management	Gökalp et al., 2017
C8	Lack of expertise in the adoption/application of modern business models	Khan et al., 2017; Raj et al., 2020
C9	Lack of digital infrastructure and internet-based networks	Bedekar, 2017
C10	Unclear financial benefit of digital investments	Marques et al., 2017
C11	Reluctant behavior toward Industry 4.0	Perales et al., 2018
C12	Lack of right culture to adopt and implement Industry 4.0	Schuh et al., 2017
C13	Lack of support from government bodies	BRICS Business Council, 2017
C14	Lack of funds for investment in Industry 4.0 technologies	Nicoletti, 2018

Out of the above 14 challenges the "Lack of knowledge" is the most important one that the majority of organizations are experiencing, while executing Industry 4.0 transformation (Almada-Lobo, 2015). Organizations are in a complete dilemma to overcome these challenges. In absence of any proven roadmap and well-developed procedures for adopting Industry 4.0, companies are trying to solve the issue using traditional methods or they are giving "hit n trial" in pursuit of some solutions, but most importantly these are not at helpful.

A few researchers have tried to deal with these problems and have suggested some solutions in their studies. According to Akkaya and Tabak (2020), organizations that wish to survive in these difficult circumstances, must step away from conventional strategies and continue to apply the improved or modern approaches more creatively. Li et al. (2018) mentioned in their study that, both managerial and operational capabilities are needed to successfully drive digital transformation. According to Kohli and Melville (2019), in this digital age, it is important to build sensing capabilities, such as entrepreneurial alertness and environmental scanning, in order to recognize innovative concepts, critically analyze design, change and ultimately deliver new business models. In this study, authors primarily try to examine the root cause of the Industry 4.0 implementation problem, the role of leadership in the Industry 4.0 era, and look for an ideal solution to the implementation problem by leveraging the leadership concept.

6.2.2 ROLE OF LEADERSHIP IN INDUSTRY 4.0

According to a study by Russell Reynolds Associates (2019), 62% of organizations in India are not willing to leapfrog to the Industry 4.0 technologies, until they completely exhaust the benefits of Industry 3.0 (Russell Reynolds Associates, 2019). So, such organizations will not transform even if all the supports are being given to them; be it management, government, and industrial support. This is a clear example that shows, how intent affects our actions and decision-making. Being an individual or an organization or a nation, without clear thought and vision one does not even start to act. Action can only be taken/performed when we intend to do that.

In the organizational context, the "strategic intent" determines the vision, mission, and strategies (to attain those futuristic positions) of the organization. People at the top level/strategic level/corporate also termed as leaders of the organization are held responsible for setting the intent of the organization. So, who is a leader actually? In general, the leader is the person who has the capacity to solve different problems in an organization or between people (Zaccaro et al., 2000), who creates, manages, and sometimes changes the organizational culture (Schein, 2010), brings people together for a purpose, and has an influence on people thanks to his/her motivation and personal characteristics (Vicentini et al., 2017). A leader's job is to decide a company's mission and vision, formulate suitable plans, motivate, and influence workers positively, empower employees, and build a corporate culture. They evaluate internal and external variables for the vision of the organization (Islam, 2012).

Strategic intent contemplates the desired leadership role and sets the parameters that companies can use to map their development (Smith, 1994). If leaders in such positions are reluctant to change, then how the organization will transform and adopt the emerging technologies.

The new industrial era does not only result in the advancement of technologies but also is a result of a new business mentality (Sterev, 2017). Industry 4.0 has challenged traditional leadership (such as bureaucratic and transactional leadership approaches, that seem less effective and adversely affecting the organizations) and demands a compatible leadership culture and approach to stay ahead and competitive enough in today's VUCA environment. This makes it necessary to understand the Industry 4.0 revolution and the leaders who will comprehend this revolution (Yüksel and Genç, 2018).

The leadership role is crucial because digitalization is not only about the new technologies, but also about their proper use (Bolte et al., 2018). Leaders are in the best position to influence the organization's culture, as mentioned earlier, and they need to have the confidence to do so. Positive organizational culture is a strategic advantage that cannot be easily duplicated by rivals and because it is one of the most difficult issues that need persistent attention, this can only be solved by leaders. It is the job of a leader to understand what the culture is, how to change it if necessary, and how to leverage that culture toward the outstanding performance of customer, client, and stakeholder services.

As far as the most appropriate style of leadership is concerned in the context of Industry 4.0, there is no mutual consensus on that and not much work has been done on this. And some believe that there is no single best leadership style that fits all situations. Leaders that use only one style of leadership are labeled as inflexible and face challenges in situations where it is not possible to apply the preferred style.

Kelly (2018), in her study, showed the leadership styles popular in different industrial revolutions. In Industry 1.0 charismatic leadership style was popular, in Industry 2.0 directive leadership style was prevalent and in Industry 3.0 transactional as well as transformational leadership style was popular. However, for Industry 4.0 Kelly mentioned that this era requires both traditional as well as new leadership characteristics. An ideal style of leadership for Industry 4.0 is digital leadership. A digital leader knows about the importance of innovations, technologies and people (Oberer and Erkollar, 2018). With a strong innovation emphasis, digital leadership (also known as leadership 4.0) is quick, multifunctional, human oriented, and with a supportive attitude.

However, the authors believe that the formal ILC at all the levels inside the organization is necessary to drive the transformational change toward Industry 4.0. The next section will highlight—What this solution is? And how it can be effectively implemented?

Leaders play a key role in the paradigm change and in fostering the innovation culture toward Industry 4.0. That is why, a leadership culture based on Industry 4.0 needs to be developed in enterprises (Oberer and Erkollar, 2018).

6.2.3 KEY LEADERSHIP COMPETENCIES FOR TODAY'S ERA

The fourth age of the industrial revolution needs leaders who can handle VUCA and adapt to the conditions (Akkaya and Tabak, 2020). Because of their structure, successful leaders do not see their role as adequate, but they are continually evolving and developing and enhancing their potential abilities. These are some primary leadership competencies that have stand out in the new era of Industry 4.0 for high-performing leaders:

- **Disruptive:** Build disruptive innovation and unique concepts, big picture thinker, traditional method problems, curious, strategic, focus on innovation, and differentiation.
- **Risk-taking:** Act deliberately and pragmatically, thrives on uncertainty, embraces difficulty, stays agile, not afraid to make errors when pushing boundaries, and seizes opportunities with enthusiasm.
- **Heroic:** Ambitious, stands firm in the difficulties, leads from the front, persistent devotion to higher expectations, commands persistence, and commitment to build followership.
- **Well-balanced:** Inspires through influence and charisma, empowers others and links the team both internally and externally, concentrates on team growth, keeps team motivated and productive.
- **Technologist:** Technically sound; carry a detailed understanding of use cases of technology and how to apply them for optimum benefit.
- **Commercial champion:** Good at numbers, bring customer insights to the table and can relate all operational decisions to the end market.

In general, the success of business leaders in the Industry 4.0 period depends on their skills and competencies (Zakaria et al., 2019). However, we need a shift in the skillset and the mindset to the organization level because no single leader is complete (Ancona et al., 2007).

6.3 ILC: KEY TO SUCCESSFUL INDUSTRY 4.0 TRANSFORMATION

ILC is about injecting the culture of leadership in the root of the system such that the leadership from a cautious act becomes a subcautious act of the system. It is about creating an environment of leadership in an organization where everyone works for the common purpose of serving the organization, and where everyone knows how he is going to serve that purpose. ILC is a concept that think leadership beyond the traits of a single leader, it is about a culture where everyone in the organization is constantly putting their efforts to move the organization in the future. Creating an internal environment of leadership that every stakeholder can sense, see, and get inspired from when they get in touch with the system. There is always a flow of energy, in and out of the organization. One of the main tasks of ILC is to make the flow of this energy go smoothly, which gets blocked at different paths due to negative energies. These bottlenecks should be removed to make the organization flow smoothly.

6.3.1 *WHY WE NEED ILC?*

The answer is sustainability of the organization. We need organizations to exist, and not exit in the long run. No organization wants to work for years and then be get wiped out of the market. Unfortunately, we have seen many organizations going out of the radar after few years of success. The statement holds true in the context of Indian business organizations too, witnessing a downfall and mergers across different sectors (RBI, 2018; Krishna et al., 2017). Business needs to be rolling in the future all the time for its incessant survival. It is something like, the heart needs to keep beating all the time to make us live, it cannot stop. But as change is natural and eternal, it affects the business cycle too, challenging the fate of the business.

Making the organization shine like a never-ending sun in the solar system of this business world will require managing the two main drivers of the business, the "*fuels*" of this system of ILC: (i) managing the current operations of the business-solving the complexities in the business, and (ii) keep visioning and rotating the wheel of the organization in the future. Both the drivers of ILC, managing difficulties and leading the organization in the future have its own requirements and entails different kinds of skills set. As said by Kotter (2001), management is about planning, organizing, and controlling, whereas leadership is about directing, aligning, and motivating. We will require people with both these skills to establish ILC. But, having them separately in an individual is a costly affair for any organization. Even more, it will be a waste of the learnings developed from the past experiences, which says that we can develop *leaders–managers*. Looks like a complex task! but no task in this world is without complexity, so does ILC. Installing leadership into the root of the organization, which is itself an adaptive change will require regulating distress, maintaining disciplined attention, the involvement of every employee, and constructive feedback (Heifetz and Laurie, 1997).

6.3.2 ESTABLISHING AND IMPLEMENTING ILC PROCESS

As mentioned by Kotter (2001), getting leadership in the culture of the organization calls for establishing the two major norms. One, *getting the right people on the bus* and two, *managing their career patterns*. However, we need to establish two more norms to maintain organizational vitality, *the culture of self-leadership and the culture of reverse mentoring*. In humans as well as in animals, being competent is a visible psychological phenomenon. It can be seen in a baby snapping a spoon for self-feeding, or in a calf attempting on the first day to stand on his feet. However, as we begin to experience setbacks in our careers, we appear to slow down this self-development process, we develop mental challenges, and build an environment of tension and anxiety around us. To get back on the road of progress, these hurdles need to be eliminated. Establishing self-leadership and reverse mentoring as a culture has the potential to make an organization run in an auto-performance mode. Primarily, *the concept of self-leadership* has got both theoretical and empirical support on improving organization vitality from the academicians and corporates (Bracht et al., 2018; Fowler, 2018; Manshi and Mishra, 2019; Neck and Houghton, 2006).

6.3.2.1 GETTING RIGHT FIRST (ACQUIRING PEOPLE WITH HIGH EMOTIONAL INTELLIGENCE (EI))

Recruiting People with High EI: The first step in the process of ILC is selecting people with high EI among those who have proved their Domain Expertise and the Intelligent Quotient (IQ). It is a two-level selection process.

As found out by Goleman (1996) in his study on personal capabilities, EI stood twice as important when compared to the other two skills for the job at all level and is a key skill, which differentiates between a good and great leader. Moreover, it starts showing its importance as the person moves up in the hierarchy. All of us have hierarchies, and we need them in the process of ILC, which authors have discussed later. Right now, authors would like to highlight that in a professional organization every employee gets equal chance of moving up in the hierarchy, and if that is true then EI becomes the most important qualifier skill to be evaluated because people low in EI will not be able to take well to great.

Check People Background on Incivility: Incivility is the rudeness of behavior, and people whose score high on incivility should be avoided even if they are competent. Incivility has the power to make the power failure in an organization by infecting organization culture. Moreover, research shows that incivility adds to the cost of the organization in the form of decreased work efforts, time spent at work, quality produced, lost work time, performance, commitment, and customer relations (Porath and Pearson, 2013).

Organization Personality and You: We all are individuals, and together we form groups and teams within an organization. The leaders' personality will define how the department or the organization will move and therefore we can say an organization also possess a personality. After performing a personality test on self, using the Myerr Brigg Type Indicator, the second author in the article found that he is more of an INTJ, which is Introvert, Intuitive, Thinking type, and Judging. When he analyzed the organization's personality type with, he is working for the past 11 years, he found that the organization personality is an ISFJ, which is Introvert, Sensing, Feeling, and Judging. He also evaluating his boss personality and came to the conclusion that the boss is ESFJ, that is, he is an extrovert, loves what is old, and moves inside the road decided, which clashes with two at times, the author's personality as well as with the organization type. Some people lie in their interview to get the position of choice, but fail because they do not perform well. One of the major reasons of their failure is that they did not fit well in the job. When people perform out of their basic personality they will not be satisfied. Choosing people based on their personality type (Bajic, 2015), organization's personality and its needs and wants, can solve many performance issues in the organization.

6.3.2.2 *MANAGING AND DEVELOPING THE CAREER PATTERN OF WORKFORCE*

Fueling the organization with a right kind of workforce, *a qualified workforce good in IQ, Domain Expertise and high on EI and Civility score)* is 20% of the job done in achieving ILC. Rest 80% of ILC is about aligning them with the organization goals, managing their career patterns, and providing the necessary support system by

- Establishing a goal-oriented purposeful corporate ladder and sensitizing employees to the associated sacrifices they must make during their journey with the organization.
- Providing a necessary support system to let them perform,
- Developing leaders–managers Profile by differentiating between managers and leaders. *"What makes a great executive and a great leader?"*
- Using active rather than the passive approach in developing employees.
- Spotting people with great leadership roles early in their career, and then nurturing them.

6.3.2.3 *INCORPORATING SELF-LEADERSHIP CULTURE*

For maintaining the organization's vitality and implementing the vision effectively down the hierarchy the leaders need followers who are self-leaders. Every contribution in the organization should be effective enough so that the assigned goals can be achieved successfully and become part of the success story of the organization. As researched by Blanchard et al. (2018) in their work on self-leadership, individuals at any level of the organization may feel the heat of failure at some point of time in their career due to personal or professional challenges, because of which even the best may develop constraints around them. It is important to question these constraints or experiences that restrict development to set the way forward for the successful execution of the vision of the leader. An organization needs more self-leaders who can replace the conventional wisdom of the Top-Down approach with the Bottom-up approach of getting what is needed to grow, instead of presuming that it is only the leader's duty to make them grow. Another core property of self-leadership is that it believes in activating different points of powers inside us to achieve both professional and personal goals. Some of these powers come to us naturally either by position, skills, personality, or behavior. But some powers can only be attained by connecting with other human resources in the organization who possesses the required skills to achieve the goals of the organization leaders. For example, an individual contributor may be in power by position in an organization but does not possess the required expertise required to make the project go great. In that case, an individual contributor should get in touch with the expert who may be present within the organization (say a project engineer having a good amount of experience but located at a different factory) or may be present external to the organization, says a friend, have a good experience in handling projects. Self-leaders are intelligent enough to realize their limits and do the best by engaging with others with the required power to go great. They are sufficiently involved to diagnose their own level of growth on the objectives set and determine the type of leadership style needed for them to grow. In short, in self-leadership, the emphasis is on supporters who challenge themselves for their developmental needs. For the global business climate, the postpandemic age has created new challenges and opportunities that will involve more self-leaders than leaders for the vitality of the enterprise. Organizations either need to find these self-leaders or need to develop a culture of self-leadership in an organization, as the concept of leading is rooted in motivation felt or perceived, which itself is rooted in the psychological satisfaction of human needs of self-development well defined in the theory of self-determination (Fowler, 2018).

6.3.2.4 INCORPORATING REVERSE MONITORING CULTURE

Industry 4.0 poses numerous challenges and demands paradigm shift or change and consequently, organizations are trying their best to bring the desired change in their entire value chain, in order to stay competitive or even to survive in the market. In today's digital era, it became imperative for organizations to acquire and retain technically and digitally sound talent. Now, the critical challenge is to train the senior and elder employees (who have been associated with the organization for a long time). Here, the concept of reverse mentoring gained momentum (Meister and Willyerd, 2010). This concept was given by the former CEO of GE, Jack Welch. Unlike conventional mentoring, in reverse mentoring senior employees are mentored by a younger or junior employee. According to Satterly et al. (2018), reverse mentoring helps the old generation in acquiring technical knowledge, understanding younger generations, and learning about current trends and concepts. On the other side, the young generation finds reverse mentoring as an effective way of developing leadership skills in them, because mentorship lets them step into a leader's shoes. Moreover, reverse mentoring offers numerous advantages to both individuals (the mentor and the mentee) and the organization such as—knowledge sharing, building strong engagement, developing leadership, and creating intergenerational relations (Gadomska-Lila, 2020). Reverse mentoring glimpse can be seen in Volvo group. Although many of their employees are working with the organization for 20–30 years and a few have just been there for a few years, they believe it is critical to have a diverse age group. This generates an environment in which many ideas and solutions emerge (Larsson & Nilsson, 2019).

6.3.2.5 ENSURING EFFECTIVE EXECUTION

Even the most visionary plans are useless if they are not executed properly. In today's VUCA environment, execution of plans is of immense importance, that sets aside the outperformed and underperformed organization. And for that matter, Raffoni (2003), outlined three keys for managers that ensure effective execution of plans. These are—Communicate the key points, develop tracking systems that facilitate problem solving and setting up formal reviews. Hence, it can be said that it is "execution" upon which the outcomes of the ILC process will depend.

6.4 CLOSING REMARKS

The fourth revolution is inevitable and economies around the world cannot afford to overlook it. Hence, adopting technology is no longer an option though its necessity. Challenges posed by this revolution are hindering the organizations from implementing Industry 4.0 technologies. Based on this study, leadership appears to be the most significant factor in driving Industry 4.0 transformation. Industry 4.0 has challenged traditional leadership (such as bureaucratic and transactional leadership approaches, that seem less effective and adversely affecting the organizations) and demands a compatible leadership embedded within the organizational culture to stay ahead and competitive enough in today's VUCA environment. By institutionalizing the right leadership culture at each organizational level (top/corporate level, middle/SBU level, and bottom/operational level) and stimulating the leadership competencies within each employee; organizations can drive the successful Industry 4.0 transformation. And, acquiring the right people, managing their career pattern, establishing a culture of Self-leadership, reverse monitoring, and effective execution, are the five building block of the proposed ILC process. So, the major takeaways of this chapter are as follows:

- Industry 4.0 transformation is posing numerous challenges to the industries that are hindering companies to adopt Industry 4.0. *"Lack of awareness of Industry 4.0 and its repercussions," "Lack of qualified and technically skilled talent," "Lack of clear digital strategic intent (vision, mission, and strategy)," "Lack of right culture to adopt and implement Industry 4.0"*—are just to name a few challenges.
- Industry 4.0 has challenged traditional leaderships styles and demands a compatible leadership embedded within the organizational culture to stay ahead and competitive enough in today's VUCA environment.
- An ideal style of leadership for Industry 4.0 is digital leadership. With a strong emphasis on innovation, digital leadership (also known as leadership 4.0) is quick, supportive, multifunctional, and human oriented. He knows the importance of innovations, technologies, and people and possess both traditional as well as new leadership characteristics. However, as no single leader is complete, leadership needs to be a culture in an organization.
- ILC is about injecting the culture of leadership in the root of the system, to create an environment of leadership in an organization where everyone works as a leader.
- Getting leadership in the organizational culture requires four major norms—*Getting the right people, managing their career patterns, culture of self-leadership, and reverse mentoring.*
- Self-leaders are intelligent enough to realize their limits and do the best by engaging with others with the required power to go great. They are sufficiently involved to diagnose their own level of growth on the objectives set and determine the type of leadership style needed for them to grow.
- Inculcating a culture of reverse monitoring is paramount. In today's digital era, organizations strive to acquire and retain technically and digitally sound talent. But the critical challenge is to train the senior and elder employees. In reverse mentoring, senior employees are mentored by a younger or junior employee. This helps the old generation in acquiring technical knowledge, understanding younger generations.

- Above all, ensuring the consistent exercise/practice of established ILC is of vital that sets aside the outperformed and underperformed organization. It is "execution" upon which the outcomes of the ILC process will depend. And eventually the successful implementation of Industry 4.0 will be affected.

6.4.1 IMPLICATIONS OF THE STUDY

This chapter maps the significant contribution of leadership in managing transformational challenges. Theoretically, the chapter examines the concept of "ILC," "Reverse Mentoring," "Self-Leadership," and Key leadership competencies in relevance to Industry 4.0. The chapter contributes to the knowledge base and understanding of future leadership. HR personnel may look for desired leadership competencies while acquiring talent or planning for training and development, and top-level managers can utilize the information for establishing better workplace culture to implementing Industry 4.0 in their organization.

6.4.2 FUTURE RESEARCH DIRECTIONS

ILC emerges as a new concept that is gaining the attraction of academicians as well as practitioners. Despite its perceived benefits and importance, not much has been written on this and rarely acknowledged. Therefore, the concept has immense potential and needs more exploration and empirical validation. The authors recommend that the ILC process proposed in this chapter can be further verified practically and empirically in different organizations and industries with respect to different conditions.

KEYWORDS

- **Industry 4.0**
- **leadership**
- **self-leadership**
- **institutionalization of leadership**
- **VUCA**
- **strategic intent and reverse monitoring**

REFERENCES

Akkaya, B.; Tabak, A. The Link Between Organizational Agility and Leadership: A Research in Science Parks. *Acad. Strateg. Manage. J.* **2020,** *19*(1), 1–17.

Almada-Lobo, F. The Industry 4.0 Revolution and the Future of Manufacturing Execution Systems (MES). *J. Innov. Manage.* **2015,** *3*(4), 16–21.

Ancona, D.; Malone, T.; Orlikow, W.; Senge, P. In Praise of the Incomplete Leader. *Harvard Business Review*, February 2007. https://hbr.org/2007/02/in-praise-of-the-incomplete-leader

Arklan, Ü. Information Society and Communication: Mass Media and Internet in the Process of Dissemination of Information. *Selcuk Contact* **2008,** *5*(3), 67–80.

Bajic, E. How The MBTI Can Help You Build A Stronger Company. September 28, 2015. https://www.forbes.com/sites/elenabajic/2015/09/28/how-the-mbti-can-help-you-build-a-stronger-company/?sh=589bf57ed93c

Bauernhansl, T.; Ten Hompel, M.; Vogel-Heuser, B., Eds. *Industry 4.0 in Production, Automation and Logistics: Application-Technology-Migration*; Springer Vieweg: Wiesbaden, 2014; pp 1–648.

Bennett, N.; Lemoine, G. J. What VUCA Really Means for You Harvard Business Review. *Harvard Business Review (January-February)*, 2014, p 27.

Blanchard, K.; Fowler, S.; Hawkins, L. *Self Leadership and the One Minute Manager*; Harper Collins, USA, 2018.

Bolte, S.; Dehmer, J.; Niemann, J. Digital Leadership 4.0. *Acta Tech. Napocensis Ser. Appl. Math. Mech. Eng.* **2018,** *61*(4).

Bracht, E. M.; Junker, N. M.; van Dick, R. Exploring the Social Context of Self-Leadership—Self-Leadership-Culture. *J. Theor. Soc. Psychol.* **2018,** *2*(4), 119–130.

BRICS Business Council. Skill development for Industry 4.0. A White Paper by BRICS Skill Development Working Group, BRICS Business Council, India Group. [Online] 2017. http://www.globalskillsummit.com/Whitepaper-Summary.pdf (accessed Oct 22, 2017).

Cezarino, L. O.; Liboni, L. B.; Oliveira Stefanelli, N.; Oliveira, B. G.; Stocco, L. C. Diving Into Emerging Economies Bottleneck: Industry 4.0 and Implications for Circular Economy. *Manage. Decision.* **2021,** *59*(8), 1841–1862.

Erol, S.; Schumacher, A.; Sihn, W. In *Strategic Guidance Towards Industry 4.0–a Three-Stage Process Model.* International Conference on Competitive Manufacturing, Jan 2016, *9(*1), 495–501.

Fowler, S. Toward a New Curriculum of Leadership Competencies: Advances in Motivation Science Call for Rethinking Leadership Development. *Adv. Develop. Hum. Resources* **2018,** *20*(2), 182–196.

Gadomska-Lila, K. Effectiveness of Reverse Mentoring in Creating Intergenerational Relationships, *J. Org. Change Manage.* **2020,** *33*(7), 1313–1328.

Gökalp, E.; Sener, U.; Eren, P.E. In *Development of an Assessment Model for Industry 4.0: Industry 4.0-MM*, International Conference on Software Process Improvement and Capability Determination; Springer: Cham, 2017; pp 128–142.

Goleman, D. *What Makes a Leader? (Harvard Business Review Classics)*; Harvard Business Press: Boston, USA, 2017.

Heifetz, R. A.; Laurie, D. L. The Work of Leadership. *Harvard Business Rev.* **1997,** *75*, 124–134.

Hofmann, E.; Rüsch, M. Industry 4.0 and the Current Status as Well as Future Prospects on Logistics. *Comp. Ind.* **2017,** *89*, 23–34.

Horney, N.; Pasmore, B.; O'Shea, T. Leadership Agility: A Business Imperative for a VUCA World. *Hum. Res. Plan.* **2010,** *33*(4), 34.

Islam, T.; Aamir, M.; Ahmed, I.; Muhammad, S. K. The Impact of Transformational and Transactional Leadership Styles on the Motivation and Academic Performance of Students at University Level. *J. Educ. Soc. Res.* **2012,** *2*(2), 237–237.

Kelly, R. *Constructing Leadership 4.0: Swarm Leadership and the Fourth Industrial Revolution*; Springer: USA, 2018.

Kerin, M.; Pham, D. T. A Review of Emerging Industry 4.0 Technologies in Remanufacturing. *J. Cleaner Product.* **2019,** *237*, 117805.

Khan, M.; Wu, X.; Xu, X.; Dou, W. In *Big Data Challenges and Opportunities in the Hype of Industry 4.0.* 2017 IEEE International Conference on Communications (ICC), May 2017, pp 1–6; IEEE: USA.

Kharb, A. Industrial Revolution–From Industry 1.0 to Industry 4.0. *J. Adv. Comput. Intellig. Commun. Technol.* **2018,** *2*(1), 1–2.

Kohli, R.; Melville, N. P. Digital Innovation: A Review and Synthesis. *Inf. Syst. J.* **2019,** *29*(1), 200–223.

Kotter, J. P. What Leaders Really Do. *Harvard Bus. Rev.* **2001,** *79*(11).

Krishna, A.; Agrawal, A.; Choudhary, A. In *Predicting the Outcome of Startups: Less Failure, More Success.* 2016 IEEE 16th International Conference on Data Mining Workshops (ICDMW), Dec, 2016, pp 798–805; IEEE: USA.

Kumar, R. Sustainable Supply Chain Management in the Era of Digitalization: Issues and Challenges. In *Handbook of Research on Social and Organizational Dynamics in the Digital Era*, 2020, pp 446–460.

Larsson, L.; Nison, J. Facilitating an Industry 4.0 Implementation, 2019. https://kth.diva-portal.org/smash/get/diva2:1334053/FULLTEXT01.pdf.

Lawrence, K. Developing Leaders in a VUCA Environment, 2015. UNC Executive Development 2013.

Li, L; Su, F.; Zhang, W.; Mao, J. Y. Digital Transformation by SME Entrepreneurs: A Capability Perspective. *Inf. Syst. J.* **2018,** *28*(6), 1129–1157.

Marques, M.; Agostinho, C.; Zacharewicz, G.; Jardim-Gonçalves, R. Decentralized Decision Support for Intelligent Manufacturing in Industry 4.0. *J. Ambient Intell. Smart Environ.* **2017,** *9*(3), 299–313.

Mburu, S. N.; Thuo, J. K. Understanding the Concept of Strategic Intent. *Int. Refereed J. Eng. Sci. (IRJES)* **2015,** *4*(6), 57–60.

Meister, J. C.; Willyerd, K. Mentoring Millennials. *Harvard Business Rev.* **2010,** *88*(5), 68–72.

Molly, R. Industry 4.0: A Great Future, January 13, 2020. https://www.manufacturingtodayindia.com/sectors/6089-industry-40-a-great-future

Neck, C. P.; Houghton, J. D. Two Decades of Self-Leadership Theory and Research: Past Developments, Present Trends, and Future Possibilities. *J. Manag. Psychol.* **2006,** *21*(3), 270–295.

Oberer, B.; Erkollar, A. Leadership 4.0: Digital Leaders in the Age of Industry 4.0. *Int. J. Org. Leadership* **2018.**

Perales, D. P.; Valero, F. A.; García, A. B. Industry 4.0: a Classification Scheme. In *Closing the Gap Between Practice and Research in Industrial Engineering*, 2018, pp 343–350.

Porath, C., Pearson, C. The Price of Incivility. *Harvard Business Rev.* **2013,** *91*(1), 115–121.

Raffoni, M. Three Keys to Effective Execution. *Harvard Manage. Update*, **2003,** *8*(2), 1–4.

Raj, A.; Dwivedi, G.; Sharma, A.; de Sousa Jabbour, A. B. L.; Rajak, S. Barriers to the Adoption of Industry 4.0 Technologies in the Manufacturing Sector: An Inter-country Comparative Perspective. *Int. J. Product. Eco.* **2020,** *224*, 107546.

Reserve Bank of India. *Report on Trend and Progress of Banking in India 2017–2018.* RBI. 2018. https://rbidocs.rbi.org.in/rdocs/Publications/PDFs/0RTP2018_FE9E97E7AF7024A4B94321734CD76DD4F.PDF

Rüßmann, M.; Lorenz, M.; Gerbert, P.; Waldner, M.; Justus, J.; Engel, P.; Harnisch, M. Industry 4.0: The Future of Productivity and Growth in Manufacturing Industries. *Boston Consult. Group*, **2015,** *9*(1), 54–89.

Russell Reynolds Associates. *Transforming Manufacturing in the Era of Industry 4.0:A Perspective from India,* 2019. https://www.russellreynolds.com/insights/thought-leadership/transforming-manufacturing-in-the-era-of-industry-4point0-a-perspective-from-india

Sae-Lim, P.; Jermsittiparsert, K. Is the Fourth Industrial Revolution a Panacea? Risks Toward the Fourth Industrial Revolution: Evidence in the Thai Economy. *Int. J. Innovat. Creativity Change.* **2020,** *5*(2), 732–752.

Satterly, B. A.; Cullen, J.; Dyson, D. A. The Intergenerational Mentoring Model: An Alternative to Traditional and Reverse Models of Mentoring. *Mentor. Tutor. Partnership Learn.* **2018,** *26*(4), 441–454.

Schein, E. H. *Organizational Culture and Leadership* (Vol. 2); John Wiley and Sons: USA, 2010.

Schuh, G.; Anderl, R.; Gausemeier, J.; ten Hompel, M.; Wahlster, W. Industrie 4.0 Maturity Index. In *Managing the Digital Transformation of Companies*; Munich: Herbert Utz., 2010.

Schwab, K. *The Fourth Industrial Revolution*; Optimist Publishing Group, 2016.

Smith, C. E. The Merlin Factor: Leadership and Strategic Intent. *Bus. Strategy Rev.* **1994,** *5*(1), 67–84.

Soylu, A. Industry 4.0 and New Approaches in Entrepreneurship. *Pamukkale Univ. J. Soc. Sci. Inst.* **2018,** (32), 43–57.

Sterev, N. Marketing Leadership: The Industry 4.0 Need of Next Generation Marketing. *Trakia J. Sci.* **2017,** *15*(1), 99–103.

Vicentini, F.; Brunetta, F.; Capo, F. Building Organizational Insight: Strategy and Organization. In *Global Business Strategies in Crisis*; Springer: Cham, 2017; pp 43–56.

Yadav, G.; Luthra, S.; Jakhar, S. K.; Mangla, S. K.; Rai, D. P. A Framework to Overcome Sustainable Supply Chain Challenges Through Solution Measures of Industry 4.0 and Circular Economy: An Automotive Case. *J. Cleaner Product.* **2020,** *254*, 120112.

Yadav, M.; Mishra, S. K. Self-Leadership as a Tool for Enhancing Performance at Workplace. *GIS Business,* **2019,** *14*(6), 76–88.

Yüksel, M.; Genç, K. Y. In *Industry 4.0 and Leadership.* 2nd International Symposium on Innovative Approaches in Scientific Studies, pp 338–341, Nov 2018 .

Zaccaro, S. J.; Mumford, M. D.; Connelly, M. S.; Marks, M. A.; Gilbert, J. A. Assessment of Leader Problem-Solving Capabilities. *Leadership Quart.* **2000,** *11*(1), 37–64.

Zakaria, N.; Nasir, A.; Akhtar, A. Are the Leaders Ready to Embrace Industry 4.0. *J. Adv. Res. Dynam. Control Syst.* **2019,** *11*(5), 543–549.

CHAPTER 7

Virtual Leadership: An Integral Phenomenon of Industry 4.0

MALABIKA TRIPATHI[1*], NALANDA ROY[2], MANVI SODANI[2], and SWAHA BHATTACHARYA[3]

[1]*Department of Applied Psychology, Maulana Abul Kalam Azad University of Technology, Haringhata, India*

[2]*Amity University, Kolkata, AIPAS, India*

[3]*Calcutta University, Kolkata, India*

[*]*Corresponding author. E-mail: malabikatripathi1@gmail.com*

ABSTRACT

The Fourth Industrial Revolution or Industry 4.0, in extremely simple terms, incorporates the advanced manufacturing techniques along with "the internet of things." This embodiment creates a system of manufacturing that is in turn connected with the physical world. Industry 4.0 includes processes like communication, analysis, and the utilization of information for further actions, in accordance to its technological pillars that have shaped this whole revolution. These pillars include cybersecurity, cloud computing, big data and analytics, robotics, augmented reality, additive manufacturing, simulation, horizontal and vertical integration, and internet of things. In all these pillars, leadership would vary immensely, however, there are some common aspects involved in virtual leadership. Essentially, leadership is the ability to guide either a group or an individual toward a specific goal or direction. Alternatively, virtual leadership would vary from traditional leadership in a way that while guiding their team, leaders would have an extremely different management style. Thus, it is extremely important to understand how

Agile Leadership for Industry 4.0: An Indispensable Approach for the Digital Era, Tanusree Chakraborty, Tahir Mumtaz Awan, Ashok Natarajan, and Muhammad Kamran (Eds.)

leadership would evolve in the Fourth Industrial Revolution. This chapter will discuss, in depth, the concept of virtual leadership and its need for the advancement of the business industry. The methodology involved in this chapter would be review of literature that would explicitly identify, select, and appraise relevant researches. The chapter will focus on understanding specific research questions like (1) Parameters of virtual leadership styles used in Industry 4.0; (2) How virtual leadership is challenging traditional leadership style; and (3) How effective utilization of this change in leadership can help organizations grow exponentially. The analysis of literature can deduce that virtual leadership is the most important aspect in Industry 4.0 to help organizations flourish in the current technologically advanced world. This chapter also provides scope for further research in this field.

7.1 INTRODUCTION

Each industrial revolution has marked the onset of a new era—be it by the invention of steam engines or digitization and artificial intelligence. The introduction of these new industries marks significant changes in the levels of job complexity and associated skill sets which are required by individuals in order to adapt to society (Selamat et al., 2017).

Digitization of industries have led organizations to build expertise in information networking, business analytics, artificial intelligence, synthesize physical reality and digital transactions, and integrate the use of artificial intelligence (Oberer and Erkollar, 2018). Industry 4.0 (I4.0) is the Fourth Industrial Revolution which is rapidly digitizing the traditional business industry. This transition is responsible for instilling heightened competitiveness among various organizations, in terms of the quality of workforce being hunted for—their skills, behaviors, motivation, and attitudes (Gehrke et al., 2015).

Adaptability of organizations to these changing and challenging times is what marks the extent of success of a business. Majority of this adaptability rests upon the shoulders of the employees of a particular company, which is why their skill sets and ability to develop and learn new techniques is what predicts the degree of successful integration of Industry 4.0 (Maisiri and Van Dyk, 2019). Digital innovations in businesses have made remarkable changes in the way businesses are done (Chakraborty and Saha, 2018). It is for this reason that new-age leaders need to effectively manage human resources (Paine, 2009) because they are responsible for strengthening a talented and indispensable workforce based on which the industry will flourish (Accenture Consulting, 2017).

7.1.1 INDUSTRY 4.0

The term Industry 4.0 was first used in Germany, 2011. It outlined the techniques and skill sets required by companies to maintain market superiority and how its' leaders could effectively integrate technology and automation into company operations (Buliga and Voigt, 2018). It can be defined as the individualization of the manufacturing process as per customer requirement, which includes a recent level of organization that controls the entire value chain of products (Lorenz et al., 2014). The focus is on the development of an open, keen assembling stage for mechanical organized data application (Bahrin et al., 2016).

7.1.2 PILLARS OF INDUSTRY 4.0

There are nine major pillars in Industry 4.0 that assist in the transformation of the organizations into highly digitalized and intelligent mode (Ambad et al., 2018). They make manufacturing a completely integrated process that is not only automated but also facilitates an optimized flow of production. This increases the efficiency of organizations and changes the traditional relationship of producers, suppliers, and customers along with the obvious change between humans and machine interaction (Lorenz et al, 2014). This includes big data analytics, the internet of things, autonomous robots, simulation, cyber security and cyber-physical systems (CPS), system integration, augmented reality, the cloud, and additive manufacturing.

7.1.3 NATURE OF JOBS IN INDUSTRY 4.0

Each industrial revolution brings about more complex skill sets required by employees to fulfill the changing organizational goals (Selamat et al., 2017). Since the world has changed from a manufacturing-based industry to now a service-based industry, jobs that are semiskilled or unskilled no longer play a primary role (Maisiri et al., 2019). Automation, artificial intelligence, and digitization have successfully taken over the physical laborious aspects of an industry. Hence, job requirements today comprise of those skills which cater to overseeing the functioning and implementation of automation. Maisiri et al. (2019) highlighted the three categories on the basis of which Industry 4.0 has affected the nature of jobs—"working practices, employees' opportunities, and jobs and skills." The most important prerequisite that employees must possess today is the ability to be adaptive and flexible in their work patterns.

7.1.4 VIRTUAL LEADERSHIP IN INDUSTRY 4.0

The concept of leadership is fundamental in industries for an effective and sustainable advancement, cultural development, and creativity. It requires the ability to have a positive impact on the team and help them to flourish together (Lussier and Achua, 2013). Virtual leadership involves managing teams through a remote working environment.

The Path Goal theory of leadership states that if required, leaders may achieve their goals and establish innovative policies more efficiently by initiating a change in the current style of leadership based on situation (House, 1971). By linking this to the current scenario, change in leadership is one of the most important necessities of organizations today because of changing industrial environment. Unconventional leadership styles with new skills and innovative problem-solving techniques are needed for businesses to thrive in Industry 4.0 (Peshawaria, 2018).

In this context, Figure 7.1 (Kolditz et al., 2017; Oosthuizen, 2017) explains the key qualities needed by leaders in Industry 4.0 in order to ensure organizational success.

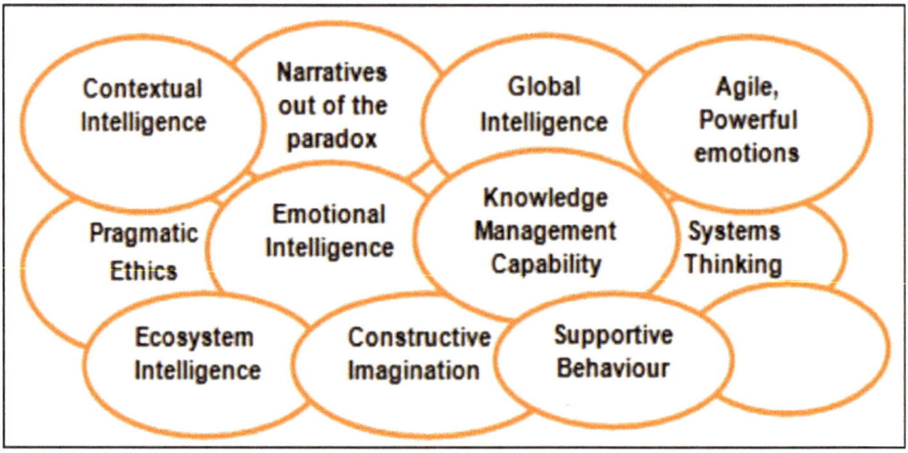

FIGURE 7.1 Industry 4.0 leaders key skills.

Source: Kolditz et al., 2017; Oosthuizen, 2017.

In a traditional office setting, the leaders constantly see the employees, they observe others' behavior, as well as interact and communicate with employees and create an environment of innovation, positivity, and productivity (Mawson, 2020). As we transcend into a big data and augmented reality world, the leaders of Industry 4.0 should be more "connected, responsive, collaborative, and experienced with networks" to execute their jobs effectively (Kelly, 2015). Advanced web-based technologies like Zoom, Microsoft Teams, Skype, and Google Meet among others are ways via which the presence of leaders may be felt even in their physical absence. Virtual leadership is the style of leadership which caters best to these needs of the organizations affected by Industry 4.0. It is bound to help overcome challenges and integrate necessary changes within the company's framework (Xu et al., 2018). Virtual leadership ensures its' prioritization, where traditional leaders fail to keep up with the integration of the latest smart technology (Peshawaria, 2018). Thus, the changing mode of production in Industry 4.0 requires a highly strategic and organized leadership approach in every dimension (Kasapoğlu, 2018). Erkollar and Oberer (2018) highlighted the ideal type of leadership in Industry 4.0 under their "4.0 Leadership Matrix," which is the Digital Leader style. They synthesize the traditional leadership knowledge with the latest technological techniques, and are able to assess the impact of innovation, employees, and digitalization on the wider society.

The present chapter focuses on analyzing three research propositions—the parameters of virtual leadership styles used in Industry 4.0, how virtual leadership is challenging traditional leadership style, and how effective utilization of this change in leadership can help organizations grow exponentially.

7.2 RATIONALE OF THE STUDY

The world has seen a multitude of growth and advancements in technology. Where certain organizations were soaring high, others were trying hard to stay afloat because of their traditional practices. There was a dire need for the practices and policies of virtual leadership and Industry 4.0, not only to sustain their livelihoods, but also to ensure their existence in the world of intelligence-based augmented reality. The traditional leadership styles were not valid in this technology-based industry. A thorough review of literature was conducted to evaluate the importance of virtual leadership in the fourth industrial revolution that can promote organizational growth to bridge the gap and assist organizations in moving forward.

7.3 METHODOLOGY

The current research is based on the review of literature, which facilitates the readers to gain an in-depth understanding through exploring the specific research questions. Additionally, this study also provides an analytical rumination and assists in the scope for future research. A literature review is the analysis of sources including articles, books, journals, and more which are pertinent to the area of study in this chapter providing a critical assessment (Fink, 2014). We conducted a search on Google Scholar, SSRN, and other online journals with the terms "Industry 4.0" and "Virtual Leadership." The researches were critically evaluated to establish and construct the chapter. The current study directs toward the following research propositions:

1. Parameters of virtual leadership styles used in Industry 4.0.
2. How virtual leadership is challenging traditional leadership style?
3. How effective utilization of this change in leadership can help organizations grow exponentially?

7.4 FINDINGS

7.4.1 RQ 1. PARAMETERS OF VIRTUAL LEADERSHIP STYLES USED IN INDUSTRY 4.0

Being seen and heard in an age where physical presence can be substituted digitally is likely the biggest challenge to leadership styles in the 21st century. With remote working environments thriving this decade, inspiring and motivating one's employees take precedence over all else. Therefore, virtual leadership depends largely on its degree of influence and effectivity in establishing open, digital dialogues in order to increase transparency between workers and the company.

Since Industry 4.0 comprises of cyber-physical systems, cloud computing and automation of manufacturing processes, adaptation to the current climate is the first and foremost goal of the business industry. Transitions in areas pertaining to safety and security, standardization of operations, aggregated communication systems, and continuous professional development are the pivots hinged on which organizations are running (Kagerman et al., 2013). They need to adapt to the advanced manufacturing processes, deal with novel challenges, focus on innovation and creativity, and establish new ways of communication inside and outside the organization (Erkollar and

Oberer, 2018). Therefore, this change in tide brought about by Industry 4.0 has raised the necessity of a new model of leadership which will address the aforementioned key aspects of the evolved business industry. The skills of a knowledge-oriented leadership style in adjunction with transactional and transformational leadership styles are combined within virtual leaders—as they facilitate easy integration of technology into a company (Sivathanu and Pillai, 2018). Together, these three styles within virtual leadership account for honing employee-talent within an organization and generating new ideas (Naqshbandi and Jasimuddin, 2018; Schwab, 2017). The most important unique role of virtual leadership is to be one-step ahead of the rapid change mobilized by Industry 4.0. Therefore, constant upgradation of specialized skills and the proliferation of novel and creative ideas is a necessity (Hensellek, 2019).

Therefore, the parameters of virtual leadership styles which are imperative in Industry 4.0 include which cater to the antecedent needs of the digitalized business industry. They must understand how the different elements of an organization fit together and must incorporate aspects such as mindfulness leadership and fundamental integration of different structures of an organization (Tareque and Islam, 2021). By internalizing this expertise, leaders can adapt to the changes brought about by Industry 4.0 effectively. Extending from these, we have listed the following set of parameters of virtual leadership used in Industry 4.0.

7.4.1.1 INCREASING COLLABORATION AND PROMOTING CREATIVITY

The conjoined variables of collaboration and creativity predict the efficiency with which technology will be integrated into a certain organization (Naqshbandi and Jasimuddin, 2018)—which is integral to its adaptability to Industry 4.0. With emerging trends related to mobile technologies, big data analytics, and 5G, virtual leadership needs to promote interconnectivity and collaboration within the organization so as to deal with technological challenges. The more collaborations are done, the chances of coming up with creative solutions and innovative ideas increases because of interaction between multiple people (Kelly, 2018). New and diversified perspectives are taken into account, which enable leaders to take better decisions and help employees think of novel ideas—thereby enhancing creativity within the company. This is an extremely important quality for virtual leaders to have as the continuous practice of promoting innovation and creativity through collaboration becomes the key to a successful organization (Ashkenas and Manville, 2018).

7.4.1.2 OPEN AND TRANSPARENT COMMUNICATION

Another major parameter of virtual leadership style to adapt to Industry 4.0 is the ability to communicate effectively (Maisiri and Dyk, 2020). Without the benefit of face-to-face interaction, the chances of miscommunication between colleagues increase, thereby hindering optimum organizational growth. Hence, virtual leaderships must focus primarily on establishing effective, open, and transparent dialogues within all strata of an organization. Proactive communication with employees regarding relevant and important company information and decision-making plans is essential to help them learn about the visions and goals that they need to work toward. To enhance communication effectivity, virtual leaderships must improve their soft skills, conflict management techniques, and emotional intelligence in order to understand the needs of the employees and establish content of communication accordingly (Carter, 2017; Krot et al. 2018). The ability to communicate a "clear vision in a complex environment" is what is required from an ideal virtual leadership (Ameer, 2020).

7.4.1.3 INCREASING EMPLOYEE MOTIVATION AND INCLUSIVITY

Virtual leadership relies significantly on its ability to increase employee participation in company proceedings and promote an overall culture of inclusivity. They must be consultative and participative, so as to ensure that employees feel valued and interested in their work, despite remote working setups (Korzynski, 2013). By integrating employee participation in different areas of a project (from planning to decision-making, etc.), the chances of yielding helpful insight, securing employee-loyalty, and establishing trust between virtual leaders and workers can be increased.

7.4.1.4 FLEXIBILITY AND RAPID DECISION-MAKING SKILLS

Adapting to today's rapidly changing environment filled with competitive businesses using the latest technology, is the primary threat that organizations in Industry 4.0 face. Hence, virtual leaderships need to be characterized by fast decision-making skills and flexibility in terms of tackling new challenges of the digitized industry (Ameer, 2020). Compared to traditional leadership styles, virtual leadership is more open and prioritizes knowledge-oriented leadership (Zakaria, 2019).

FIGURE 7.2 Model representing key parameters of virtual leadership.

Source: Developed by authors.

Based on the extensive research conducted, there are a few more important parameters that play a crucial role to enhance the virtual leadership practices in organizations that can be seen in Figure 7.2. In the current remote-working conditions and Industry 4.0, it has become necessary for the leaders to create an environment where the employees can not only converse openly but can also promote creativity and flexibility among employees and leaders.

7.4.2 RQ 2. HOW VIRTUAL LEADERSHIP IS CHALLENGING TRADITIONAL LEADERSHIP STYLE?

In a time when mechanical operations would dominate the business and manufacturing industry, traditional styles of leadership fit hand-in-glove with the needs of organizations. Leaders have traditionally always been seen through the smokescreen of being in a position of power, authority, and as being a reinforcer of hierarchy. The reason why the business community during Industry 1.0 and 2.0 was best suited to traditional leadership style was because of the cultural aspect of society during the 1800s to the 1900s. People needed focus, direction, authority, and control because of the uncertain periods that the world had gone through—World War I (1914–1918); the Great Depression (1929); World War II (1933–1945); the Cold War (1947–1991). However, with technology on the rise from late 1900s to the early 2000s, and now with the onset of Industry 4.0, focus of industries have shifted from craving stability to integrate creativity and growth. Virtual leadership provides an alternative to the traditional leadership styles that would require physical, face-to-face interaction with employees. Organizations in the present context would require a positive organizational culture (Chakraborty and Ganguly, 2019), will need positive approaches to understand employees by appreciative questioning (Chakraborty and Mishra, 2019; Chakraborty et al., 2019). Organizations are more agile after a disaster, so the role of human resource professionals in managing after effects of a crisis is very crucial (Chakraborty and Saha, 2017). In essence, it challenges the functioning of leadership styles that do not adhere to digital platforms.

With the mode of communication, production, and other activities being digitized, neither the traditional leadership styles are relevant in Industry 4.0, nor the typical skills that employees would previously possess (Thomas, 2018). This change in the industry has created scope for new opportunities, development of skills, and diversified ways via which businesses can function. Hence new forms of leadership are required to mirror these transformations and ensure the efficient functioning of organizations (Vlasov and Chromjaková, 2018). In a time when the unpredictability of the business industry can potentially prove to be a massive threat to organizations, the ways by which virtual leadership challenges traditional leadership styles must be understood in order to quicken its integration into business models mentioned below.

7.4.2.1 INITIATING THE GROWTH OF CREATIVITY AND INNOVATION

Traditional leadership styles operate on the idea of growing businesses pre-Industry 4.0. New strategies and ideas for organizational growth are restricted due to their limited knowledge of the technological aspects of the current global scenario. This is why virtual leadership are rapidly turning into the only viable alternative to traditional leadership styles because it keeps assessing the latest technology and how new artificial intelligence can be integrated into business models, so that organizational goals can be achieved (Senduk, 2017). Virtual leaders entrust employees with more independence to facilitate creative growth. Since remote working depends on the trust built between leaders, managers, and employees, it invariably results in employees receiving the freedom to work on the basis of their own innovations while adhering to company guidelines. University of Michigan (1950s) further highlights how virtual leadership triumphs over the previously dominant employee-centered leadership style by focusing more on guiding employees to provide innovative and novel proposals, rather than by only enforcing a passive, supportive workplace environment. Therefore essentially, virtual leadership moves one step ahead by addressing multiple variables that may increase employee satisfaction and performance levels—which, in turn, affects their creativity (Taherkhani, 2015).

7.4.2.2 CHANGE IN COMMUNICATION PATTERN AND PRIORITIZATION OF STRUCTURE

The Ohio State Model of Leader Behavior (1940s) will be analyzed to provide a better understanding of the differences in communication patterns between virtual leadership and traditional leadership styles. It also incorporates their difference in prioritization of roles structure in an organization, and how this difference makes virtual leadership more suitable for Industry 4.0.

As per the Ohio State Model of Leader Behavior (1940s), there are four leader behavior styles based on two variables—*consideration* and *initializing structure*, where the former refers to the leader's attitude regarding the welfare of employees, and the latter refers to the degree to which a leader defines the roles in an organization and enforces one-way communication. The four behavior styles are "high consideration, high structure," "low consideration, low structure," "high consideration, low structure" and "low consideration, high structure," and is generally used to describe traditional leadership styles based on the priorities of the leaders.

Traditional leaderships can therefore choose the axis of their leadership style—whether they will prioritize downward communication, structure, or concern for employees. Virtual leadership challenges this model of traditional leadership is because there remains no option for a virtual leader to choose one variable over the other. They need to establish a balance between both promoting employee participation and maintaining a two-way, open communication with them. Catering the needs of employees (consideration) involves constructing ways by which employee engagement, satisfaction, and motivation can be increased. On the other hand, ensuring that their roles are defined well enough (initializing structure) so as to avoid possible miscommunication and overlapping of tasks in the future. The reason why virtual leadership needs to maintain an equilibrium between both aspects is because the risk of employee disengagement, demotivation, and social isolation is higher in the remote working condition of Industry 4.0 (Fedakova and Ištoňová, 2017). This further increases the chances of social loafing and decreased performance levels (Wojcak et al., 2016). Thus, by establishing an open, honest, and simplified dialogue, by initiating inclusive activities, virtual leadership ensures that workers feel valued and trusted. This trust is crucial, as it reduces negative effect of social loafing (Robert, 2020). Since virtual leadership maintains both aspects of transparent communication and employee welfare, instead of choosing between the two, it is more suitable than traditional leadership styles in Industry 4.0.

7.4.2.3 COMPATIBILITY WITH TECHNOLOGY

Leaders who follow traditional styles are those who are afraid of change (Martin, 2017). Today's industry is characterized by new innovations, the latest technology (Stock et al., 2018), creativity, agility, and boldness (Ameer, 2020), and therefore leaders must also possess these same qualities in order to remain compatible with Industry 4.0. Traditional leadership styles still focus largely on the physical and manual aspects of organizations and its functioning and are therefore not suitable to run a successful integration of humans and automation (Zakaria, 2019) in a digital industry (Peshawaria, 2018).

Virtual leadership, on the other hand, requires in-depth understanding and acceptance of the latest technology, including internet of things, big data analytics, artificial intelligence, automation, etc. (Avolio et al., 2014). Possessing the ability to manage the changing dynamics between technology and human skills, is unique to virtual leadership—which is why it can be regarded as the most compatible style of leadership in Industry 4.0.

FIGURE 7.3 3C Model of virtual leadership.

Source: Developed by authors.

7.4.3 RQ 3. HOW EFFECTIVE UTILIZATION OF THIS CHANGE IN LEADERSHIP CAN HELP ORGANIZATIONS GROW EXPONENTIALLY?

Kolditz et al. (2017) stated that the inability of leaders to efficiently adapt to Industry 4.0 may be responsible for the failure of organizations. Hence, change in leadership to one that prioritizes adaptation to the latest technology and relevant skill sets is necessary—that is why virtual leadership style should be integrated to secure the exponential growth of the business industry. The well-known key ingredients of a basic, successful business are efficient management, finance, and technical pillars (Mishra and Mishra, 2020). Additionally, the businesses in Industry 4.0 rely on certain parameters to help set it apart from its competitors. These include innovation efforts, talent development, effective communication, and employee motivation (Silva, 2017). As virtual leadership is able to fulfill all these requirements, its utilization ensures the success of an organization. Exponential growth of an organization today relies significantly on the steps that it takes to both maintain and go beyond the three staunch pillars of management, finance, and technology.

In terms of the three pillars highlighted by Mishra and Mishra (2020), this change to virtual leadership may ensure *management* success in terms of developing a "systematic approach," defining company goals and priorities and ensuring "flexible" activity (Merkevičius, 2015). Maintenance of digital database and dashboards, all employees of a company will be able to receive updated, real-time data regarding company operations and employee performances from remote working zones.

According to the second pillar, that is, *finance*, virtual leadership is likely to ensure that operations under its leadership also are digitized and that the integration of artificial intelligence and robotics are used in manufacturing and delivery processes, thereby enforcing cost-effective operations. They reduce waste production, decrease the possibility of human errors, and standardize all operations.

Catering to the final pillar of *technology*, virtual leadership provides the personalized benefit of being the best suited to bring about the integration of the latest technology and digital strategies for business development. One of the most significant methods by which this change in leadership ensures technology integration is by incorporating features such as mandatory mobile technology and meetings and performance appraisals in various online platforms.

The interconnected aspects of innovation efforts, enhancing talents (Silva, 2017), effective communication, and employee motivation are also areas where utilization of virtual leadership ensures exponential growth.

Regarding *innovation efforts*, correct utilization of virtual leadership may have a significant impact (Wrede and Dauth, 2020). Since it has roots in adaptability to rapid change, virtual leadership is forced to challenge itself to be constantly agile and innovative in order to face the unpredictable nature of Industry 4.0 at the blink of an eye. This gives businesses their competitive edge. Adopting the innovative strategies by the virtual leaders will increase the chances of the employees too will be given the same opportunity. Therefore, by giving them the space to cultivate their creativity, virtual leadership ensures *talent-development* among the employees—which will benefit their own personal growth as well as the organization's exponential growth.

Tokuyama (2018) stated that when one sense loses its ability to function, "the unaffected senses take up the responsibilities of the affected sense." Connecting this statement's relevance to facilitate business *communication* in Industry 4.0, the affected sense could be considered as physical interaction among leaders and employees. Since operations are now digital, (more facilitated rapidly by the COVID-19 pandemic) this "sense" is no longer functional. Virtual interaction could be considered as the "unaffected senses" which will take up the responsibility of the affected sense. Essentially, this means that virtual interaction would become more detailed, transparent, and frequent so as to make up for the loss of physical interaction and the benefits that it yielded. It is for this reason that virtual leadership can be utilized for effective communication, which is now a requirement for organizations due to digitized operations. As a domino effect, effective communication will cause *employee motivation* to rise because trust will be built between leaders and workers. This support from leaders is essential to job loyalty and employee satisfaction (Farrukh et al., 2019).

Overall, by maintaining employee satisfaction, loyalty, open dialogue, and fulfilling the requirements of the management, financial and technology pillars of the organization, virtual leadership ensures a necessary transition in organizational culture—one that adapts more suitably to an era of digital workplaces.

7.5 PROPOSED MODEL FOR FUTURE RESEARCH IN CREATING MORE RESILIENT WORKING ENVIRONMENTS THROUGH VIRTUAL LEADERSHIP

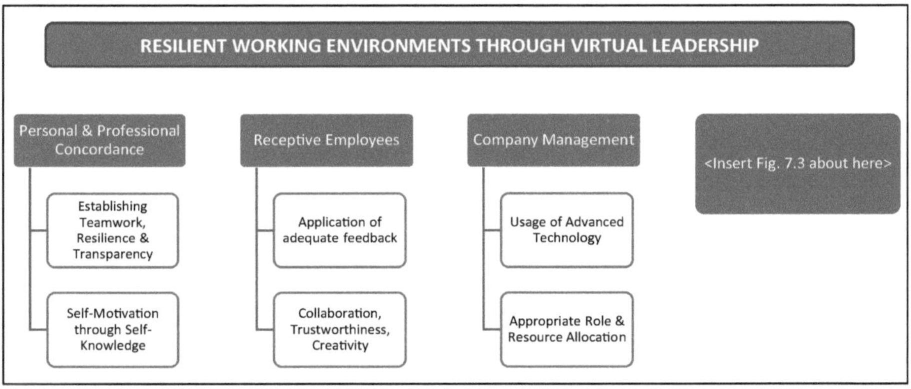

FIGURE 7.4 Proposed model for creating more resilient working environments through virtual leadership.

Source: Developed by authors.

Based on the challenges that businesses face due to the rapid transition into Industry 4.0, creating receptive and resilient working environments has become vital. This Model provides scope for future research to explore the smooth transitioning from traditional leadership styles to virtual leadership styles.

7.6 CONCLUSION

In order to adapt to the current times, virtual leaders must acquire the qualities of promoting creativity, establishing transparent communication, being flexible in decision-making, and creating a culture of inclusivity. From the above explanations, it can be concluded that, to establish organizational growth in the advancing technological era, businesses should transit efficiently from traditional leadership styles to virtual leadership styles and promote the same. The integration of virtual leadership ensures maintenance of the management, financial, and technological branches of organizations, while also, focusing on talent development, innovation efforts, and employee motivation. Care should be taken to remove psychological stress among employees for better functioning (Deb et al., 2008). The proposed model for future research aspires to elucidate on the requirements needed by virtual leaders to promote a resilient working environment which can ensure more growth and prosperity.

KEYWORDS

- **Industry 4.0**
- **virtual leadership**
- **jobs**
- **creativity**
- **inclusivity**
- **decision-making**

REFERENCES

Accenture Consulting. Manning the Mission for Advanced Manufacturing, 2017. https://www.accenture.com/il-en/_acnmedia/PDF-50/Accenture-Manning-Mission-Advanced-Manfacturing.pdf

Almada-Lobo, F. The Industry 4.0 Revolution and the Future of Manufacturing Execution Systems (MES). *J. Innov. Manag. JIM* **2015,** *3*(4), 16–21.

Ambad, P.; Bhosle, S.; Vaidya, S. Industry 4.0 – A Glimpse. *Procedia Manuf.* **2018,** *20*, 233–238. https://doi.org/10.1016/j.promfg.2018.02.034

Ameer, T. The "Basic" Framework: Leadership Imperatives for Industry 4.0. https://apcoworldwide.com/blog/leadership-imperatives-for-industry-4-0/ (accessed Jan 17, 2020).

Ashkenas, R.; Manville, B. The Fundamentals of Leadership Still Haven't Changed. https://hbr.org/2018/11/the-fundamentals-of-leadership-still-havent-changed (Nov 7, 2018).

Avolio, B. J.; Sosik, J. J.; Kahai, S. S.; Baker, B. E-Leadership: Re-Examining Transformations in Leadership Source and Transmission. *Leadersh. Quart,* **2014,** *25*(1), 105–131. DOI: 10.1016/j.leaqua.2013.11.003.

Bagheri, B.; Yang, S.; Kao, H. A.; Lee, J. In *Cyber-Physical Systems Architecture for Self-Aware Machines in Industry 4.0 Environment*, IFAC Conference, 2015, *48*(3); pp 1622–1627.

Bahrin, M. A. K.; Othman, M. F.; Nor, N. H.; Azli, M. F. T. Industry 4.0: A Review on Industrial Automation and Robotic. *Jurnal Teknologi (Sciences & Engineering)* eISSN 2180–3722, 2016; pp 137–143.

Baldoni, J. *The Next Big Thing: Virtual Leadership*. Forbes. https://www.forbes.com/sites/johnbaldoni/2020/05/12/the-next-big-thing-virtual-leadership/?sh=36b02ce53e5e (accessed May 12, 2020).

Bauernhansl, T.; Ten Hompel, M.; Vogel-Heuser, B., Eds.; *Industry 4.0 in Produktion, Automatisierung und Logistik- Anwendung, Technologien, Migration*; Springer Vieweg: Wiesbaden, 2014; pp 1-648.

Bienzeisler, B.; Schletz, A.; Gahle, A. K. Industry 4.0 Ready Services Technologie Trends 2020; Fraunhofer IAO: Stuttgart, 2014.

Blake, R. R.; Mouton, J. S. *The Managerial Grid*; Gulf Publishing Company: Houston, 1964.

Brettel, M.; Friederichsen, N.; Keller, M. How Virtualization, Decentralization and Network Building Change the Manufacturing Landscape: An Industry 4.0 Perspective. *Int. J. Mech. Aerosp. Indus. Mechatron. Manuf. Eng.* **2014,** *8*(1), 37–36.

Carter, D. Creativity in Action–the Information Professional is Poised to Exploit the Fourth Industrial Revolution: The Business Information Survey 2017. *Bus. Inform. Rev*. **2017,** *34*(3), 122–137. https://doi.org/10.1177/0266382117722440

Chakraborty, T.; Saha, R. Digital Innovations: Breakthrough Opportunities to Build Novel Business Models. *Bus. Perspect. Aims Scope* **2018,** *1*, 46.

Chakraborty, T.; Saha, R. Proactive and Reactive Role of Human Resource Professionals in an Aftershock Management. *Train. Dev. J.* **2017,** *8*(1), 1–11.

Chakraborty, T.; Mishra, N. Appreciative Inquiry: Unleashing a Positive Revolution of Organizational Change and Development. *Int. J. Econ. Commer. Bus. Manag.* **2019,** *6*(2), 32–37.

Chakraborty, T.; Ganguly, M.; Natarajan, A. Predicting Entrepreneurial Satisfaction: The Role of Non-Financial Incentive Factors and Quality of Life Among Women Digital Entrepreneurs. *J. Global Bus. Adv.* **2019,** *12*(3), 328–355.

Chakraborty, T.; Ganguly, M. Crafting Engaged Employees Through Positive Work Environment: Perspectives of Employee Engagement. In *Management Techniques for Employee Engagement in Contemporary Organizations*; IGI Global, 2019; pp 180–198.

Chang, W.; Lee, C. Virtual Team e-Leadership: The Effects of Leadership Style and Conflict Management Mode on the Online Learning Performance of Students in a Business-Planning Course. *Br. J. Educ. Technol.* **2013**, *44*(6), 986–999. DOI: 10.1111/bjet.12037.

Davis, R. Industry 4.0. Digitalization for Productivity and Growth; European Parliamentary Research Service, PE 568.337, 2015. http://www.europarl.europa.eu/RegData/etudes/BRIE/2015/568337/EPRS_BRI (2015)568337_EN.pdf.

Deb, S.; Chakraborty, T.; Chatterjee, P.; Srivastava, N. Psychological Stress of Traffic Police Officers Causal Factors Behind the Same and their Coping Strategies. *J. Indian Acad. Appl. Psychol.* **2008,** *34*(1), 19–28.

DeRue, D.; Ashford, S. Who Will Lead and Who Will Follow? Social Process of Leadership Identity Construction in Organizations. *Acad. Manag. Rev.* **2010,** *35*(4), 627–647.

Erol, S.; Jäger, A.; Hold, P.; Ott, K.; Sihn, W. In *Tangible Industry 4.0: a Scenario-Based Approach to Learning for the Future of Production*, 6th CLF - 6th CIRP Conference on Learning Factories, Procedia CIRP, 2016, 13–18.

Farrukh, M.; Kalimuthu, R.; Farrukh, S. Impact of Job Satisfaction and Mutual Trust on Employee Loyalty in Saudi Hospitality Industry: A Mediating Analysis of Leader Support. *Int. J. Bus. Psychol.* **2019,** *1*(2), 30–52.

Fedáková, D.; Ištoňová, L. Slovak IT-Employees and New Ways of Working: Impact on Work-Family Borders and. *Ceskoslovenska Psychol.* **2017,** *61*(1), 68–83.

Fraunhofer IOSP. Industry 4.0 - Information Technology is the Key Element in the Factory of the Future; Fraunhofer Institute of Optronics, System Technologies and Image Exploitation, PR, 2013; pp 1–5.

Ganschar, O.; Gerlach, S.; Hämmerle, M.; Krause, T.; Schlund, S.; Spath, D. Production Work of the Future - Industry 4.0. *Fraunhofer IAO*, 2013.

Gehrke, L.; Kühn, A.; Rule, D.; Moore, P.; Bellmann, C.; Siemes, S.; Standley, M. A Discussion of Qualifications and Skills in the Factory of the Future: A German and American Perspective, 2015,.

Grosvenor, R.; Qin, J.; Liu, Y. A Categorical Framework of Manufacturing for Industry 4.0 and Beyond, Changeable, Agile, Reconfigurable & Virtual Production. *Procedia CIRP* **2016,** *52*, 173–178.

Guzmán, V. E.; Muschard, B.; Gerolamo, M.; Kohl, H.; Rozenfeld, H. Characteristics and Skills of Leadership in the Context of Industry 4.0. *Procedia Manuf.* **2020,** *43*, 543–550. DOI: 10.1016/j.promfg.2020.02.167.

Henke, M. Management der Industry 4.0 –Management of Industry 4.0 - Challenges and Fields of Action for the Success-Oriented Management of Companies. *Present. Zukunftskongress Logistik* **2014,** *32*(5).

Hensellek, S. Digital Leadership. *J. Media Manag. Entrepreneurship* **2019,** *2*(1), 55–69. https://doi.org/10.4018/jmme.2020010104

House, R. J. A Path Goal Theory of Leader Effectiveness. *Adm. Sci. Quart.* **1971,** 321–339.

Judge, T.; Ilies, R.; Bono, J.; Gerhardt, M. Personality and Leadership: A Qualitative and Quantitative Review. *J. Appl. Psychol.* **2002,** *87*(4), 765–768.

Kelly, R. *Constructing Leadership 4.0: Swarm Leadership and the Fourth Industrial Revolution*. Palgrave Macmillan: Cham, Switzerland, 2018.

Kipper, L. M.; Iepsen, S.; Dal Forno, A. J.; Frozza, R.; Furstenau, L.; Agnes, J.; Cossul, D. Scientific Mapping to Identify Competencies Required by Industry 4.0. *Technol. Soc.* **2021,** *64*, 101454. DOI: 10.1016/j.techsoc.2020.101454.

Koch, V.; Kuge, S.; Geissbauer, R.; Schrauf, S. *Industry 4.0. Opportunities and Challenges of the Industrial Internet*; Strategy and PwC: Munich, 2014.

Kolditz, T. A.; Casas i Klett, T.; Strackhouse, J. Are You a Leader of the Fourth Industrial Revolution?, 2017.

Korzynski, P. Online Social Networks and Leadership. *Int. J. Manpow.* **2013,** *34*(8), 975–994. DOI: 10.1108/ijm-07-2013-0173.

Krot, K.; Mazgajczyk, E.; Rusińska, M.; Woźna, A. *Strategy of Improving Skills of Nnovation Managers in the Area of Advanced Manufacturing Technologies*, International Conference on Intelligent Systems in Production Engineering and Maintenance, Sept 17–18, Wroclaw, Poland, 2018.

Lee, J.; Kao, H. A.; Yang, S. In *Service Innovation and Smart Analytics for Industry 4.0 and Big Data Environment, Product Services Systems and Value Creation*, Proceedings of the 6th CIRP Conference on Industrial Product-Service Systems, Procedia CIRP, 2014, 3–8.

Lussier, R.; Achua. C. *Leadership. Theory, Application and Skill Development*; South-Western Cengage Learning, 2013.

Maisiri, W.; Dyk, L. V. Industry 4.0 Skills: A Perspective of the South African Manufacturing Industry. *SA J. Hum. Res. Manag.* **2021,** *19*. DOI: 10.4102/sajhrm.v19i0.1416.

Maisiri, W.; Van Dyk, L. Industry 4.0 Readiness Assessment for South African Industries. *South Afr. J. Ind. Eng.* **2019,** *30*(3), 134–148. https://doi.org/10.7166/30-3-2231

Martin, W. *Leadership: Outdated Theories and Emerging Non-Traditional Leadership* (Doctoral dissertation, Northcentral University, 2017); ProQuest LLC, 2017.

Mawson, A. *The Perfect Virtual Leader*. Forbes. https://www.forbes.com/sites/amawson/2020/07/01/the-perfect-virtual-leader/?sh=6ed0e3967d62 (accessed July 1, 2020).

Merkevičius, J.; Davidavičienė, V.; Raudeliūnienė, J.; Buleca, J. Virtual Organization: Specifics of Creation of Personnel Management System. *E+M Ekon. Manag.* **2015,** *18*(4), 200–211. DOI: 10.15240/tul/001/2015-4-014.

Mintzberg, H. *The Nature of Managerial Work*; Harper and Row: New York, 1973.

Mishra, A.; Mishra, D. The Three Pillars concept for a Successful Business: Management, Finance & Technical. *Int. J. Arts, Manag. Hum.* **2020,** *9*, 34–37.

Moore, K. *Virtual Leadership: The Essential Skills|monday.com Remote Blog*. Monday.Com Blog. https://monday.com/blog/remote-work/3-essential-virtual-leadership-skills-that-create-a-thriving-remote-culture/ (accessed Jan 12, 2021).

Müller, J. M.; Buliga, O.; Voigt, K. I. Fortune Favors the Prepared: How SMEs Approach Business Model Innovations in Industry 4.0. *Technol. Forecast. Soc. Change* **2018,** *132*, 2–17. https://doi.org/10.1016/j.techfore.2017.12.019

Naqshbandi, M. M.; Jasimuddin, S. M. Knowledge-Oriented Leadership and Open Innovation: Role of Knowledge Management Capability in France-Based Multinationals. *Int. Bus. Rev.* **2018,** *27*(3), 701–713.

Neugebauer, R.; Hippmann, S.; Leis, M.; Landherr, M. In *Industries 4.0- Form the Perspective of Applied Research,* 49th CIRP Conference on Manufacturing Systems, 2016, 2–7.

Oberer, B.; Erkollar, A. Leadership 4.0: Digital Leaders in the Age of Industry 4.0. *Int. J. Organ. Leadersh.* **2018,** *7*(4), 404–412. https://doi.org/10.33844/ijol.2018.60332

Paine, S. *The Emerging Role of HR Practitioners: Expectations, Challenges and Trends*; Unitec: New Zealand, 2009.

Peshawaria, R. Minimum Supervision, Maximum Performance. *Lead. Lead.* **2018,** *2018*(87), 42–47.

Rennung, F.; Luminosu, C. T.; Draghici, A. In *Service Provision in the Framework of Industry 4.0,* SIM 2015/13th International Symposium in Management, Procedia - Social and Behavioural Sciences, 2016, 372–377.

Robert, L. P. Behavior–Output Control Theory, Trust and Social Loafing in Virtual Teams. *Multimodal Technol. Interact.* **2020,** *4*(3), 39. DOI: 10.3390/mti4030039.

Rüßmann, M.; Lorenz, M.; Gerbert, P.; Waldner, M. *Industry 4.0: The Future of Productivity and Growth in Manufacturing Industries*, 2015, 1–14.

Salina, R. *5 Must-Have Virtual Leadership Skills* [+Tips & Examples]. Ariel Group. https://www.arielgroup.com/virtual-leadership/ (accessed July 28, 2020).

Schwab, K. The Fourth Industrial Revolution: Crown Business, 2017.

Selamat, A.; Alias, R.; Hikmi, S.; Puteh, M.; Tapsi, S. Higher Education 4.0: Current Status and Readiness in Meeting the Fourth Industrial Revolution Challenges. *Redesigning High. Educ. Towards Ind.* **2017,** *4*, 23–24.

Senduk, A. Leadership in the Digital Era: How to Grow Digital Leaders? https://medium.com/@andrewsenduk/leadership-in-the-digital-era-how-to-grow-digital-leaders-72f6143a9b95 (accessed Nov 30, 2017).

Silva, K. Culture: A Key Ingredient in Business Success. *Strategic HR Rev.* **2017,** *16*(4), 166–170. DOI: 10.1108/shr-03-2017-0021.

Sipsas, K.; Alexopoulos, K.; Xanthakis, V.; Chryssolouris, G. In *Collaborative Maintenance in Flow-Line Manufacturing Environments: An Industry 4.0 Approach*, 5th CIRP Global Web Conference Research and Innovation for Future Production, Procedia *CIRP 55,* 2016, 236–241.

Sivathanu, B.; Pillai, R. Smart HR 4.0 – How Industry 4.0 is Disrupting HR. *Hum. Res. Manag. Int. Dig.* **2018,** *26*(4), 7–11. DOI: 10.1108/hrmid-04-2018-0059.

Stackman, R.; Devine, K. Leadership and Emotional - Rational Coherence: A Start? *Acad. Manag. Perspect.* **2011,** *25*(1), 42–44.

Stock, T.; Obenaus, M.; Kunz, S.; Kohl, H. Industry 4.0 as Enabler for a Sustainable Development: A qualitative Assessment of its Ecological and Social Potential. *Process Saf. Environ. Prot.* **2018,** *118*, 254–267.

Taherkhani, E. Review of Relation Between Creativity and Job's Satisfaction of Physical Education Teachers. *Sci. J.* **2015,** *36*(3).

Tareque, M. A.; Islam, N. Digital Leadership: The Perspectives of the Apparel Manufacturing. *Int. Bus. Res.* **2021,** *14*(2), 124. DOI: 10.5539/ibr.v14n2p124.

Thoben, K. D.; Wiesner, S.; Wuest, T. Industries 4.0 and Smart Manufacturing- A Review of Research Issues and Application Examples. *Int. J. Autom. Technol.* **2017,** *11*(1), 4–16.

Thomas, M. E. *Modeling Instruction in High School Science: The Role of School Leadership*, Chicago State University, 2018.

Tokuyama, D. *Supersensors: How the Loss of One Sense Impacts the Others*, Wu Tsai Neurosciences Institute, 2017.

Vlasov, V.; Chromjakova, F. The Effect of the Fourth Industrial Revolution Economies and Management. **2018,** *10*. 1007/978-3-319-74216-8_54.

Witkowski, K. In *Internet of Things, Big Data, Industry 4.0 - Innovative Solutions in Logistics and Supply Chains Management*, 7th International Conference on Engineering, Project, and Production Management, Procedia Engineering, 2017, 182763–182769.

Wojcak, E.; Bajzikova, L.; Sajgalikova, H.; Polakova, M. How to Achieve Sustainable Efficiency with Teleworkers: Leadership Model in Telework. *Procedia Soc. Behav. Sci.* **2016,** *229*, 33–41. DOI: 10.1016/j.sbspro.2016.07.111.

Wrede, M.; Dauth, T. A Temporal Perspective on the Relationship Between Top Management Team Internationalization and Firms' Innovativeness. *Manag. Decis. Econ.* **2020,** *41*(4), 542–561. DOI: 10.1002/mde.3119.

Wyld, D. What Matters More in Growth Companies: Leader or Idea? *Acad. Manag. Perspect.* **2009,** *23*(2), 95–96.

Xu, M.; David, J. M.; Kim, S. H. The Fourth Industrial Revolution: Opportunities and Challenges. *Int. J. Financial Res.* **2018,** *9*(2), 90.

Zakaria, N.; Nasir, A.; Akhtar, A. Are the Leaders Ready to Embrace Industry 4.0? *J. Adv. Res. Dyn. Control Syst.* **2019,** *11*(05), 7th ser.

Zhang, S.; Fjermestad, J.; Tremaine, M. In *Leadership Styles in Virtual Team Context: Limitations, Solutions and Propositions*, Proceedings of the 38th Annual Hawaii International Conference on System Sciences, 2005, 48c–48c. DOI: 10.1109/HICSS.2005.382.

Zuehlke, D. Smart Factory – Towards a Factory-of Things. *Annu. Rev. Control* **2010,** *34*, 129–138.

CHAPTER 8

Employee Engagement Perspectives in Agile Organizations: Managing People in Industry 4.0

NANDITA MISHRA[1*] and TANUSREE CHAKRABORTY[2]

[1]*Chetana's Institute of Management and Research, Mumbai, Maharashtra 400051, India*

[2]*Administrative Staff College of India, Hyderabad, Telangana 500034, India*

**Corresponding author. E-mail: nanditamishra06@gmail.com*

ABSTRACT

As Industry 4.0 starts permeating into the economy, there is an increased complexity, disruptiveness, and uncertainty, leading to changes in the way business is done. To manage the changing demands, the organizations and the leaders must innovate products and services, and on a more frequent basis. Firms need to get used to changes in the business environment and develop processes that focus on new trends, identify improvements, look at new manufacturing systems, all while assessing their impact on the management of their organizations. Leaders must learn to embrace this change and adapt to moving quickly with the fast-paced changing business dynamics. The new dextrous organization requires a totally different leadership approach. The HR functions under agile leadership also must be dynamic and agile in nature. To this effect, the current chapter focuses on the role of certain HR agile practices and tries to establish how they lead to employee engagement. Discussing agility in business and major HR agile strategies

Agile Leadership for Industry 4.0: An Indispensable Approach for the Digital Era, Tanusree Chakraborty, Tahir Mumtaz Awan, Ashok Natarajan, and Muhammad Kamran (Eds.)

and functions, this chapter brings to light the importance of agile HR systems and consequential employee engagement.

8.1 INTRODUCTION

Organizational agility and agility in leadership have become the two most discussed topics among scholars pertaining to the changes that organizations have to take up today. Organizational agility is about the adaptability of the company to the changes it faces from different spheres of operation. Agility in organizations is generally embedded with speed and stability. Earlier the workplace scenarios were not as we see it today. The employees of those days were more governed by the loyalty and organizational commitment. Employees exhibited higher levels of loyalty to the organization, expressed commitment and intended to engage in lifetime employment with their organizations. The nature of business and organizations are changing today. This change and ambiguity in businesses have led to the need for organizational agility that is directed toward success and sustainability. In order to achieve their objectives, organizations are required to adopt a more agile approach, that is, an approach defined by empowerment among employees, engagement among employees, a culture to embrace disruption, and finding out newer and strategic ways how to be competitive. This will help agile organizations to do everything in order to assist managers and employees employ their best efforts. Either individually or as a team, employees use agile methods to sustain, advance, and drive business performance.

Today, business operations have become global, and today's leaders look for greater engagement and competencies among their subordinates. Dexterity and deftness of the employee has taken precedence over loyalty. Loyalty is now a hygiene condition. The employment contract also became much different. Adroit talents left organizations, performances have varied kind of expectations, and the growth and development needs of employees also became dynamic. This needed for something new-fangled and fresh. To sustain the performance and effectiveness of organizations in the face of the ever-changing business environment, fresh approach toward employee engagement is the cry for the day. Today, employees take greater risks at work, decision-making has become decentralized and leadership demands versatility and dynamism. This requires an agile approach in the leadership team with respect to organizational structure, organizational processes, and the people who are the backbone of the organization, giving rise to the need for a metrics in the process and a continuous check on how things are responded to.

With this background, the present chapter discusses the concept of organizational agility and throws light on how employee engagement is an essential element to organizational agility. Further, we discuss how the strategic leadership fosters organizational agility, how agile job resources and employee engagement are interrelated. The following sections will discuss the changed way of performance management and training and development in agile organizations. Finally, it explains how employee engagement impacts agile career development and talks about the role of leadership style in fostering employee engagement in agile industries.

Examples of agility in business:

1. In no time, the automakers leaped into the making of ventilators.
2. In no time, the makers of drinking alcohol jumped into making hand sanitizers.
3. The rise of Netflix in a short time span over Blockbuster.
4. Changing paper documents to online forms.
5. Developing flexible job descriptions to meet organization needs.
6. Creating an environment that adheres to change as needed.
7. Allowing cross-functional and different ideas in training, and education that can develop individual capacity, capability, and cooperation.
8. Valuing results and impact, no longer rewarding longevity or seniority.

8.2 ORGANIZATIONAL AGILITY AND EMPLOYEE ENGAGEMENT

Majority of the companies now work under continuous pressure of meeting their target, albeit in a changing business landscape that influences the business practices. The Economist Intelligence Unit' Survey stated that at a time when rapid development is taking place for knowledge management, the organizations' sustainability is dependent on its adaptability and core competency. There could be three different approaches to fulfil the organization's avidity, such as organization's responsiveness (reactive agility), organizational effectiveness (proactive agility), and organization's resourcefulness (innovative agility) (Saha et al., 2017). According to Alhadid (2016) and Najrani (2016), the knowledge of organizational agility is in the form of impact of expertise in manufacturing, promptness in understanding the market setting, understanding the changing patterns in dealing with human resources and adopting the right technology. To address the organizational agility (OA) in a business, it is imperative that competitive challenges turn into business opportunities. Organizational agility (OA) is nothing but the ability to think, respond, and create competitive advantage.

Deloitte Report (2020) mentions the changing crisis and agility of an organization in three phases. Handling a typical crisis can be explained over three times frames. The current situation is the respond phase to deal with the present crisis and to manage continuity. In the recovery phase, the company has learnt the lessons well and adapted to changes to become stronger. Resilient leaders view recovery as a journey for the organization, the employees, the teams, and all stakeholders. For an organization to remain agile, it is important to have a mindset shift, to identify and address the uncertainties and embed trust in the employees to enable the organization to recover. During the final stage, the agile company prepares for the future and develops new strategies. This phase is called the thrive phase and is very important for agility. The changing dynamics of the post-COVID-19 workplace has affected the individuals and organizations differently. Recovery of any organization will never be static, there will be interruptions and interventions from time to time. However, agile organizations handle disruptions more efficiently and effectively. An agile organization engages the employee differently. At the corporate level, a balance is sorted between survival instincts versus sense of purpose. Agile organizations rethink globalizations as making indelible global footprints. The focus is shifting from just-current business to sustainable business.

For an agile organization, the employee engagement is in terms of employee centric profiling and employee experience. A new workplace should have more collaborative space, Internet of things, and technical resilience for the employee to be more innovative, efficient, and adaptable to the workplace. The vision and mission statements of such agile organizations clearly focus on purpose, potential, and perspectives.

Sambamurthy et al. (2003) highlighted the alertness and responsive capability as main components of agility. The same is discussed and supported by the Deloitte Report 2020.

Organizations must study the deficiencies and the changes needed to adapt to the changing industry requirements. Dove (2005) says that fast and intelligent decision-making capacity is required to respond to business needs. Only a leader with skills like flexibility and adaptability will be able to achieve the shift from resource-based perspective to process-based perspective.

8.3 STRATEGIC LEADERSHIP TOWARD AGILITY

In developing leadership roles, organizations may look at strategies, such as empowerment and care. Organizations need leaders who can care and lead the staff to change. Creating, sharing, or distributing equal opportunities might not be liked by aspirational leaders. In many cases, they perceive themselves as leaders and want to avail all opportunities of growth. Once the leadership context changes, it will change the content too, which will represent leadership requirements for the future. Today's organizations need strategic executor to make the organization agile. Strategic leadership means looking for the future of business and developing insights. In context to that an executor will work only on the current challenges. The challenge is that employees are either inclined toward strategy or toward implementation. A good approach will be to design a strategy, evaluate how it unfolds, and implement the strategy keeping in mind the immediate and the future requirements. Similarly, it is not advisable to spend time only strategizing and not focusing on the execution. Strategic leadership is all about strategizing the present, keeping in mind the future, and solving today's problems for a better tomorrow. For instance, updated technical skills is a future need and understanding people's needs and leading them is an immediate need. Many leaders driving technological advancements are not equipped to understand the human part and adjustments every employee does. This is also applicable for a people specialist, who do not understand the impact of technology. A strategic leader's role is to help develop along with people development and thereby offer better future to the people. Strategic leaders should also preserve the past, because the company core should not change but also focus on moving to the future and create always fresh ideas. That is what is known as innovation in strategic leadership. Innovation is to experiment with fresh ideas and encourage others to do so. Strategic leadership is democratic in nature, taking a 360° approach make decisions based on variety of inputs. Ability of a leaders to admit their mistakes and learn from the mistakes brings confidence throughout the organization. Leaders in an organization should maintain integrity and build trust in all interactions. Strategic leaders accrue support, negotiate to form coalitions, and overcome resistance to maintain progress. Strategic leaders are at the cross-section of a company's main operations and its growth opportunities. Strategic leaders are responsible to bring stability within the organization by making the working conditions conducive. Strategic leadership style is also transformational. Employees might have predetermined business functions and growth opportunities that they complete as per schedule assigned, but employees are constantly used out of their comfort zone by the leader.

What makes strategic leaders different from the others is their focus on organizational limitations and image, which they think are transformable. In a study by Rooke and Tolbert (2005) 10 CEOs in six different industries were interviewed. Each of the participants filled out a Leadership Development Profile, which showed that five of them were strategy-oriented while and five were action oriented. All five strategist leaders succeeded in generating one or more organizational transformation over a 4 year period, their companies' market share, reputation, and profitability all improved. But among the nonstrategist leaders, only two of them succeeded in transforming their organization, despite help from consultants. Strategists have three distinct levels of social engagements, their personal relationships with the team, organizational relationship as per organization's structure and national and international developments. The strategist works to create ethical principle and practices beyond the self-interest.

8.4 AGILE JOB RESOURCES AND EMPLOYEE ENGAGEMENT IN INDUSTRY 4.0

Personal and job resources are two very important contributors to employee engagement. There are certain job resources which have always been established as strong predictors of employee engagement. They are social support, feedback for significant others at work, growth and development, task independence, skill variety, etc. (Bakker and Demerouti, 2007). Today, job resources do not refer only to hard skills, they embrace the aspects of physical, social, or organizational issues that reduce the job demands and also minimize the physiological and psychological costs. Agile workplace calls for job demands to be adequately functional and goal-directed and also include resources that promotes learning and self-development (Schaufeli and Bakker, 2004). Job resources also impact employee motivation. Workplaces that have all resources for effective and efficient performance will demand for dedication and performance at work from the employees (Meijman and Mulder, 1998). An agile work culture, an agile work place, and agile leadership can foster employee engagement on the whole. In agile work environment, work is accomplished successfully and work goals are attained smoothly. A supportive work team and providing feedback on performance increases the workplace success. As Schaufeli and

Salanova (2007) states, if basic needs are met and work goals are satisfied, employee engagement is likely to be augmented. With respect to Industry 4.0, the growth witnessed by businesses is exponential. This has called for much higher levels of skill sets, physical and mental ability, resilience, and competence. Industry 4.0 revolution has showered the employment sector with incredible opportunities and enormous economic impact. This has implications on the job resources as well. Conservation of resources (COR) theory (Hobfoll, 2001) states that if people value certain resources, they tend to acquire, retain, and save them. They collate this with stress and assert that depending on the resources lost, people experience the extent of stress.

Hobfoll et al. (2000) have posited that people are apt to bring in resources so that their resources are not at stake. In this connection, the agile organizations call for those job resources that do not allow employees to become obsolete. People with high resource pool are less vulnerable to resource loss and strong resource pool always helps to seek more opportunities, so that they can expect increased resource gains. Agile organizations demand that today's employees might be required to perform tomorrow jobs that do not even exist today, they might be removed, retrained, or more open to accept changes. Industry 4.0 has brought that rapid pace of technological changes that it requires shaping the workforce agility in order to emphasize leveraging the best combination of people and technology. It calls for developing job resources for a longer term. Leadership teams should be able to proactively recognize individuals whom to nurture and develop into future job roles and further leadership roles. Agile leadership should also look at the placement of their not so high performers so that their existing job resources can also be met with the requirements. This shows that when employees are supported, they find relevant lateral moves in the agile business so that they can align with the goals of other individuals. Agile organizations should encourage employees achieve their career aspirations along with fulfilling their personal goals. This has a connection with employee engagement, as an agile team will be ready to grab opportunities. Due to the global digital revolution and is changing business landscape, business leaders are uncertain about what is waiting for them and so they need to be doubly prepared for investing in agile workforce than they required to do before. Human Resource Department can in fact design or administer different organizational systems that can be implemented in other departments in the process.

8.5 AGILE PERFORMANCE MANAGEMENT AND EMPLOYEE ENGAGEMENT

Agility in organizations calls for an increasing emphasis on augmenting performance measurement systems and agile ways of performance management practices. Performance measurement as a concept gained greater popularity with the work of researchers in the field of human resource management for over some time (Johnson and Kaplan,1987; Bertalanffy, 1968; Neely, 1999; Weiner, 1948). Performance measurement, to some extent, is influenced by theories of organizational and management control which have emerged from general system theory. Performance management is a tool to appreciate how employees perform in an organization with respect to organizational control and management control. Performance management has a different approach in conventional organizations, but for agile organizations, three changes are extremely crucial to organizational success. It needs organizational goals to be aligned to individual goals and setting the priorities of such goals. For performance management, agility in the operations has become the norm of the day. Agility is observed in terms of goal setting, in place of delegating specific responsibilities or assigning tasks directly to employees. Organizations are pressing on collaborative approaches to goal accomplishment; analysis of one's potentials also makes the system flexible. Employees are given a wider scope to choose and decide on their goals, this gives them freedom as well, enhancing their morale. Managers as mentors handhold their subordinates rather than doing traditional performance evaluation processes. Organizations are also open to remote working now. This change began much before the pandemic had set in. There are various types of performance management softwares now to track employee performance, their progress on various projects, and to help them with real-time feedback. Telecommunicating has also become popular in this regard. Agile performance management has also adopted flexible work hours and it respects that every individual does not have same productive work hours. While morning hours appear to be productive for some people, others do better during evenings. Women have responsibilities of families, and after work, family sometimes appears to be a second work shift for them (Desai et al., 2011). Within a logically chosen slot, employees are given the opportunity to be present at work at variable time of the day. This helps employees decide their working hours, this flexibility has also resulted in a robust performance output. Another remarkable change in performance management is the shift in focus about employee results;

there are companies, such as Netflix and Linkedin, who do not count the number of days employees take off (Heam, 2017). They just focus on results. They do not offer fixed number of holidays to their employees. For this, they have policies of continuous communication with managers, check target accomplishment at regular times and emphasize on genuine results. Needless to mention, job sharing is another milestone achieved in terms of agile performance management practices. Job sharing means balancing core work to work-life balance among today's employees. Job sharing is a grand alternative for working mothers. Today's working mothers who are considered as high achievers are retained as valued resource in the organizations through those flexible performance management practices. The use of enterprise performance management systems are very useful for agile performance management. They aim at providing apt resources and actionable decision-making information which are extremely useful above any data source. EPM solutions are significant for managing various types of organizational actions by connecting operational and financial metrics and linking them to managerial insights. Flexibility at work improves employee engagement (Anderson and Kelliher, 2009). In order to enhance employee engagement at work (Setiyani et al., 2019), organizations are also focusing on training their employees, engaging with them even before needed, develop a better performance management system and create notifications as feedback to employees (Performyard, 2021). Organizations are also introducing mechanisms that activate automatic reviewing processes in the performance management systems. There are tools to examine the consistency and coherence that are vital for performance elements. Organizations are also trying to keep the performance management system clean and along with that they are defining meticulous working groups for reviewing the processes and implementing them smoothly (Alfaro-Saiz et al., 2013). To enhance employee engagement and positive employee feelings about work, it is important to have an appreciative approach toward employee capabilities (Chakraborty et al., 2019).

8.6 AGILE EMPLOYMENT DEVELOPMENT PRACTICES

Traditional as well as today's agile companies realize that the human resource is an invaluable asset. Organizational success depends not only on finances, assets, physical resources but also on human capital. Hence, training and development makes the major HR function in today's agile organizations. With agility in business processes, HR managers have been required to rethink their approach and roles in an organization and are incessantly creating new plans and policies to keep up with the needs to augment the professional development and employee performance (Hubauer, 2020). Creating a positive work environment marked by growth and development increases employee engagement (Chakraborty et al., 2019). In addition, organizations need to look into those practices of employee development that ensure growth and development (Chakraborty et al., 2020). Organizations being agile today need to look into the employee development functions more flexibly. Employee development needs to collaborate with various groups and internal teams must develop effective solutions for resolving the challenges. It does not only imply that it has to be completely technology-dependent, but it needs a more customized training design for each employee. These days, employees who undergo the right development programs provided by the organization are the game changers in the organization (Sendawula et al., 2018). Nowadays, companies are providing training in the form of classroom sessions blended with online workshops. Flexibility in training is the utmost necessity of the day as the pandemic has also brought a pressure on the budget and movement of people to different training venues. With the development of technology, the cost of training is reducing and the use of smartphones has also paved the way for an extensive and wider reach of training targets and resources. So, virtual training has given employees the scope and time to learn more (Rose, 2021). It has also become imperative to manage employee development programs with empathy and clear communication. Empathy and perceived work-life balance would ensure employee engagement. To make the employee development programs flexible and engaging, individual online learning has been the choice. Employees also engage in some group processes to facilitate knowledge sharing and learning. Digital language trainings are also helping employees learn new skills. With online sessions, a huge number of participants have been trained at one time. Soft skill development is integral while working on employee development. This fosters interpersonal relationships better and helps employees to avoid the disasters of overly digitized communication. With too much shift to digitized communication, employee relationships are suffering (Chakraborty et al., 2021). Thus, the attempts of organizations to engage in agile employee development practices create engagement among employees and also make them more committed.

8.7 EMPLOYEE ENGAGEMENT AND AGILE CAREER DEVELOPMENT

A comprehensive agile organization touches every facet of the organization, starting from people, process, strategy, structure, and technology. Organizations in the technology sector are born-agile. However, most organizations must undergo a transformation. Organizations must build new capabilities required to sustain agility. Top leadership should be aligned with the organization's change and strategy. Leadership concentrating more on strengths and challenges rather than weakness and threats are more likely to succeed. (Mckinsey Research, 2019) It is crucial to define the career path in flatter organizations to sustain the transformation and to design roles to enact the new operating model. Agile team members and employees are versatile to work from any location and make ways of working and collaborating different teams and departments within the organization or outside the organization. Most organizations require the existing staff to take on new role or responsibilities and expect them to be equipped with the necessary skills from day one. The practicality is not so simple. A typical career path goes through different phases. First, identifying agile coaches and recruiting them for training the employees identified. Second, the career paths and new agile roles should be defined so that the opportunities are realized. Third, the organization should support life-long learning and improvisation across the organization.

The culture team is an essential component of an agile transformation. Agile transformation acknowledges that one may not know everything but can discover as one progresses and subsequently take the learnings from there.

Careers in an agile organization are significantly different from what employees have been doing earlier. Employees often leave their existing organizations without sufficient knowledge and experience for better prospects. It is every organization's responsibility to develop employees according to demand-driven systems in the industry. Solution to career development can be achieved by partnering with employees and upskilling them.

Career path can be developed and designed through (1) creating an environment conducive to learning, this is very important as it includes entrepreneurial developments too, (2) specific training for high performance managers and mediocre performers, and (3) incorporating business relevant modules in existing business situation to equip employees with advanced career paths.

8.8 LEADERSHIP STYLE AND EMPLOYEE ENGAGEMENT IN AGILE INDUSTRIES

Organizations must begin by studying the competencies that made leaders successful in the past. Post that, organizations must analyze what new skills are required to have a change mindset and transform the way it will work for building new capabilities. Leadership styles have therefore changed for agile organizations, and especially post-pandemic, there has been a quantum jump. Some leadership styles generally practiced are:

8.8.1 COMMONLY EFFECTIVE

This is exactly what is known as democratic leadership. The leader usually makes a decision considering all the inputs from the team members. Even if the leader takes the final call in everything, every employee has a say in the decision.

8.8.2 NOT VERY EFFECTIVE

Country club leadership is nonintruding and noninterventional. Leaders, practicing it, afford all authority to their employees. A start-up company founder may allow complete freedom in the organization, while they focus on overall workings of running the company. Sometimes, this style might limit the growth of both the employees and the company. Therefore, leadership style should be kept in check under all circumstances.

8.8.3 RARELY EFFECTIVE

Bureaucratic leadership style in agile organizations is hardly effective. Similarly autocratic leadership will not work in agile organizations. Employees may lack freedom and people may feel stifled in the roles. This will quickly shut down innovation, as people would be shy away from taking risks. Any organization with predetermined target and goals growth should never go for bureaucratic or autocratic leadership style.

8.8.4 SOMETIMES EFFECTIVE

Transformational and transactional leadership styles are sometimes effective, depending upon the company's convention. Transactional leaders allocate roles and responsibilities for each employee and even reward the employees to motivate employees. But such plans should auger well with company's vision and should be used in addition with other gestures of appreciation or goal achievement levels. Transformational leadership style is very common in high-growth companies. But transformational leaders might miss seeing the growth perspective of all team members. What is required is the right mentoring and guidance to shoulder new responsibilities.

8.8.5 MOST EFFECTIVE

Coach style leadership is most effective when organizations are getting ready to transform. Coaches focus on strategies to work and collaborate better. These coaches are more aligned to strategic and democratic leadership styles. An agile leadership style is not about focusing on similar skills and goals for all employees. These leaders focus on developing each employee's different strengths and working on their weaknesses. A good leader would always be able to take the team together even in a divided group. Difference in opinion should never become indifference within the team. Strategic leaders develop the organization, as well as develop each individual as per their forte. Coach style leaders are aware of their team's personal and professional needs and balance the two accordingly in helping the team, as well as the organization's growth.

8.9 CONCLUSION

Lack of institutional agility is often linked to ineffective procedures in the organization, such as lack of proper guidelines, no defined end time and no real positive impact (Temmmerman, 2021) Adapting quickly to new ideas, competently dealing with unexpected pressures, both internal and external and streamlining the process to succeed, when confronted with change, is what agile leadership is. The focus as discussed in this chapter is on diagnosing what changes are needed in the existing structure to make an organization agile. Diverse people may have divergent points of view working together, but still finding common grounds in order to move forward. It is the leadership at the top, which truly impacts the governance of an organization. An agile organization therefore should promote innovation and encourage open communication. Today's agile nature of organization by itself demands the practices in the organization to be agile and dynamic. Thus, this chapter has some takeaways which are extremely essential for organizational success, effectiveness, and employee engagement. They are (1) for organizations to become defter, they must respond quickly to changes within the market, (2) the various strategies with respect to human resource practices needs to agile, (3) Leadership styles enforce greater agility in organizations, (4) employees should always stay tuned and keep working on their job resources to complete in the agile environment. The business world is going through a VUCA phase. So preparations to meet with the agile requirements are needed both by the managers and those who are managed.

KEYWORDS

- **employee engagement**
- **agile leadership**
- **people management**
- **human resource management**

REFERENCES

Alfaro-Saiz, J. J.; Verdecho, M. J.; Rodriguez-Rodriguez, R. In *How to Achieve Dynamic and Flexible Performance Management Systems for Collaborative Processes*, Working Conference on Virtual Enterprises, Sept 2013; Springer: Berlin, Heidelberg, 2013, pp 639–647.

Alhadid, A. The Effect of Organization Agility on Organization Performance. *Int. Rev. Manage. Bus. Res.* **2016,** *5*(1), 273–278.

Anderson, D.; Kelliher, C. Flexible Working and Engagement: The Importance of Choice. *Strategic HR Rev*. **2009,** *8*(2), ISSN: 1475-4398.

https: /www.universityworldnews.com/post.php?story=20210419105006889 (accessed Aug 16, 2021).

Bakker, A. B.; Demerouti, E. The Job Demands Resources Model: State of the Art. *J. Manage. Psychol.* **2007,** *22*(3), 2007.

Bakker, A. B.; Demerouti, E. Job Demands-Resources Theory. In *Work and Wellbeing*; Chen, P. Y., Cooper, C. L., Eds.; Wiley Blackwell: USA, 2014; pp 37–64.

Bertalanffy, L.V. *General System Theory: Foundations, Development, Applications*; Braziller: New York, NY, 1968.

Chakraborty, T.; Ganguly, M. Crafting Engaged Employees Through Positive Work Environment: Perspectives of Employee Engagement. In *Management Techniques for Employee Engagement in Contemporary Organizations*; IGI Global: USA, 2019; pp 180–198.

Chakraborty, T.; Gohain, D.; Saha, R. What Comes in the Way of Engagement? Moderation Analysis of Stress on Women Marketing Executives Work Life Balance. *Int. J. Hum Res. Dev. Manage.* **2020,** *20*(3–4), 349–368.

Chakraborty, T.; Mishra, N. Appreciative Inquiry: Unleashing a Positive Revolution of Organizational Change and Development. *Int. J. Econ. Commerce Bus. Manage.* **2019,** *6*(2), 32–37.

Chakraborty, T.; Tripathi, M.; Saha, S. The Dynamics of Employee Relationships in a Digitalized Workplace: The Role of Media Richness on Workplace Culture. In *Critical Issues on Changing Dynamics in Employee Relations and Workforce Diversity*; IGI Global: USA, 2021; pp 175–205.

Desai, M.; Majumdar, B.; Chakraborty, T.; Ghosh, K. The Second Shift: Working Women in India. *Gender Manage.***2011,** *26*(6), ISSN: 1754-2413.

Dove, R. Agile Enterprise Cornerstone, Knowledge, Values and Responsibility. In *Business Agility and Information Technology Diffusion*; Baskerville, R. L., Mmathiassen, L., Pries-Heje, J., Decrop, J. I., Eds.; Springer: New York, 2005.

Gould-Williams, J.; Davies, F. Using Social Exchange Theory to Predict the Effects of HRM Practice on Employee Outcomes. *Public Manage. Rev.* **2005,** *7*(1), 1–24.

Heam, S. *Introducing Work Flexibility Through Performance Management Systems.* Clear Review. [Online] January 20, 2017. https://www.clearreview.com/flexibility-way-forward-performance-management-systems

Hobfoll, S. E. Conservation of Resources: A New Attempt at Conceptualizing Stress. *Am. Psychol.* **1989,** *44*(3), 513.

Hobfoll, S. E.; Shirom, A.; Golembiewski, R. Conservation of Resources Theory. In *Handbook of Organizational Behavior*; 2000; pp 57–81.

Hubauer, J. *Managing employee learning and development with flexible working.* Speexx. [Online] October 7, 2020. https://www.speexx.com/speexx-blog/managing-employee-learning-and-development-with-flexible-working/

Johnson, H. T.; Kaplan, R. S. *Relevance Lost*; Harvard Business School Press: Boston, MA, 1987.

mckinsey.com/business- function/organization/our-insights/ the -journey-to -an- agile organization

Meijman, T. F.; Mulder, G. Psychological Aspects of Workload. In *A Handbook of Work and Organizational Psychology*; Psychology Press: UK, 2013; pp 15–44.

Najrani, M. The Endless Opportunity of Organizational Agility. *Strategic Direction* **2016,** *32*(3), 37–38.

Neely, A. The Performance Measurement Revolution: Why Now and What Next? *Int. J. Operat. Product. Manage.***1999,** *19*(2), 205–228.

Performyard. *16 Key Features of a Performance Management System.* [Online] February 21, 2021. https://www.performyard.com/articles/features-performance-management-software.

Rooke, D.; Torbet, W. R. Seven Transformation of Leadership; Hbr.org/2005/04/seven-transformation- of- leadership, 2005.

Rose, M. *Workplace Training Now Needs to Be More Flexible and Adaptive Than Ever.* eLearning Industry. [Online] May 12, 2021. https://elearningindustry.com/workplace-training-now-needs-flexible-adaptive-ever.

Saha, N.; Gregar, A.; Saha, P. Organizational Agility and Employee Engagement. *Int. J. Org. Leadership* **2017,** *6*, 323–334.

Sambamurthy, V.; Bharadwaj, A.; Grover, V. Shaping Agility Through Digital Options: Reconceptualizing the Role of Information Technology in Contemporary Firms. *MIS Quarter.* **2003,** *27*(2), 237–263.

Schwartz, J.; Pelster, B.; Kwan, A.; Neveras, N.; Erickson, R.; Szpaichler, S. Talent 2020: Surveying the Talent Paradox from the Employee Perspective. *Deloitte Insights* [Online] January 2, 2013. https://www2.deloitte.com/us/en/insights/topics/talent/talent-2020-surveying-the-talent-paradox-from-the-employee-perspective.html (accessed Aug 16, 2021).

Sendawula, K.; Nakyejwe Kimuli, S.; Bananuka, J.; Najjemba Muganga, G. Training, Employee Engagement and Employee Performance: Evidence From Uganda's Health Sector. *Cogent Bus. Manage.* **2018,** *5*(1), 1470891.

Setiyani, A.; Djumarno, D.; Riyanto, S.; Nawangsari, L. The Effect of Work Environment on Flexible Working Hours, Employee Engagement and Employee Motivation. *Int. Rev. Manage. Marketing* **2019,** *9*(3), 112.

Temmerman N. *Cultivating the Agile University Requires Good Leadership.* University World Press: Apr 24, 2021.

James, T. K. Leadership in Context, kingsfund.org.uk/publications/articles/ leadership-context lessons- new- leadership- theory- and current- leadership, 2011.

Weiner, N. *Cybernetics*; Wiley: New York, 1948.

CHAPTER 9

Leadership Style in Times of Crisis: Traditional Mentoring to Remote Monitoring

TRISHA BARUA[1*] and IPSITA C. PATRANABIS[2]

[1]*Deloitte India (Offices of the US), Hyderabad, Telengana 500081, India*

[2]*Techno International New Town, Kolkata, India*

**Corresponding author. E-mail: trisha.barua6@gmail.com*

ABSTRACT

The world, for the past one year, is experiencing a pandemic that has caused uncountable human suffering and death, alienated personal relationships, and in many situations has robbed people of livelihoods. The coronavirus pandemic has also challenged the medical systems and uncovered dreadful inequity of the society. The pandemic has also reinforced the belief that unprecedented events lurk around us and the ones who are proactive in designing prevention mechanisms are the ones who will be a winner ultimately. This chapter discusses about the various challenges and scopes that leadership is experiencing today amidst a global crisis. This chapter attempts to highlight the challenges of this novel intervention and the transition from normal work environment to a futuristic hybrid work environment.

9.1 INTRODUCTION

Unprecedented advent of events is a major challenge because it hits hard when the world is least ready to retaliate. The year 2020 will be etched in the

---

Agile Leadership for Industry 4.0: An Indispensable Approach for the Digital Era, Tanusree Chakraborty, Tahir Mumtaz Awan, Ashok Natarajan, and Muhammad Kamran (Eds.)

pages of history for being a harbinger of chaos and uncertainty owing to the COVID-19 outbreak worldwide. The world economy and human emotions, arguably, have been impacted as much more than the other events affected by the pandemic. A leader is someone who successfully collaborates with his team to achieve greater ends. An effective leader seems to understand the goals and purposes of the employees and help them channelize their competence. He must seem to be able to satisfy them, understand the consequences of his actions and have clarity about his decisions. The best leaders are always ready to be accountable in a crisis and face challenges beyond their control. They work toward generating team focus and develop appropriate measurement criteria to monitor and manage performance. Under any kind of external crisis, for example, natural, financial, or even health emergencies the effective leaders will respond promptly and take control of the situation. The leaders who are the custodian of all forms of crisis and are capable to minimize loss and firmly regain control of difficult situation are the ones hailed as heroes.

The current research focuses on the increasing demand of leadership skills that needs to be exercised while mentoring and monitoring teams virtually. Unlike the traditional, collocated teams, virtual teams are challenged by space, time, geographical, and organizational boundaries. The effectiveness of the performance of such virtually scattered teams has often been brought under scrutiny and questions have been raised about the roles and responsibilities that the team leaders should play and adapt and types of interventions that managers can use to launch and sustain these teams. The term remote working is not a novel idea that has been introduced to the world. It was already existing, being explored all throughout by various industries. But remote working would become an integral part of every household at the blink of an eye could not been foreseen earlier. In relation to this, this chapter tries to find answers to various pertinent questions with the help of virtual discussions with leaders from different domains that would help in the process of predicting similar situations of crisis in future and help employees sail through isolation, dislocation, and disengagement.

9.2 LITERATURE REVIEW

At first, the study brings to light what leadership truly means, with the help of past research. In their work (Probert and James, 2011), they highlight two different crises. The first one refers to a crisis in the leadership development literature. According to the authors, the organizations must focus on the governing framework that organizations have inculcated in their culture about leadership. The second type of crisis refers to the challenges the organizations are facing. Any form of sudden disruption provides opportunities to promote changes in the leadership concept, as the situation becomes dynamic and malleable. Leadership is a concept that touches every segment of the society. Be it a school prefect to being a business leader, the one who is looked up at and admired wins the game. One such elaboration was made by the medical professionals cum authors who raise some important issues that the world is facing today with the onset of the pandemic. They have cited the important traits and practices that effective leader demonstrate during a time of disaster. They have spoken largely about the core leadership principles that one needs to adhere to during a crisis. To combat crisis, different offices, such as the clinical enterprise, educational enterprise, and research enterprise to work together. Talking from a healthcare sector point of view, they are sanguine that crisis leads to opportunity to react and proact and it will greatly benefit in planning for crisis with renewed commitment for achieving long-term goals (Kaul et al., 2020). Preparedness can only apply with proactive training. Any disaster management requires proper training because a proper disaster management training builds the appropriate workforce who remain prepared to respond at all time and at all levels of disasters (Nazli et al., 2013). Talking of the future implications of virtualization, this study highlights that as working from office had enabled breadth of a task, at home, the tasks have gained greater depth. The learning and gaining have been too much outside the normal office set up and hence, the future of work will be a hybrid one. Flexibility of work environments has created new challenges for managers and leaders. It requires new skills and perspectives to handle a hybrid workplace. Sometimes, intervals of crisis can become moments of opportunities (Steenbarger, 2020). Any new intervention comes with its pros and cons. Virtualization and remote working has been interpreted both in the light of being advantageous and disadvantageous. The business needs to evolve to adapt to a post-pandemic world and it can start by revising the organizational policies where employees will work majorly from home and only will come to office to collaborate on projects (Teismann, 2020). Managing a diverse workforce remotely brings in a plethora of concerns that the managers and leaders need to keep in mind while managing a hybrid workforce. Molding a company culture according to the given situation of crisis should be of primary importance to the managers to make the workforce as much productive as possible. Secondly, there should be an unbiased balance in managing both remote and traditional workers with the help of specific policies so that either of them do not feel neglected by the management. Thirdly, using a technology that supports the organization work culture will serve as the backbone of any organization (Sambandam, Managing a Hybrid Workforce, 2019). The hybrid workplace has thrown fresh new challenge to the managers and who are managing employees located both in office and remotely. The author puts emphasis on how the managers need to design processes and implement practices which will ensure a connect among in-office employees and employees working from home and thus a transparency in the workflow. The biggest strength that the leaders and managers need to utilize is their ability to empathize with the employees and allowing them to feel included in the system even if they are working remotely. One is to follow the open-door policy even in a remote workspace so as not to make employer–employee relationship strained in times of crisis and the other is to constantly mentor and coach the teams virtually so that they do not feel out of place and have a good rapport built up with the management (Kelly, 2020). Talking about virtual mentorship, it talks about the importance of effective mentorship programs that can be widely introduced in the new workplace, mostly, remotely. But what it needs to assess more is to segregate between who needs mentoring most and who can serve as the mentor. The author opines how mentorships in the current circumstances may be more of a way to calm the fears of the remote employees. The author is of the opinion that now that the mentor and the mentee are remotely situated is not an obstacle to success, but rather in some cases may make the relationship easier. But, the bigger challenge, however, is the question of commitment from the mentees, because in most cases, the commitment wears off after few initial sessions of the mentorship program (Weldon, 2020). The paradox of maintaining the connection while social distancing is a challenge. What is of more importance now is the need for remote mentoring that delivers positive and powerful effects for the displaced employees. Investing in these mentoring programs can boost morale, performance and motivation, which are even more tangible in this period of anxiety and uncertainty. Organizations need to tailor few necessities before the onset of the mentorship program, that is, managing expectations from the mentoring program, adapt to the new work environment to suit the mentorship program, create a backup plan for tech issues and most importantly, maintain connection, trust, and confidentiality (Goredema). The move to virtual working requires a change in the mindset for both leaders and employees. It requires trust, ownership of processes, credibility, innovation, and care. There is a pressing need to have a relook at the existing leadership model to meet the need for a digital and virtual work environment. The virtual collaboration in the times of crisis has provided insights into the understanding of the true meaning of leadership. The responsibility of a leader is to make the environment congenial for the teams to work toward organizational shared goals along with mutual support. It is also expected of the leaders to engage the teams in trying out new things to keep the remote workforce strong and motivated (Ruff and Schoenwald, 2020). The review and preview of the relevant literature for this chapter focuses on changes in work practices, economic and sociopsychological impacts, and various demographic factors to generate disparate effects. In this chapter, the authors present the implications, issues, and insights for way forward. The broad scope of the study provides a consolidative approach toward the implications of COVID-19 for organizations and also identifies scope for a proactive approach (Kniffin, 2020).

9.3 OBJECTIVES OF THE STUDY

To highlight the importance and responsibility of an accurate leadership amidst a crisis, that has left the whole world cripple.

To conduct focus group discussions with middle- and high-level managers who are in leadership positions in MNCs.

To analyze the responses in the light of the current pandemic.

To recommend the major takeaways from the same that can help build a resilient future for the companies in similar situations.

9.4 RESEARCH METHODOLOGY

For the current research, a focused interviews were designed with middle- and high-level managers to align the research on the platform of discovering common areas that have affected the leadership skills and the metamorphosis that the existing skills had to undergo to adapt to the challenges thrown by the unprecedented pandemic. The methodology involved gathering respondents form esteem backgrounds to discuss the current topic of research. In this form of qualitative research, the respondents were asked a set of standard questions to collect their views and opinions about the same. The choice of respondents was made across different industries, such as training, learning and development, oil and natural gas, finance, IT, and service sectors with the aim of getting a heterogenous response to the questions. As it was difficult to conduct group interviews during the lockdown, Google Meet and Zoom sessions were the possible options as the respondents could not come up with mutual convenient time.

The focus group for the study comprised of the following respondents:

Mr. Rahul Bose, Business Consultant, IBM—An experienced Learning Lead with a demonstrated history of working in the information technology and services industry. He is skilled in Business Process, Operations Management, Global Delivery, IT Strategy, and Leadership. With a total 34 years of experience, he has managed teams of various hierarchies and sizes throughout his entire career.

Miss Swati Gupta, Head Learning and Development, Better.Com—A leading Talent Development professional and a seasoned facilitator with over 15 years of experience. She holds a remarkable track record of constructing behavioral turnarounds in the organizations that she has worked with through her keen insight into Organization and Talent Development life cycle.

Mr. Angshuman Bhattacharya, DGM, Paradip Refinery, IOCL—He has 21 years of total experience across verticals, such as recruitment and selection, salary and wage administration, contract management and compliance, general management, and welfare. He also holds expertise in General HR, Talent acquisition, Contract Management, IR&Legal, CSR, and SAP-HCM.

Mr. Arijeet Talapatra, CEO, Transsion Holdings—An expert in building and executing Go-To Market strategies across Urban and Rural markets. His experience spans growth markets as well as highly competitive and penetrated industries. Over a career of 20 years, he has worked in Global Blue-Chip companies in sectors as diverse as FMCG, Telecom, and Mobile handset industry.

Mr. Adhideb Ghosh, Executive VP, MullenLowe Lintas Group—Advertising and Marketing professional with over 16 years of experience. Skilled in Advertising, Integrated Marketing, Brand Management, Client Relation and Business Development.

Mr. Prithwish Ghosh, General Manager Operations, WNS Global Services—With a total of 22 years in service industries, he has been in Finance domain all throughout and had an initial experience of audit of various sectors, such as Public and Private Sectors and has an experience of Bank Audit.

Mr. Avelo Roy, Managing Director, Kolkata Ventures—A tech entrepreneur, investor, four times TEDx speaker and TV Host. Avelo started his first company at the age of 19 around his patent pending technology while still studying at the Illinois Institute of Technology. His mentorship through Kolkata Ventures in the past 3 years has resulted in 400+ revenue generating start-ups, two exits and 10 crores in funds raised (Table 9.1).

TABLE 9.1 Representation of the Sample Design.

Sl. No.	Name	Organization	Position	Experience (in years)
1	Rahul Bose	IBM	Business Consultant	34
2	Swati Gupta	Better.Com	Head Learning and Development	15
3	Angshuman Bhattacharya	IOCL	DGM	21
4	Arijeet Talapatra	Transsion Holdings	CEO	20
5	Adhideb Ghosh	MullenLowe Lintas Group	Executive VP	16
6	Prithwish Ghosh	WNS Global Services	General Manager Operations	22
7	Avelo Roy	Kolkata Ventures	Managing Director	15

The sample was distributed keeping in mind not only the heterogenous sectors to which the respondents belong to, it was also to highlight their years of experience which is one of the primary reasons behind their understanding of what leadership is, their individual leadership skills that has developed throughout their years of services, their experience in handling different team sizes, and their encounter with organizational crises. The aim was to understand how they applied their skills and their past experiences in developing new set of leadership skills to build a resilient future for their respective organizations.

The questionnaire was kept open-ended with the intention to induce more information as compared with other forms of questionnaire like, close-ended or multiple-choice questions. Since the focus group belonged to a heterogenous community with demographic variations, it served as a feasible choice of research instrument for carrying the study further. Open-ended questions also serve as the best platform for easy analysis unlike quantitative data analysis.

Some of the salient questions enlisted in the questionnaire are:

As a leader, during crisis, what are some of the big questions that a leader needs to be asking oneself?

How does your company ensure a clear and transparent communication with all stakeholders during a crisis? Kindly give details about the same. Please mention whether the organization brought in a change in the communication system in the face of a novel crisis?

How resilient is your business model under the current impact of the crisis? How do you plan to manage future potential crises?

Please throw some light on the ways you handled the teams remotely and the ways you both mentored and monitored them.

Are you in support of the opinion that a hybrid workforce model will dilute an organization's culture? As a leader what are the innovative ways that you have used to keep the organizational culture intact even remotely?

How aligned is your organization, both internally and externally right now? How has the pandemic affected the way your organization operates and maintains communication?

The questions are drafted based on the current scenario and the major challenges faced by the organizations. The collected response has helped in getting a clear picture of how different organizations have faced unique challenges and the unique ways they overcame the same. The study has also drawn references from the secondary data, such as online journals, articles, and blogs. Therefore, the research design has been curated with a mixed method of data collection. The focus of the study is to find a similarity in the pattern of how different organizations have faced the challenges of COVID-19, how have they planned the new normal for their business, how did they manage their remote workforce, how did they keep them aligned and motivated and how, amidst all the uncertainties, have they kept their businesses thriving. The primary focus of the study is to understand the leadership techniques applied during the crisis and how the leaders successfully managed to mentor and coach the scattered workforce remotely while keeping the culture of the organization intact (Fig. 9.1).

9.5 SUITABLE SUGGESTION OF MODEL

Virtual leadership throws a lot of challenges to the leaders all over the world with regard to implementation. The world was indeed not prepared for such a raging pandemic given the advancement it has seen in the fields of science, technology, and medicine. But being a surprise phenomenon that it is, this pandemic has also brought along many first-time lessons that were long overdue for the entire world to be learnt. Among many such unprecedented advent of events was the virtual set up of work space that brought the entire world closer but moved people away from one another. Leading and managing

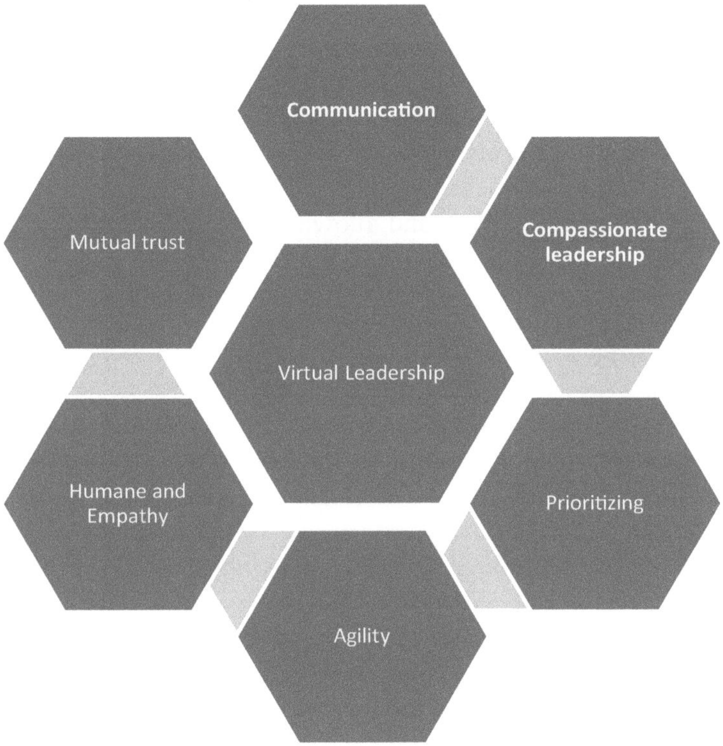

FIGURE 9.1 Key dimensions of effective remote monitoring *(self-compiled)*.

a diverse and dispersed population was and still not a child's play. Though many big companies had had the infrastructure to shift the entire population of the organization remote, many small companies struggled hard to even make ends meet which led to poor business and eventual lockdown.

Humans being organizational resources deserve to be checked upon, spoken to, heard, seen and emotionally felt. And leaders being the pivotal head of the pecking order need to cater to such needs. The pandemic taught the world that resources, even if located diversely, can be treated with equal empathy and attention and this where the importance of virtual leadership arose.

Leadership on the virtual platform can be strengthened with the following key dimensions:

Prioritizing events and issues with equal importance as it would have been done in the physical office environment.

Agility of the organizations to take crucial decisions in testing times.

Treat all the resources with empathy and with a humane outlook.

Building a mutual trust between the employees and the organization as this is the time when both of them are in equal need of one another.

Keeping the channel of communication smooth and clear and devoid of any barriers that might prevent the employees to reach out to their employers in dire need of circumstances.

Keeping compassion alive and allowing the entire organization act as a family, even remotely.

9.6 FINDINGS

Just when the world was ready to welcome the new opportunities that came with a new year, making resolutions to prepare for a better year than the previous ones, it was hit by an unprecedented guest, not known to many and hence was named as the novel coronavirus. The COVID-19 virus may have throttled our lives, but things have not come to a stop. The world continues to move on with it albeit in very different ways. And this is what will be discussed in broader ways in this section where analysis will be drawn based on the responses from industry experts and leaders who have found this pandemic an opportune moment to redesign not only the organizational structure but also to sharpen the leadership skills as well.

The major key findings in the study are as follows:

Communication is the key to successful team management.

Compassionate leadership is need of the hour to make the work population feel like a family.

Prioritizing work, roles and responsibilities by the leadership will develop a sense of ownership among the employees.

Challenging times are not to be sympathetic but to be empathetic.

Leaders need to be humane in their dealing with employees, especially in a time when people are experiencing a splurge of unforeseen emotions.

Trusting one another has become imperative in a challenging time like COVID-19 and only mutual trust can help overcome both the situation and the emotions associated with it.

The respondents being middle- and high-level managers with experience ranging from 15 to 30 years, have shared their unique experiences on how the pandemic has paved newer path of exploration and has made them more resilient as leaders. According to all of them, there has been crises of some form or the other that has at some point in time hit their career hard, the

2008 financial crisis being the major one for many. Apart from that, there have been other short-term crises at business or organizational level that with immediate interventions were controlled.

An organizational crisis, on the one hand, can pose a threat to the sustainability of the business and will have deep impact on its relationships with the stakeholders, whereas a global financial crisis highlights the tension between external forces and a challenging macroeconomic environment on the other. According to the respondents, the challenges this time was even bigger because the crisis was not only limited to an organizational or financial level but also caused a global health crisis that had put human lives at stake too. Therefore, the bigger challenge that as leaders they faced were to manage their human capital amidst all the uncertainty and cacophony surrounding layoffs and retrenchment, the leaders had to don the cape of a savior, with the sole mission of holding the family together from disintegrating and albeit, they were successful unlike many big and small companies.

Good leadership must be exercised whether in a crisis or not. Communication is the key, as reiterated in a respondent's answer, is even more important during a crisis. It is one of the salient responsibilities of the leader to take everyone on the same page, make them understand the gravity of the situation, the reason for the decisions taken and the consequent response to the same. Digital innovations in businesses have made extraordinary modifications in the way businesses are done (Chakraborty and Saha, 2018).

Not always will the leaders find their employees bring up their issues for an open discussion. There are times, especially in the rise of a crisis that the leaders need to hear and understand words that are left unsaid. As Peter Drucker has said, the most important thing in communication is to hear what is not being said. Creating a sense of unity in crisis is the greater communication the organization can practice holistically and foster a sense of loyalty in the employees. Leaders need to think of innovative ways to keep the communication going varying from big talks to little talks, with the later getting greater preference. During organizational crisis, compassionate leadership is what most leaders should be adhering to because besides a pandemic, the world is also grappling with the virtual world, which to many is as novel as the virus.

As a leader during a crisis, one deals with a lot of immediate issues and existential crises for the company. These are the testing times when leaders must be asking themselves few significant questions that might help them deal with similar situations in future. Leaders have found themselves asking questions on how to keep the organization trained with regard to business continuity plan considering challenges of short-term and long-term orientation. Prioritizing is another key word that needs to highlight upon factors like what to prioritize, whom to prioritize, and when to prioritize. Sharing his experience, one of the respondents narrated how his mentor in an international company followed the method of dehumanization in order to sustain the Business Continuity Plan and laid off 1500 employees during COVID-19 from the existing 2100 employees and continued business with the rest 600 employees with the justification that the existing workforce will help him make business and keep it running whereas retaining all 2100 would bankrupt him. He also added that if he can sustain his business during the apparent crisis, he will be able to bring the severed workforce back into the organization when things go back to normal. Unpopular and straight decisions like this upset the mental models of the employees on the one hand, but on the other hand, it raises big questions for the leaders and that is, whether he is good enough to take these decisions all by himself. But what is more important in the face of adversity is to think of measures which will make the organization grow and leverage the strength of the adversity that has hit it.

Millions of professionals around the world are engaging themselves in mass experimenting on how to bring more innovative ways to keep the remote work teams engaged and yield maximum productivity inspired by the competency model of Innovation Leadership established by Dr David Gliddon. According to the philosophy of this model different leadership styles prove influential for employees to produce creative ideas, products and services eventually giving birth to a creative leader. This pandemic has stood witness to the fact that virtual teams and remote work accelerate the transformation of existing leadership models. Collaboration in the virtual mode has provided important insights into the understanding of leadership and has in fact, in many ways imposed a forced transformation of the existing leadership models to meet the present demand in the novel work processes. Now that the entire world has moved into a hybrid workspace, it is even more cardinal to have the remote teams closely connected since the basic human touch has been missing from the work culture. Since regular communications in the form of chance meetings has been replaced, communication via digital media should be more frequent, must be more frequent and above all more regular. From weekly zoom luncheons to weekend fitness drive sessions, the companies have adapted to broader engagement activities, albeit digitally. Besides relying upon digital platforms, such as WhatsApp, Google Meet, zoom calls to ensure communication is at place all the time, the daily health tracker in-house app of one of the companies has also come handy in tracking the health status of the employees. The in-house well-being team has also been active in communicating on a regular basis with the employees to help them fight mental health issues, hit worst by the pandemic. Engagement programs, such as Stay Happy, Stay Fit were devised under which various activities were organized to keep employees physically fit and mentally healthy while they are remotely placed. This, according to the respondent, has been able to generate more trust and as a result, the attrition rate has also been much less during the pandemic. Virtual collaboration reiterates the demand for a face-to-face colocation with an efficient digital leadership style that must have reliability, integrity, and confidentiality to foster trust and respect. The establishment of predominant structures positively encourages the dynamics of the team and also creates synergy within the company.

Working from home has always existed and was exercised in the IT sector. But it will be the new normal one day, was never thought of then. While there were respondents who spoke extensively on how well their organization has adapted to the change and could shift the entire set up in remote locations within a few days, all thanks to their organizational agility, there were also companies who had to start the set up from the scratch and yet had to stick to the Business Continuity Plan. Once the workforce shifted the base, there were other issues that started affecting work and that is finding a new environment to work in. Employees had to deal with challenges, such as background noises, interrupted work, childcare, chores, looking after elderly parents, weak internet connection, and taking rejuvenation breaks. These inframeasures were not initially addressed quite well by the companies but gradually they were given allowance to upgrade their workspace. According to one respondent, some people in his organization utilized the allowance to upgrade to better connection to avail uninterrupted internet connection. Another advantage of the pandemic for a company was when it proposed a corporate spin-off to manage the organization's infrastructure services. Employees were also allowed to travel to Tier-2 and Tier-3 cities, preferably their hometown, to find the desired peace and time to finish the work on time, otherwise hindered by city paraphernalia. Technology as a business model has always been a blessing to IT companies as they were always savvy with it. What most of the companies also did was to introduce the empathy scheme and extended the human touch to the distressed employees to give them assurance of their job so that they are not constantly exposed to the fear of losing it. Organizations in the present context would necessitate a positive organizational culture and people orientation (Chakraborty and Ganguly, 2019), there is need for positive approaches to understand employees (Chakraborty and Mishra, 2019; Chakraborty et al., 2019).

Organizations are more agile after a disaster, so the role of human resource professionals in managing consequences of a crisis is very crucial (Chakraborty and Saha, 2017).

Of all the many things that the pandemic normalized for the entire world, a hybrid workforce model topped the list. That it is here to stay even longer, was what most of the respondents agreed to. However, essential services, such as oil and gas, insurance, and banking were barred from the luxuries of remote working. The immediate urgency of their respective services will always stand contradictory to a remote workspace. One of the major upsetting events of the pandemic was the mass scale exodus of the migrant laborers that made headlines all around the world. Sharing his experience, one of the respondents replied that with the announcement of the lockdown, their organization planned for the major maintenance shutdown after operating for 3 years at a stretch. While migrant laborers in other parts of the country were forced on an arduous and perilous journey home, their organization nestled them in safe home with lodging and food for completing the manual industrial labor required during the shutdown activities. The eventual outcome of this decision demonstrated the virtue of humane approach rather than a thoughtless and mechanical decision. While there were uncountable positive cases followed by a couple of death by the virus, yet the organization did not stop its operation and stood testimony of peace, prosperity, and a wholehearted embrace of the larger society surrounding them.

The paradigm shift concerning mentoring to monitoring, albeit digitally, needed a quick yet flawless response for the scattered workforce. The initial set back of the pandemic was the huge pandemonium around losing jobs. While many companies were losing their workforce, the respondents to my research retained their employees. There were also events of job rotation where the companies ensured growth for the existing workforce. For instance, sales personnel, debarred from traveling, were given the role of leading the sales team digitally during which their on-the-field experiences were utilized for pushing business online to the customers. What did affect the business during the pandemic was, companies were unable to announce promotions or allow increments for the financial year. Nevertheless, they kept the spirits of the workforce high by providing them with various learning and development opportunities. Recreation wise, the teams had their weekly calls which was not only about work discussion but also about discussion that was beyond work and talking about how they are coping up with working remotely. Leaders reached out to those employees more frequently who were away from family and staying on their own with limited resources at their disposal. They also did a lot of virtual fun activities to keep the employees engaged and motivated and even celebrated festivals digitally. One essential aspect of remote mentoring is to trace burnout among employees. Leaders need to be attentive to the behavioral changes among the employees and help them get professional help for the same. Another important aspect of online monitoring was building mutual trust where it was ensured that the organization still operated as a single unit, like a family. The bigger lesson inculcated into the minds of the employees is together we rise, or together we fall since psychologically the workforce went into a self-preservation mode, cutting themselves off from one another, suspecting each other and creating mental cages around themselves. Therefore, more than anything else, the leaders had to monitor the mental health of the employees while mentoring them for generating business. Talking about the drawbacks of remote mentoring, one of the respondents replied by saying that it took him quite some time to understand the person as an individual because virtually, they had very limited chances to interact unlike a normal office set up. The respondent deployed a buddy for the newly inducted, freshly into remote working employee, so that the former can share instances of his prior experience as a remote worker (as IT companies had always had the culture) and make him feel at ease to ask for assistance, if need be, for a new working environment. Remote working had proved beneficial in many ways, to name a few, talent development, increased self-esteem in employees and deeper engagement are now an indispensable part of the newly acquired hybrid work culture.

An organization's culture is its backbone that includes an organization's expectations, experiences, philosophy, self-image, inner workings, and future expectations. When the entire world shifted its base to remote working with no clear understanding of the timeline as to when can regularity will be restored, there was an immense ambiguity as to how the companies will handle the intangible but the powerful tools of businesses, such as meetings, socializing, and perks, such as free entertainment, allowances, and many more. A survey during the pandemic has proved that the impact of culture around the world has not been that detrimental and that supportiveness and kindness have increased than got deteriorated. The standard of communication upped, and employees are getting greater response than ever. There's also been a huge shift in flexibility, with companies acknowledging the complex lives of their employees in terms of responding to the needs of children, aging parents, health concerns, and so on and so forth which the pandemic has pushed to the fore. With reference to this, a respondent replied by saying that he has allowed more flexibility in the work timings for the employees rather than binding them into strict deadlines. So, the employees in his organization now enjoy the freedom to finish their task at their own feasible time, albeit the importance of sticking to daily work deadline is still uncompromised.

A crisis is a time when relationships start disintegrating. To combat the same, businesses need to be well aligned to the internal and external policies so that it is a storehouse of information for not only its employees but also to its external stakeholders too. Companies, especially in telecom and service sectors took the initiative to reach out to its customers and offered their assistance without being asked for the same. Earlier when for availing the same service, a client company had to travel long distance to the office of the service provider, was now available just a phone call away. When the entire world was under the impression that everything came to a standstill, few companies continued with their businesses and assured full assistance to their stakeholders. These companies kept in mind that any lag in communication will make way for the customers own version of internal and external problems and priorities. Therefore, to nip in the bud, the companies did a thorough mind mapping of the employees' current needs and requirements.

9.7 SUGGESTIONS

It is always preferable to proactively implement remote work policies and training in advance for sudden changes that may occur anytime in future. Though it might not always end up being feasible, but managers can engage and improve the productivity of the remote employees.

Managers need to understand and recognize factors that can make remote work demanding and productive at the same time. If not implemented, then the high performing employees may experience a decline in productivity while working remotely.

Ensuring regular check-ins in the form of video-conferencing can prove to be the key to managing remote workforce. For managers and employees new to remote working, this can be a great deal to connect, share, and collaborate.

Communicating the desired outcome with regard to regular tasks and responsibilities can prove to be an imperative aspect of regular communication. Lack of communication in the normal workplace is itself a challenge and now when the entire workforce has shifted remotely, communication is paramount.

Focusing on outcomes rather than activity can be a great way to allow employees enjoy freedom at work and exercise ownership. Clearly, defining desired goals and results and then allowing employees to develop a plan of action enhance creativity and ownership.

It is the right time to mentor than only manage. Employees these days are well informed and understand the subtle nuances and difference between management and leadership. Employees are in more need of encouragement, motivation, training, and opportunities than ever.

It is indeed true that the world is stuck in a situation of volatility, uncertainty, complexity and ambiguity but that does not mean it brings a halt in the process of development of both the teams and the managers. It is about time for mutual growth and development and the sun is shining just right for the hay to be made.

9.8 CONCLUSION

Despite its far-fledged implications for organizations and its stakeholders, research on crisis management still remains fragmented. We have integrated research from multiple perspectives to bring coherence to this literature, and have developed a holistic framework for better understanding of the crisis process. The pandemic has proved that the employees will be expecting a lot from their present leaders in the future because the new experience of remote working may lead to a paradigm shift which will demand new-generation leaders with relevant competencies. In the new hybrid work model, leaders will be judged by their capability to activate the ecosystem and facilitate new competencies with the help of innovative learning opportunities, inclusive work culture, and opportunities for self-development. Leaders will be expected to promote digital literacy in the new workplace. It is essential to build momentum for projects and engage employees located away from one another. Traits, such as building relationships to foster fun and enthusiasm, nomadism to organize remote work, and emotional intelligence to foster conscious stress management and resilience building is bound to challenge the status quo and drive evolution. The hybrid workplace has been a positive chance to experiment for many organizational leaders throwing them to the edge of pushing their own boundaries and creating an environment of democratic leadership. When hit by the unprecedented crisis, the organizations took resort in reactive approaches to initiate changes. What needs to be explored further is how the organizations will be proactive in their approaches and what new structural changes be implemented for continuing with the novel work culture. The organizations have discovered one very crucial observation and that is they are flexible in operations and open-minded to adapt to change. However, it just needed a global pandemic to unveil certain hidden qualities.

KEYWORDS

- **leadership theory**
- **management style**
- **motivation**
- **team formation**
- **working conditions**

REFERENCES

Chakraborty, T.; Ganguly, M. Crafting Engaged Employees Through Positive Work Environment: Perspectives of Employee Engagement. In *Management Techniques for Employee Engagement in Contemporary Organizations*. IGI Global: USA, 2019; pp 180–198.

Chakraborty, T.; Ganguly, M.; Natarajan, A. Predicting Entrepreneurial Satisfaction: The Role of Non-Financial Incentive Factors and Quality of Life Among Women Digital Entrepreneurs. *J. Glob. Bus. Adv.* **2019,** *12*(3), 328–355.

Chakraborty, T.; Mishra, N. Appreciative Inquiry: Unleashing a Positive Revolution of Organizational Change and Development. *Int. J. Econ. Commerce Bus. Manage.* **2019,** *6*(2), 32–37.

Chakraborty, T.; Saha, R. Proactive and Reactive Role of Human Resource Professionals in an Aftershock Management. *Train. Dev. J.* **2017,** *8*(1), 1–11.

Chakraborty, T.; Saha, R. Digital Innovations: Breakthrough Opportunities to Build Novel Business Models. *Bus. Perspect. Aims Scope* **2018,** *1*, 46.

Goredema, O. (n.d.). Remote Mentoring: How to Cultivate a Virtual Connection. Chronus.

Kaul, V.; Shah, V. H.; El-Serag, H. Leadership During Crisis: Lessons and Applications from the COVID-19 Pandemic. Gastroenterology **2020,** *159*(3), 809–812.

Kelly, T. *Engaging Remote Employees Through Coaching and Mentoring*; Training Industry. February 20, 2020.

Kniffin, K. F. *Covid-19 and the Workplace: Implications, Issues and Insights for Future Research and Action*; Harvard Business School: Boston, June 5, 2020.
Nazli, N. N.; Nazli, N. N.; Sipon, S; Radzi, M. H. Analysis of Training Needs in Disaster Preparedness. *Procedia Soc. Behav. Sci.* **2013,** *140*, 576–580.
Probert, J.; James, K. T. Leadership Development: Crisis, Opportunities and the Leadership Concept. *Leadership* **2011,** *7*(2), 137–150.
Ruff, S. K.; Schoenwald, L. *Covid-19 Demands a New Era of Authentic Leadership.* [Online] May 26, 2020. https://www.capgemini.com/2020/05/covid-19-demands-a-new-era-of-authentic-leadership/
Sambandam, S. *Managing a Hybrid Workforce*. Entrepreneur India, February 19, 2019.
Shah, V. K. *Leadership During Crisis: Lessons and Applications from the COVID-19 Pandemic*; Elsevier: Amsterdam, 2020.
Steenbarger, B. *Preparing for Tomorrow's Hybrid Workplace*. Forbes, July 15, 2020.
Teismann, C. *What's the Future of Work? A Hybrid Workforce*. Entrepreneur India, September 19, 2020.
Weldon, D. *The Benefits and Challenges of Remote Mentorship Programs*. CMS Wire, May 8, 2020.

CHAPTER 10

The Power of Empathy: Rethinking Leadership Agility during Transition

TUSHARIKA MUKHERJEE

Work and Organizational Psychologist, Justus Liebig University, Giessen, Germany
E-mail: tusharikamukherjee@gmail.com

ABSTRACT

The world in 2020 was a different world. While it has projected health to be of paramount importance globally, it has also led to unprecedented changes in the way the world functions, especially for economic purposes. As we transition into a changing work culture in Industry 4.0, leadership plays a pivotal role in moving the organization and the individuals-within forward. One key competency that has increasingly been deemed critical to success among leaders in today's times is the vicarious sharing of emotions. Contemporary work structures often render blurring of boundaries between work and non-work domains, causing a significant impact on both mental and physical health, including burnout. Similarly, multicultural and globalized business environments pose considerable performance pressure on the workforce. Such emergent work conditions create a constant demand for greater flexibility and understanding on the part of the organization. Contrary to the traditional perception of leadership, the evidence provides that empathy in a leader operates against employee burnout while enhancing resilience, job performance, and organizational citizenship behavior through effective leader–member exchange. As communication becomes crucial in managing changes or crises, leader empathy enables sustain trust in the organization

Agile Leadership for Industry 4.0: An Indispensable Approach for the Digital Era, Tanusree Chakraborty, Tahir Mumtaz Awan, Ashok Natarajan, and Muhammad Kamran (Eds.)

and leader decision. In this chapter, we take an evidence-based approach to empathic leadership with a focus on its values in the times of adversity and transition, and explore empathic leadership as a capacity building strategy for creating a secure workplace for the modern workforce.

10.1 INTRODUCTION

Organizational life and leadership have changed drastically during the pandemic which brought with them more turmoil and crisis than ever before in living memory. In this rapidly evolving situation, as adversities continue to emerge, they also bring with them new outlook and need to discover latent capacities. The workforce of the present and future necessitates leaders who embody skills considerably different from what was traditionally regarded as leadership. Leadership models highlighting functions, such as planning, coordinating, controlling, and organizing demands a renewal based on qualities, such as self-awareness, self-control, and empathy (Goleman, 2013). Goleman expressed that "executives who can effectively focus on others emerge as natural leaders regardless of organizational or social rank." Daimler (2020) asserts that post-2020 era will witness a paradigm shift in the way leadership competencies will be described. Daimler claims that while the emotional intelligence will scale up emotional agility, time management will make way for context management, communication skills will be renewed via empathy. Industry 4.0 conceptualized from a competency-based approach underscores eight competencies, termed as the "big eight," include leading and deciding, supporting and cooperating, interacting ad presenting, analyzing and interpreting, creating and conceptualizing, organizing and executing, adapting and coping, enterprising and performing (Prifti et al., 2017). Competencies that have been shown to bear close association with empathy.

Emotions have been the subject matter of organizational behavior, work psychology, and management studies for a long time. However, Lewis et al. (2010) claim that much of the advances in behavioral sciences in the workplace have circumscribed learning, cognitions, and perception, and feelings have been largely undervalued or ignored for long. Muchinsky (2000) argues that the focus has been on the "cognitive explanation of emotions" and rarely the reverse "emotional explanation of cognitions," and suggests that feelings must be incorporated within the scientific inquiry if we are to comprehensively understand psychological constructs such as job satisfaction and job stress. Emotions in organization manifest at five discernible, occasionally coinciding levels, namely, intraindividual, interindividual (between persons), interpersonal interaction, groups, and the organization itself (Ashkanasy and Humphrey, 2011).

In this chapter, we reflect on leadership from an emotion-based perspective, we scrutinize empathy as leadership value and strength much needed in the times of adversity and transition, and explore empathic leadership as a capacity building strategy for creating a secure workplace for the modern workforce. The term "empathic leadership" is used in this chapter to steer our gaze on leadership style(s) where empathy is a dominant expression in leader–subordinate interaction and is perceived as empathic behavior.[1] Here, we approach empathic leadership from a multidimensional perspective (based on the five-level model of emotion in organization; Ashkanasy and Humphrey, 2011)—as an emotional response, as attitudinal and behavioral outcome, as a perceived variable in interpersonal relation, as collective behavior, and as an overall climate or culture, and a marker of organizational success.

10.2 THE EMERGING LEADERSHIP LEXICON—EMPATHY

Since the Hawthorne studies by Elton Mayo and Fritz Roethlisberger (1924–1932), the Ohio State studies by Timothy Judge and colleagues, and the human relations movement thereafter precipitated the sense that work behavior is more than merely a by-product of fiscal stimulus and authoritative control, and kindled a new line of exploration in leadership in the context of the emotional needs of the workers. Moving further from the preparadigmatic stages of interpreting the anomalies within the existing theoretical structures, systematic inquiry specifically targeted to understanding emotions in organizational research, that is, "organizational behavior affective scholarship" evolved only recently with the advanced researchers focusing on affect as a social construct (Barsade et al., 2003). Only during the past two decades has exploration of emotions been part of mainstream organizational behavior research agenda, and is subjected to investigation from a multilevel position (Ashkanasy and Humphrey, 2011).

---

[1] Reference to the term "empathetic" appears in the context of empathetic leadership model (Kock et al., 2019), and in this chapter, we purport to deduce the two terms empathic and empathetic identically.

Derived from the German word *einfühlung* which translates to "feeling into," empathy has become a part of scientific inquiry since the late 1940s (Stueber, 2019). Defined as the "the ability to comprehend another's feelings and to re-experience them oneself" (Solovey and Mayer, 1990, p.194), empathy is one of the key components of emotional intelligence and building social relationships (Goleman et al., 2017). One of the pioneering works that described empathy as an emotional event came from Stotland, who defined empathy as "an observer's reacting emotionally because he perceives that another is experiencing or is about to experience an emotion" (1969, 272). Stotland held that empathy encompassed a broad spectrum of emotional responses, including annoyance, relief, distress, envy, pity, and *schadenfreude* (German word for the pleasure or satisfaction derived from others misfortunes and failures). The emerging landscape of empathy research appraises empathy as distinct from four congruous reactive emotions, namely, emotional contagion (reflexive triggering of similar emotions as others), affective versus proper empathy (vicarious sharing of affect with the ability to distinguish between self and others), sympathy (awareness of others despair), and personal distress (self-oriented emotional response to others sufferings), however, they have been interchangeably assessed in the empirical observations (Steuber, 2019).

Empathy, though frequently regarded as a single trait, is composed of three abilities. Goleman (2013) described it as the "empathy triad" encompassing cognitive empathy (understanding other's viewpoint), emotional empathy (understanding others feelings), and empathic concern (being able to sense another person's requirements). Emotional empathy additionally comprises feeling the same emotion as the other person, feeling distressed while perceiving others' strain, and feeling compassionate for others (Hodges and Myers, 2007). Defining empathy as a malleable and developable skill, cognitive approaches interpret empathy in terms of the accuracy and thoroughness in recognizing and understanding others' emotional states (Hodges and Myers, 2007).

10.3 EMPATHY IMPRINTS ON ORGANIZATIONAL LEADERSHIP

Daniel Goleman observed that emotional intelligence is twice as important as technical knowledge and cognitive abilities for leadership at all organizational levels (1998, 1995). In one of the most celebrated articles "What makes a leader?" Goleman noted that "different situations call for different types of leadership. Most mergers need a sensitive negotiator at the helm, whereas many turnarounds require a more forceful authority. I have found, however, that the most effective leaders are alike in one crucial way: they all have a high degree of what has come to be known as emotional intelligence" (1998, p. 93). He identified five basic components of emotional intelligence relevant for work, namely, self-awareness, self-regulation, motivation, empathy, and social skills. According to the author, empathy is "the ability to understand the emotional makeup of other people," and "skill in treating people according to their emotional reaction" (1998).

Leadership theories and approaches have evolutionarily accorded value to the emotional transactions or reciprocation between the leader and subordinates. For instance, contrary to the traditional conception of charismatic leader, the motivational approach to *charismatic leadership* holds that behavioral characteristics displayed by the leader in his/her interaction have a motivating effect upon the subordinates (Choi, 2006). Empathy in leadership is one of the three primary components within this paradigm (the other two are envisioning and empowerment) which builds into trust, emotional bonds, and identification with the leader. This further translates into collective identity, group cohesiveness, and organizational citizenship behavior toward the leader and co-workers (Choi, 2006). More recent models, such as transformational and ethical leadership also have evidence of association with empathy. *Transformational leadership* style emphasizes individualized consideration as one of its fundamental components, which translates into attending to subordinates' developmental needs while displaying concern for those who feel ignored (Megerian and Sosik, 1996). Similarly, three out of the four variations of empathic behavior (perspective taking, empathic concern, and empathic matching; personal distress being unrelated) have positive bearing upon transformational leadership (Skinner and Spurgeon, 2005). Typical quality of *ethical leader* like being supportive and assisting subordinates to develop flourishes with the inclination to understand others' needs and emotions. Mahsud et al. (2010) contend that empathy drives ethical leader to adequately attend to subordinates' career aspirations and job attitudes. Findings reveal that the synthesis of empathy and ethics works as a deterrent of unethical organizational practices and preventing other executives from derailing (Val Velsor and Leslie, 1995). Studies reveal that conditions, such as interpersonal problem, deficiency in developing and leading a team, and falling short of adjusting to transitions, contribute to executive derailment. It seems reasonable to assume that empathy, whether or not combined with ethical concerns, can leverage unconducive conditions.

Parallels between empathic and *servant leadership* can be intuitively drawn based on the importance accorded to subordinate's emotions. Kock et al. (2019) clarify that servant leadership is aimed at securing subordinate needs, even by risking leader or organizational needs, whereas, empathy orientation attempts to cater to emotional needs in a more unified and collaborative manner.

Secondly, empathy is reported to enhance individual likelihood of **emotional contagion** as a social influence, which in turn plays a pivotal role in perception of leadership and follower effectiveness, team dynamics, organizational citizenship behavior, decision-making, conflict and negotiations, and job performance (Barsade et al., 2018; Doherty, 1997). Consequently, empathy is intuitively contemplated to be synergistic with relation-oriented leadership styles. On the contrary, studies show that empathy is a concomitant of task-oriented leadership styles as well. Planning and organizing tasks frequently involve working through confounding information about the task, diverse people and environments, which are facilitated by empathic concerns via effective information processing and perspective taking (Kellett et al., 2006; Wolff et al., 2002). Another study on the ease of employees to return to work after a long sick leave explores effective leadership qualities. Analysis across varied leadership styles showcases 10 dominant categories perceived by both supervisors and subordinates include taking initiative in maintaining contact, being considerate, understanding employee problems, being empathetic, recognizing and appreciating, ability to judge, being communicative, to listen and be receptive, facilitate inclusion, and being able to solve problems, where most of the qualities unambiguously pertain to empathic attitude and behavior of the leader (Aas et al., 2008)

The emerging organizational behavior scholarship on empathy posits it as a consistent predictor of employee behavior. Research has linked empathy with subordinates' well-being and goal progression (Scott et al., 2010); positive emotional states in both leaders and subordinates (Boyatzis et al., 2006; Owens and Hekman, 2012); forgiveness climate and pro-social conflict resolution (Fehr and Gelfand, 2012); reduced burnout and higher compassion satisfaction (Wagaman et al., 2015). However, the singular most prominent behavioral criterion in the work context is undeniably **performance.** Extant research suggests that empathic emotions contribute to successful work performance (e.g., Ashkanasy et al., 2002; Skinner and Spurgeon, 2005). When leaders display understanding of others, **Empathetic leadership model** demonstrates that empathy results in two distinct yet interrelated support origins—emotional and instructional (Kock et al., 2019). The authors describe the process of empathetic leadership optimizing employee performance, which is summarized as subordinate empathy need -> empathy display by leader -> subordinate's appraisal of being valued -> subordinate's positive stance for the leader -> workplace stress decline and generation of positive affect -> enhanced job satisfaction -> elevated effort and risk taking readiness (innovative behavior) -> increased employee performance -> improved perception of subordinates abilities by leader (empathetic outlook) and consequent improved leading -> overall performance enhancement and success (2019, p. 222).

Effective communication behavior and competence serve as anchors in the leader–employee relationship to the extent that certain **communicative skills** facilitate leader emergence (Johnson and Bechler, 1998). While what is spoken has conventionally received a fair proportion of scientific interest, little attention has been accorded to listening as a communication tool and necessity in organizational communication. Reports indicate that listening promotes positive organizational culture, positive interpersonal relationships, supportive climate, and job performance (e.g., Bodie, 2012; Gentry et al., 2007). The limited literature in organizational listening scholarship within the leader–member exchange theory framework also reveals that listening strengthens alliance between leaders and subordinates, as well as job satisfaction, commitment, and productivity (Bodie, 2012; Parks, 2012). In addition to being an information processing mechanism, listening competence derives its caliber from personality differences including empathy. An empathic listener communicates attentively and is likely to engage in a relational or other-oriented dialogue (Parks, 2012). Highlighting the significance of active listening in business communication, Drollinger and Corner (1999) conceptualized active-empathetic listening as a process capacitating sensing, perceiving, and responding.

A pertinent argument in this context concerns practicability of **listening** in certain types of leadership circumstances. In a communication-intensive setting, such as a small group, listening competencies influence leadership emergence. Johnson and Bechler (1998) found that individuals who were perceived as efficient listeners subsequently emerged as group leaders. The findings imply attitudinal and behavioral similarities between effective listening and effective leadership, such as staying focused during discussions, showing interest in others opinion, tolerating disagreement, being non-judgmental or being able to segregate issues from person (Johnson and Bechler, 1998). Similarly, in cross-cultural teams where empathic listening can mitigate the complexities induced by cultural diversity. In the model of cross-cultural leadership Grisham and Walker (2008) elucidate that trust is at the core of cross-cultural leading, which is built upon power, transformation, communication, and empathy. The challenges of leading are compounded when multicultural teams operate virtually. Lack of physical proximity and sharing of office space can hinder developing trust and managing personality differences (Oertig and Buergi, 2006). Effective communication with empathy can aid breaking the communication barriers in teams that are not co-located. In describing emotional intelligence at work, Goleman specifies cross-cultural sensitivity as one of the hallmarks of empathy (1998).

Empathy as a leadership behavior can be decisive in **leadership perception** and development practices. Emphasizing the values of focusing attention on others, Goleman (2013) holds that leaders with empathy are easily recognized by others as natural leaders. A study comprising 38 countries showed that leader's performance (appraised by the supervisor) is linked to the subordinate's perception of the leader's ability to express empathy (Sadri et al., 2011). Contrary to the belief that leaders are not expected to be empathic in high power distance cultures due to centralized power structures, the evidences revealed that leaders' empathy was readily recognized in high power distance cultures as compared with cultures which have more linear power structures. The authors hold that empathy expressed by the leaders is perceived as protection and support from a higher authority (parent like) figure in high power distance cultures. Further, the 2020 State of Workplace Empathy Research reports that despite almost 90 percent of employees and executives believing that empathy is crucial in adverse situations and transitions, there is a substantial gap between perception of empathy in one's organization by top executives and their subordinates. Three quarters of the interviewed workers agreed to work for longer durations for an empathic employer, and also conveyed that empathy in organization is a strong motivator for them, facilitates their productivity, and reduced turnover intentions.

In addition to the dividends in leader–member exchange (Cropanzano et al., 2017), empathy increases the likelihood of **corporate philanthropy** in the form of allocating financial aid and employee volunteering programs for causes like AIDS where executive decisions are affect-infused (Muller et al., 2014). It is imperative to distinguish empathy-based decision-making as an individual choice or a social phenomenon driven by **collective empathy**. As in the case of corporate philanthropy, affect-infused decisions are targeted to events outside the organization' empathic decision unlike other decision that are intraorganizational. Muller et al.'s model of collective empathy identifies group identification, norms of sensitivity, and appropriate communication, produces collective empathic arousal, which intensifies managerial intent to make decisions not exclusively fuelled by economic or strategic (reputational) agenda. This is in line with contemporary advocacy that empathy is essential for a collective perspective within which common humanity and individual differences can be nurtured (Lussiez, 2009).

10.4 VALUES IN LEADERSHIP DURING TRANSITION

Organizations can seldom avoid encountering disruptions and crisis. From natural disasters, civil unrests, to economic shutdowns, organization in all its levels is grossly impacted. The colossal task of organizational repair and revival, in other cases continuity and progress, typically rests on the crisis management ingenuity of the leadership. As D'Auria and Smet (2020) have observed during the coronavirus pandemic of 2020, leaders can strengthen their organization and workforce in five different ways: creating a network of collaborative crisis-response teams and empowering them, raise optimism with credibility, making decisions and reflecting on them, communicating effectively with transparency, and finally, responding with empathy. In this section, the focus is on the transition agility of leaders with empathy, especially in response to a crisis. Organizations in the present context would require a positive organizational culture (Chakraborty and Ganguly, 2019), will need positive approaches to understand employees by appreciative questioning (Chakraborty and Mishra, 2019; Chakraborty et al., 2019). Organizations are more agile after a disaster, so the role of human resource professionals in managing after effects of a crisis is very crucial (Chakraborty and Saha, 2017).

10.4.1 DIGITAL TRANSFORMATION

Industry 4.0 business models are increasingly committing to digitization to achieve agility, changeability, and efficiency. Digital innovations in businesses have made remarkable changes in the way businesses are done (Chakraborty and Saha, 2018). The technological sharpness in the workforce insists on a leadership which is more malleable, adaptable, and willing to provide more support through concern and curiosity for the emotions of the workers. The 2021 "Future of Leadership Global Executive Study" by Schrage, Pring, Kiron, and Dickerson suggests that "successful digital transformation demands that leaders measurably transform themselves. Efforts to lead digital transformation are unlikely to be effective without a leader's own *affective* digital transformation—one that makes purpose, engagement, and fairness as important to workplace success as data-driven agility and productivity." The authors claim that the digitized workplace has not only blurred the boundaries between work and non-work identities, and those between personal and organizational vales, but has also made issues, such as sustainability, discrimination, inequity more readily detectable. Schrage and colleagues assert that this blending of work and home territories and the ever-growing risks of burnout stipulates that these leaders can express themselves more authentically in the digitized workspace. The pandemic has exacerbated the divide between the burgeoning transformation, current capacity in handling them, as well as the associated risks. Leading effectively during challenging times requires greater bottom-up empowerment and accessibility, and that can be achieved within a psychologically safe work environment that allows mutual sharing and respect for each other emotions.

10.4.2 ORGANIZATIONAL RESILIENCE

Evidence of emotional turmoil and psychological distress has been recurrent during economic crisis precipitated by the pandemic. The corporate stress in 2020 is similar to the global recession of 2009, albeit way too faster (Jain et al., 2021). Surviving amidst uncertainties is a challenge and requires risk management strategies and effective transformation. Elucidating crisis-response, Jain and colleagues expressed that organizational resilience enables withstanding adversity and effective coping. They also propose that while organizations prepare for economic, operational, and social resilience, it can only be achieved when psychological safety is guaranteed. Similarly, empathy is a salient contributor in organizational healing, that is, recovery from crisis to normalcy (Powley, 2013). As one of the primary mechanisms of organizational healing, leaders' attention to individual and organizational needs fosters positive relationships, initiates social support, reduces fear, and uplifts employee morale (Powley, 2013). Additionally, at the individual level, crisis and its management are associated with anxiety and stress. Research claims relative proficiency of empathic leader in handling primarily due to their ability to timely predict and gauge a crisis. Perceiving warning signs of a crisis, strategizing and making well-timed decisions, communicating about the actions responsibly to the stakeholders, and converting exigencies to normalcy, and other crisis management tasks can be somewhat adequately managed by an empathy-driven leader (Koenig et al., 2020).

10.4.3 ORGANIZATIONAL SUSTAINABILITY

The primary understanding of sustainability on the context of organizations is based on growth in terms of economic prosperity and goods and services which offers long-term competitiveness. Contemporary idea about sustainability albeit goes beyond the financial underpinnings and are motivated considerably, if not entirely, by conservation of environment, and an organizational society that guarantees health and well-being of its members. Highlighting the role of a leader in sustainability Hargreaves and Fink (2003) assert that a culture of shared responsibility, social justice, human resource development, and diversity can be secured when sustainability is at the core of leadership. Florea et al. (2013) in their study on the values that contribute to organizational sustainability suggest that empathy is a major determinant of individual progress, positive interpersonal relationship, as well as sustainable organizational pursuits. Nurturing empathy concurred with altruism, reciprocity, and private self-effacement through effective human resource management practices ensures intrinsically attainment of sustainability goals. According to Florea and colleagues, organizational practices like clear communication of expectations and goals, developmental activities like on-boarding and coaching, consultation targeted to identification of areas of improvement in skills and performance, and promotion of quality of life, serve as empathy-enhancing practices.

10.4.4 INNOVATION

Taking other people's perspective serves as a crucial link between the company and the end users and customers in the innovation process. Though innovation potentials add a competitive edge to companies, the changing needs of the customer can lead to volatility and anxiety in business that ends in distancing itself from originality and inspiration. Organizational leadership that assumes shared vision and responsibility, values and nurtures creativity, creates an environment that is based on open communication and inclusiveness where employees feel secure and accepted, helps innovation to flourish (Agbor, 2008). In the context of management innovation, the leader is instrumental in creating a climate that fosters employee readiness in voicing ideas, as has been frequently observed with transformational leaders (Vaccaro et al., 2012). Empathy-based innovation facilitation is more appreciative of the cognitions, experiences, as well as emotions of those engaged in the innovation process (Montonen et al., 2014). Montonen and colleagues suggests that innovation empathy during the idea generation phase of innovation yields greater business development and enhanced customer satisfaction.

10.4.5 DIVERSITY AND INCLUSION

Rapidly changing demographic composition of the workforce globally has forced organizations to continue to counter diversity (of cultures and skills) and inclusion concerns. Researches illustrate that leader or manager empathy can be targeted to optimize diversity initiatives. In diversity trainings for instance, perspective taking leads to more positive attitude via empathic attitude (Madera et al., 2011). By attending to others' feelings in culturally diverse contexts, global leaders can benefit from acknowledging the similarities and differences from multiethnic perspectives, sometimes referred to as inclusive cultural empathy (Pedersen and Pope, 2010).

10.5 CREATING WORKPLACE EMPATHY—IMPLICATIONS AND RECOMMENDATIONS

The basic assumption underlying creation of any leadership style is that people change and leadership is, at least in certain ways, a learned behavior. Personality predispositions, innate traits, influence how and what is learned, but complex reality and consistency of change necessitates personal and societal/organizational transformations. Evidences from cognitive neuroscience reveal that empathy, much like leadership, can be developed or enhanced such that people learn to better understand and utilize emotional capacities (Holt and Marques, 2012). Describing empathy in workplace as a tool for effective leadership, Gentry et al. (2016) advocate that leaders' empathy can be enhanced through coaching and training initiatives where leaders are supported to understand empathy and its advantages, learn active listening and holding judgments, practice perspective taking and appreciating diverse experiences, nourish compassion as a virtue for contemporary work life (2016; p. 6–8). Engaging in empathic interaction with team members facilitates trust and cooperation and aids in creating an empathic culture (Kouzes and Posner, 2011).

A point of forethought at the core of empathic understanding, and conceivably the first step toward sustainable growth is the intention or willingness to change or develop (McKee et al., 2008). The desire to change entails self-discovery in multiple directions and is attained via unraveling what brings purpose or meaning to the person or organization. Advancing further enables identifying one's strengths and weaknesses and their impact on others, leading to evaluation of what is expected and what has been obtained. Following this, one engages in outlining mechanisms to reach an outcome based on learning and motivation, finally yielding the desired behavioral change (McKee et al., 2008). This development accentuates how leaders contribute in creating emotionally healthy work environment. Leaders can strive for a climate of empathy through recognizing emotional burden of jobs, authentically expressing emotions and encouraging others to communicate adequately, reinforcing genuine display of emotions or desired behavior with reward, and arranging for skill development trainings to develop emotional competencies (Ashkanasy and Daus, 2002).

Areas of departure from the current state of affairs in empathy research are the arguments, first, is there gap between how empathy is described and how it is practiced? And, second, how much of empathy is too much? The first argument comes from the assumption of empathy as an absolute virtue (Roth, 2019). Bloom's (2016) critique specifies that empathy does not necessitate or guarantee benevolence. Further, Bloom asserts that individual misfortune is at the focal center of empathic decisions, which can facilitate emergence response while eluding distant outcomes and dissatisfying those who do not have empathic needs (Roth, 2019). To address the second concern, we direct attention on the widely accepted organizational benefits of empathy. A case in point is the relationship between empathy and burnout, which plausibly is assumed to share an inverse link. A study on helping professionals revealed that empathy can be emotionally exhausting and result in depersonalization and social disengagement (Williams, 1989). It is worth acknowledging that empathy–burnout relationship is industry-dependent, that is, more likely in service-related professions, but it does affirm a complex relationship between the two constructs. The study indicated that emotional exhaustion can be sustained to an extent, but it can lead to social avoidance in the long run (Williams, 1989). Empathy undeniably facilitates positive mental health and human flourishing, but there could be downsides to promoting empathy as an incontrovertible technique.

10.6 CONCLUSION

As the global workforce continues to grapple with uncertainty and changes of present times, much of the organization's survival and continuity rests on its leadership. The success of the leader in adapting to changes is as much an emotional process as it is a cognitive restructuring, and empathy has a profound impact on the leader's response to the needs of the organization and its people. The present chapter contributes to highlighting the benefits of empathic concerns in leader especially in the context of modern work culture of automation, diverse workforce, blurring work and home boundaries, and a host of other issues that have compounded the complexities of modern work life. In a nutshell, one of the primary steps toward organizational sustainability and resilience is to generate an environment where employees feel emotionally secure and motivated to perform, and this can be effectively achieved by leader's affective engagement with the workforce.

KEYWORDS

- **empathic leadership**
- **leadership perception**
- **empathy culture**
- **sustainability**

REFERENCES

Aas, R. W.; Ellingsen, K. L.; Lindøe, P.; Möller, A. Leadership Qualities in the Return to Work Process: A Content Analysis. *J. Occup. Rehabil.* **2008,** *18* (4), 335.

Agbor, E. Creativity and Innovation: The Leadership Dynamics. *J. Strat. Leader.* **2008,** *1* (1), 39–45.

Ashkanasy, N. M.; Daus, C. S. Emotion in the Workplace: The New Challenge for Managers. *Acad. Manag. Perspect.* **2002,** *16* (1), 76–86.

Ashkanasy, N. M.; Hartel, C. E. J.; Daus, C. S. Diversity and Emotion: The New Frontiers in Organizational Behavior Research. *J. Manag.* **2002,** *28*, 307–338.

Ashkanasy, N. M.; Humphrey, R. H. Current Emotion Research in Organizational Behavior. *Emot. Rev.* **2011,** *3* (2), 214–224.

Barsade, S.; Brief, A. P.; Spataro, S. E. The Affective Revolution in Organizational Behavior. The Emergence of a Paradigm. In *Organizational Behavior: The State of the Science*; Greenberg, J., Ed.; Laurence Erlbaum Associates: Mahwah, NJ, 2003; pp. 3–52.

Boyatzis, R. E.; Smith, M. L.; Blaize, N. Developing Sustainable Leaders Through Coaching and Compassion. *Acad. Manag. Learn. Educ.* **2006,** *5*, 8–24.

Chakraborty, T.; Ganguly, M. Crafting Engaged Employees Through Positive Work Environment: Perspectives of Employee Engagement. In *Management Techniques for Employee Engagement in Contemporary Organizations*; IGI Global, 2019; pp. 180–198.

Chakraborty, T.; Ganguly, M.; Natarajan, A. Predicting Entrepreneurial Satisfaction: The Role of Non-financial Incentive Factors and Quality of Life among Women Digital Entrepreneurs. *J. Glob. Bus. Adv.* **2019,** *12* (3), 328–355.

Chakraborty, T.; Mishra, N. Appreciative Inquiry: Unleashing a Positive Revolution of Organizational Change and Development. *Int. J. Econ. Comm. Bus. Manag.* **2019,** *6* (2), 32–37.

Chakraborty, T.; Saha, R. Proactive and Reactive Role of Human Resource Professionals in an Aftershock Management. *Train. Dev. J.* **2017,** *8* (1), 1–11.

Chakraborty, T.; Saha, R. Digital Innovations: Breakthrough Opportunities to Build Novel Business Models. *Bus. Perspect.* **2018,** *1*, 46.

Choi, J. A Motivational Theory of Charismatic Leadership: Envisioning, Empathy, and Empowerment. *J. Leader. Organiz. Stud.* **2006,** *13* (1), 24–43.

D'Auria, G.; De Smet, A. Leadership in a Crisis: Responding to the Coronavirus Outbreak and Future Challenges. Mckinsey and Company [Online] 2020. https://www.mckinsey.com/business-functions/organization/our-insights/leadership-in-a-crisis-responding-to-the-coronavirus-outbreak-and-future-challenges

Fehr, R.; Gelfand, M. J. The Forgiving Organization: A Multilevel Model of Forgiveness at Work. *Acad. Manage. Rev.* **2012,** *37* (4), 664–688.

Florea, L.; Cheung, Y. H.; Herndon, N. C. For All Good Reasons: Role of Values in Organizational Sustainability. *J. Bus. Ethics* **2013,** *114* (3), 393–408.

Gentry, W. A.; Weber, T. J.; Sadri, G. Empathy in the Workplace a Tool for Effective Leadership [Online] 2016. https://cclinnovation.org/wp-content/uploads/2020/03/empathyintheworkplace.pdf\

Goleman D. What Makes a Leader? *Harv. Bus. Rev.* **1998,** *76* (6), 93–102.

Goleman, D The Focused Leader. *Harv. Bus. Rev.* [Online] 2013. https://hbr.org/2013/12/the-focused-leader

Goleman, D.; Mckee, A.; Waytz, A. *Empathy (HBR Emotional Intelligence Series)*. Harvard Business Press. 2017.

Grisham, T.; Walker, D. H. Cross-Cultural Leadership. *Int. J. Manag. Proj. Bus.* **2008**.

Hargreaves, A.; Fink, D. Sustaining Leadership. *Phi Delta Kappan* **2003,** *84* (9), 693–700.

Hodges, S. D.; Myers, M. W. *Empathy: Encyclopedia of Social Psychology*; 2007; pp. 296–298.

Holt, S.; Marques, J. Empathy in Leadership: Appropriate or Misplaced? An Empirical Study on a Topic that is Asking for Attention. *J. Bus. Ethics* **2012,** *105* (1), 95–105.

Jain, R.; Nauck, F.; Poppensieker, T.; White, O. Meeting the Future: Dynamic Risk Management for Uncertain Times [Online] 2021. https://www.mckinsey.com/business-functions/risk/our-insights/mckinsey-on-risk/mckinsey-on-risk-number-10

Johnson, S. D.; Bechler, C. Examining the Relationship between Listening Effectiveness and Leadership Emergence: Perceptions, Behaviors, and Recall. *Small Group Res.* **1998,** *29* (4), 452–471.

Kellett, J. B.; Humphrey, R. H.; Sleeth, R. G. Empathy and the Emergence of task and Relations Leaders. *The Leaders. Q.* **2006,** *17* (2), 146–162.

Kock, N.; Mayfield, M.; Mayfield, J.; Sexton, S.; De La Garza, L. M. Empathetic Leadership: How Leader Emotional Support and Understanding Influences Follower Performance. *J. Leadersh. Organ. Stud.* **2019,** *26* (2), 217–236.

Kouzes, J. M.; Posner, B. Z. *Credibility: How Leaders Gain and Lose It. Why People Demand It*; Jossey-Bass: San Francisco, 2011.

Lewis, M.; Haviland-Jones, J. M.; Barrett, L. F., Eds. *Handbook of Emotions*; Guilford Press. 2010.

Lussiez, Y. W. *The Empathic Principal: The Relationship Between Empathy, Transformational Leadership, and Teacher Self-Efficacy*; The University of New Mexico, 2009.

Schrage, M.; Pring, B.; Kiron, D.; Dickerson, D. *Leadership's Digital Transformation: Leading Purposefully in an Era of Context Collapse*. MIT Sloan Management Review and Cognizant, 2021.

Madera, J. M.; Neal, J. A.; Dawson, M. A Strategy for Diversity Training: Focusing on Empathy in the Workplace. *J. Hosp. Tour. Res.* **2011,** *35* (4), 469–487.

Mahsud, R.; Yukl, G.; Prussia, G. Leader Empathy, Ethical Leadership, and Relations-Oriented Behaviors as Antecedents of Leader-Member Exchange Quality. *J. Manag. Psychol.* **2010,** *25* (6), 561–577.

Megerian, L. E.; Sosik, J. J. An Affair of the Heart: Emotional Intelligence and Transformational Leadership. *J. Leaders. Stud.* **1996,** *3* (3), 31–48.

Montonen, T.; Eriksson, P.; Asikainen, I.; Lehtimäki, H. Innovation Empathy: A Framework for Customer-Oriented Lean Innovation. *Int. J. Entrep. Innov. Manag.* **2014,** *18* (5–6), 368–381.

Muchinsky, P. M. Emotions in the Workplace: The Neglect of Organizational Behavior. *J. Organ. Behav.* **2000,** *21* (7), 801–805.

Muller, A. R.; Pfarrer, M. D.; Little, L. M. A Theory of Collective Empathy in Corporate Philanthropy Decisions. *Acad. Manage. Rev.* **2014,** *39* (1), 1–21.

Oertig, M.; Buergi, T. The Challenges of Managing Cross-Cultural Virtual Project Teams. *Team Perform. Manag.* **2006,** *12* (1/2), 23–30.

Owens, B. P.; Hekman, D. R. Modeling how to Grow: An Inductive Examination of Humble Leader Behaviors, Contingencies, and Outcomes. *Acad. Manag. J.* **2012,** *55*, 787–818.

Pedersen, P. B.; Pope, M. Inclusive Cultural Empathy for Successful Global Leadership. *Am. Psychol.* **2010,** *65* (8), 841.

Powley, E. H. The Process and Mechanisms of Organizational Healing. *J. Appl. Behav. Sci.* **2013,** *49* (1), 42–68.

Prifti, L.; Knigge, M.; Kienegger, H.; Krcmar, H. In *A Competency Model for "Industrie 4.0" Employees*, Proceedings Der 13. Internationalen Tagung Wirtschaftsinformatik (WI 2017), Leimeister, J. M., Brenner, W., (Hrsg.); St. Gallen, Switzerland, 2017; pp. 46–60.

Sadri, G.; Weber, T. J.; Gentry, W. A. Empathic Emotion and Leadership Performance: An Empirical Analysis Across 38 Countries. *Leadersh. Q.* **2011,** *22* (5), 818–830.

Scott, B. A.; Colquitt, J. A.; Paddock, E. L.; Judge, T. A. A Daily Investigation of the Role of Manager Empathy on Employee Well-Being. *Organ. Behav. Hum. Decis. Process.* **2010,** *113* (2), 127–140.

Skinner, C.; Spurgeon, P. Valuing Empathy and Emotional Intelligence in Health Leadership: A Study of Empathy, Leadership Behavior and Outcome Effectiveness. *Health Serv. Manage. Res.* **2005,** *18*, 1–12.

Stueber, K. Empathy. The Stanford Encyclopedia of Philosophy (Fall 2019 Edition) [Online] 2019, Edward N. Zalta, Ed.; https://plato.stanford.edu/archives/fall2019/entries/empathy/

Vaccaro, I. G.; Jansen, J. J.; Van Den Bosch, F. A.; Volberda, H. W. Management Innovation and Leadership: The Moderating Role of Organizational Size. *J. Manag. Stud.* **2012,** *49* (1), 28–51.

Val Velsor, E.; Leslie, J. B. *Why Executives Derail: Perspective Across Time and Cultures. Academy of Management Executive*; Vol. 9, No. 4, 1995, pp. 62–72.

Wagaman, M. A.; Geiger, J. M.; Shockley, C.; Segal, E. A. The Role of Empathy in Burnout, Compassion Satisfaction, and Secondary Traumatic Stress Among Social Workers. *Soc. Work* **2015,** *60* (3), 201–209.

Williams, C. A. Empathy and Burnout in Male and Female Helping Professionals. *Res. Nurs. Health*, **1989,** *12* (3), 169–178.

CHAPTER 11

Leadership: The Crux of a Buoyant Organizational Structure

SOUMITA MITRA* and SARITA KAR

Department of Humanities and Social Science IIT (ISM), Dhanbad, India

**Corresponding author. E-mail: soumita.rupa@gmail. com*

ABSTRACT

The industrial sectors have been responsible for the continuation of both production and deliverance of services, since a long time. It is not a hidden fact that the manufacturing units of a country is responsible for not only providing their services within the country but also for outside the boundaries. They play an instrumental role in analyzing the rise and fall of the GDP or gross domestic product. It also forms a part in deciding how much revenue or foreign exports a country will secure. Thereby, it can be well understood that each person in an organization holding an authoritative post, needs to be well equipped with his leadership skills. His managerial qualities should not only ensure maximum and flair production but at the same time would also maintain a satisfactory environment among the employers within the concern. In this respect, some of the leadership lessons advocated by Swami Vivekananda would benefit in achieving a dexterous position for the company. The agile characteristic in an industry can only be ensured by a leader's competitiveness.

Therefore, this work of mine would work in justifying the thoughts of Swami Vivekananda that "All the powers in the universe are already ours. It is we who have put our hands before our eyes and cry that it is dark." Drawing inspiration from those golden words can prepare the leaders working in an

Agile Leadership for Industry 4.0: An Indispensable Approach for the Digital Era, Tanusree Chakraborty, Tahir Mumtaz Awan, Ashok Natarajan, and Muhammad Kamran (Eds.)

organization to face all hurdles with valor of mind and hardwork. It would also accentuate how the presence of leadership quality acts as catalyst in ensuring the all-round development of the company and the country at large.

11.1 INTRODUCTION TO INDUSTRY 4.0 AND ITS IMPACT ON THE ECONOMY

As per the reports of the Accenture and the UN Global Compact, only a very small portion of the chief executive officers of the big companies feels that the commercial enterprises are putting their actual endeavors in order to meet the issue of providing continuous balance in the worldwide job market. This is an essential and important matter for encountering the increased pressure caused in the ratio of job opportunities as compared with the expertise utilization level of any individual. For instance, we may start feeling hopeful when we come across the news that the solar energy would become economical than the coal but at the same time, we feel tormented as we see the constant increase in the carbon release every year. Therefore, increased awareness of the gradual transformation taking place all around has given the call for leaders who would create sustainable environment. In one of the conferences organized by the Institute for Sustainable Leadership under the University of Cambridge, named "The international Academy for Business in Society's Conference" which was held in the year of 2014, accentuated the idea that leadership is a prerequisite characteristic, required for creating a better and imperishable tomorrow. But the main drawback lies on the way leadership is perceived by the management institutions. We only assume it to be an occurrence formed within the society but fail to see that it is the consequence of the natural disposition which dwell originally in an individual. The subject of leadership forms an integral part of the curriculum in almost all the B-schools. The main purpose behind teaching the emphasis of leadership is to make the pupils able to handle responsibilities bestowed upon them with utmost care. Such responsive nature makes them becoming a desirable personality, wherever they choose to work (Bendell and Little, 2015). A hardworking and diligent person is always respected and preferred among any other ordinary person. Every concern wants to bestow their trust and responsibilities upon such an individual, on whom they can have their complete faith that they would be able to combat any struggle, that might approach their way, with his leading ability.

11.2 CONSTRAINTS FACED BY THE INDUSTRIES AND ORGANIZATIONS DURING INDUSTRY 4.0

The touristry can be considered to be prime source, among many other sources to contribute and support the economic position of a nation. The name of Spain gets emphasized when the names of the places well-known for their tourists attraction is discussed. It is at this juncture, the lodging facilities concentrated at that particular place plays a crucial role. Although, during the year of 2008, the situation of credit crunch or economic downturn caused adverse impacts. It reduced the rate of travelers traveling or visiting the place, both within the country and also from other places. This in turn started affecting the rate of jobs. A steep decrease in the source of income available, exhausted their share of saved profit and also in their willingness to make attractive changes for pleasing the visitors. At this moment of difficulty, it depends on the board of manager's and their ability to restabilize the circumstance and help that they would provide to maintain a ray of hope among the workers. Thus, the whole onus of bringing back a disadvantageous condition to its original state depends on a leader. His personal traits of controlling and surviving a disaster creates a long-lasting positive impact not only upon his fellow members but they begin assuming him to be their idealized person. This assists in lighting a spark of aspiration within the workers and also pushes the productivity of that concern to new achievements (Quintana et al., 2015). The cultural background of a person leaves a far-reaching impact on his personality. It plays a crucial part in carving out the actual image of a leader from an ordinary person. The cultural identity bears an impression on the leader's way of evaluating situation, his ethical understandings, his thought process and finally on his behavioral approach. Difference existing in the societal ambience on one hand may invite innovational outcomes in terms of problem-solving and fetching desired results in any work front. But, on the other side, it might bring some adverse effects in terms of collision between the existing workers. Therefore, in order to derive maximum output from the workers of a particular concern, who belong to different places, they need to promote an all-inclusive strategy for their workers, where the need and problems of them would be given equal consideration. (Jungwirth and Barth, 2013). According to some executives, holding important positions in the governmental concerns, justifies their explanation on the topic of setting up industrial activities within a lock-up. As per them, it is one of the most viable options to refrain the offenders of law from committing the same types of offenses. It would help in keeping them engrossed in learning some new life skills, which later on would allow them to return back to the normal course of life. In another way, it would eliminate the causes of boredom from the life of the imprisoners and it might also aid in reducing the sentence of imprisonment declared for them. But, controversies have been raised on this issue like in what way these concerns of managing the training and service would be handled within a cell. In the second place, difficulties are there in estimating the proper wages or payment, which would be offered to these working population in the cell. Also, there is a lagging in charting out plans for the marketing of these produced goods at a proper place (Pryor, 2005). Thus, from the above incidence, it once again becomes evident that the leadership skills or the guiding instincts are an indispensable attribute, which cannot be compromised at any sphere of life. The vigor of coming forward and taking the lead to guide the rest of the team members is a matter of immense mental strength and confidence. Only a few possess such type of rare quality.

11.2.1 ROLE OF LEADERSHIP IN MEETING THE CONSTRAINTS OF INDUSTRY 4.0

Leadership abilities are an indispensable aspect which is required in any organization for its smooth running and for confirming its progress simultaneously. But, the presence of guiding skills in a person which assists him in directing all activities in a systematic manner, is of varying magnitude, available in every other individual. Their ways of providing guidance might also seem to vary, depending on the cases, he is giving his furtherance.

Some experts who are conducting their researches in the field of leadership have expressed the view that there are categories of leadership qualities which seem to create their specific affects. For example, the transformational leadership is assumed to be more impactful rather than any other leading attributes. While some other experts have went on mentioning that the expression of a managerial skill or the prowess of leading a team takes place under the influence of a particular environmental factors and cultural background present there. Among the challenges that create hurdles for the future leaders to fulfill their duties, the prime one is, bias followed between males and females at different places. The existence of such differentiation cause stumbling block in the path of women population from becoming eligible for dignified positions. Such situations also get extended in the form of prejudice exercised on payment of salaries they receive. The term "transformational leadership" basically deals with a process, where it includes negotiating process among the team leader and the other team members. This allows in sharing the work load equally amidst all, in order to complete that particular task within a proper timeframe. It follows a system of self-absorption for keeping the group members continuously motivated and for boosting them to work hard. Moreover, in order to refrain from committing faults in the final output, this system, prefers holding back the work given, for some time prior moving to the final stage of result. This step accentuates the importance of stimulation, encouragement, having clear motives, grooming of one's talent, and finally performing in a better way. It ultimately creates an extra room in the form of increased productivity. These types of leading qualities help carving out personalities of steady exemplar or idol, for others who would follow them. They gradually grow up as transformers of the community, who helps others to have a proper target and then work dexterously for achieving those targets. It carries out the function of preparing the young personnels into desirable organizational personalities by following a chain of four functions. The first step consists of creating positive supremacy. In the second step, it consists of motivating the pupil. The third part includes, enabling the learners to do critical thinking and decision-making on their own. And the last step, it consists of personalized effect. Its main intention is to promote the team members to upgrade their level of output. It promotes the idea of equality for all the members within a team. It also speaks about providing equal importance about listening and solving their difficulties (Gourbalova, 2015). It is necessary for any organization to not only train and prepare its employees to accomplish their objectives, within the office, but it becomes an essential obligation for them to lend an ear to the difficulties faced by the employees. It is a well-known fact that "a healthy mind rests in a healthy body." Thus, a balanced psychic and physical well-being allows a person in uncovering their talents in a more better way. The mental disposition of an individual decides his decision-making capacity, his ability to face uncertain conditions, and also his power of leading or guiding others. The quality of leading everyone in an organization plays some of the important roles. In the first place, it is responsible for causing motivation and encouragement among the other fellow persons within an organization. Secondly, it is directly linked with boosting the performance level of the employees. Increased in the productivity will ultimately benefit the overall progress of a company. But at the same time, there exists a small section of leaders who do not possess any moral sense. They even express an exploitative and avaricious views toward the world and its beings. These are psychotic persons who continue to exist in the companies and thereby in the community, very easily. They generally have good academic status, yet they secretly hold a cruel and mean nature within themselves. Such types of people are common at most organizational jobs, but they exist mostly at company jobs, where the work is related to high remunerations and incentives. At such places, these people adopt unscrupulous methods in order to earn maximum positional benefits and income. It has been studied that these company maniacs make use of various manipulations for creating influence upon their superior subordinates for fetching better opportunities, which is mostly above their possessed ability. They forbid facing any disagreements from the fellow colleagues or simply they prefer ignoring any queries by them as their primary interest is only saving their own personal motives. At the same point, they also try to save the situation from getting disclosed in front of everyone. In order to maintain their authoritative role within the organization, they resort to use of force, oppression, and also imply tyrannizing other employees. Increase in the number of such employees result in the thrive of gloomy ambience and mentalities within an office. For instance, this may give rise to scenarios where, the employees may skip their work or duties or they may be found lazily hanging over at office premises, just passing time. Such situation occurs in the presence of antipathetic and uncooperative leader, who leads the fellow members and the company, at large (Boddy, 2017).

The talent of leading and guiding others is also an important and desirable feature for the learning institutions. In the teaching and learning centers, the trace of the management and superintending qualities is a common layout. Starting from the academic head to the educators and also the learners, everyone seems to have the positivity of directing and helping others, in their best possible way. Such constructive mindset is an essential aspect not only for the personal welfare of that person but it also allows a share of happiness and a sense of security in the fellow workmates. The manners or conduct of the team members is believed to be firmly connected with the habits of the leaders. Thus, pliability in the emotional, sociocultural, and corporal factors help in developing the suppleness and adjustive character of the leader (Rajbhandari, 2015). Innumerable mortal interventions give rise to various ecological and community-based disadvantages. Thus, the talent of guiding others in a right way can be considered to bring about positive behavioral transformations in the individuals, which would help in making everyone more responsible toward the ecology, thereby adopting viable actions. In order to bring strong and effective steps toward protecting the ecosystem, assistance would be required both from the people who are associated with politics and it would also require the aid of the localities, from whom it will be easy to know the actual reasons of any environment-related troubles, especially affecting their region. Thereby, the sphere of leadership has now become extended, broad, and interdisciplinary. Although, it still holds its close association with the disciplines which deals with administration skills and managerial teachings (Evans et al., 2015). As per the theory of evolution or the law explaining "the survival of the fittest" always speaks about a condition that in the state of any emergency, a leader needs to prove his mettle or else he might get displaced of his position by a more befitting figurehead. Whereas, there are few others who holds on to a totally different idea, concerning leaders. According to them, it is completely centered on the personal temperament of that person who leads. It is for him to decide, as to how he will manage a disaster. If he succeeds in opting for a judicious resolution at most of the issues, then it can be expected from him that he will be able to do the same justice, even at times of an upheaval. Thus, as per Zaccaro (2007), the initiation of the concept related to leadership gave birth to some main points. In the first place, it describes the flair which enables a person to guide others as a special and rare skill, which at times carries the power of even alternating the existing historical stream. In the next part, it is thought to be a distinguished trait, which might have been adapted by him, from his ancestral lineage (Rizzi and Reber, 2013).

11.3 SWAMI VIVEKANANDA: HIS VISIONS AND PHILOSOPHIES TOWARD AGILITY IN LEADERSHIP

When we look for the traditional system of keeping oneself mentally motivated, the theme of "Karma Yoga" from the Bhagawat Gita becomes evident. It speaks about following the ideal of universal equality, sticking to the path of performing one's responsibility even under the face of extreme adversity, and finally, staying impartial while deciding to bestow honor on someone. Therefore, it goes on justifying that the quality of being morally good, maintaining a bond trust, and comradeship with everyone (Mulla and Krishnan, 2014). The theory of "Karma Yoga" or being a "Karma Yogi" has been equally highlighted by Swami Vivekananda. Swami Vivekananda has always spoken about the upliftment of the young generation within a country, as it ensures progress of the nation through their success. Thus, he goes on saying that a "man of action" need not follow any strict maxim to fulfill his duties, instead he just needs to carry on his work with sincerity. He has also mentioned that for becoming a dedicative person, there is no requirement for him to adopt a life of severe austerity. He can continue doing his good deeds by continuing with the earthly ways. He even lays emphasis of the proposition of Gita called "Niskamakarma." It speaks about the thought that one should have while doing any work, that is, not to keep any expectations from the actions performed by him. He says that a person is able to perform maximum when he keeps himself aloof from any impression of obtaining advantage from the work done (Lal, 2013). Staying under the constant stress of waiting for results can have a dampening effect on the zeal with which an individual wants to perform the task. Keeping oneself aloof from the burden of presupposition that the work done by them would help in earning fame, money or power, also can drown him in the sea of great despair. Thus, the individual should rather keep his focus fixed toward attaining his goal. Such dedication and attentiveness for achieving one's target, makes its fulfillment, very definite.

11.4 IMPLEMENTING THE IDEALS OF SWAMI VIVEKANANDA TOWARD PROBLEM-SOLVING IN INDUSTRY 4.0 ERA

The multidimensional role of the leadership has become very obvious in every field. It has gradually moved on from the pages of history to the new practical world of today. The quality of uniting and guiding all the people of a country to think responsibly, not only for their own but also for others. Thus, in this respect, the term called "Sāttvika Leadership," abbreviated as SL, gets featured. It basically propagates the theory of utilizing one's learning and understanding to support others in their good-motives, thereby taking care that everyone is able to get the equal share to flourish. The presence of such leading qualities helps the team members in becoming psychologically strong and ethnically stable, which would assist them in becoming a person with noble thoughts. Thus, such group members and team lead both adds in not only creating an aura of hopefulness but also sustaining that atmosphere to allow continual progress within a concern. Encouragements need to be provided to such good team spirit. It would help in avoiding many unpleasant incidences occurring like business impropriety, jealousy existing amongst fellow workers or disruptions in the work atmosphere. The significance of leadership has been mentioned even in the traditional system of "Darshanas." The word has been derived from the Sanskrit origin, "drsir preksaṇe," which means, to perceive closely. It speaks about some of the basic practices which needs to be followed. In the first, it says that whatever learning's is being discovered by a leader, must be set free and made available, for others also. In a second place, all the good virtues studied and ethical teachings learned must be practiced by all in their day-to-day life. They should not be kept limited only till studying, and not implementing those in real terms (Alok, 2017). Rather, they ought to be applied in solving their day-to-day worries. Setting up an ethereal environment within the office, helps the employees in facing with the painful and conflicting situations, in their daily working life. Practicing devotional existence in one's life helps them in building up better understanding with the other employees. It creates expansion in their level of analyzing any circumstance with an open mind. This reduces the maximum chances of collision between the office mates. In working toward strengthening one's bond with others, the practice of performing yoga serves to create effective impact. It prescribes five types of execution, also known as, "yamas," which belongs to Sanskrit origin and implies "self-discipline." It helps in discovering the sense of oneness with the other workers working with him. Generally, the term yoga is quickly assumed to be an essential aspect in training one's physical outlook by practicing some body poses. But, implementation of the yogic system in one's life involves meditating, controlling one's breath, following rules of conduct, maintaining sanitation or having faith in God. Carrying out these activities on a regular basis helps in achieving a positive change from within. It not only helps in becoming physically fit but also helps in mentally robust. The main motive behind performing such practices is to enable an individual in maintaining a permanent or constant position of harmony, happiness and allowing oneself to remain active in the service and care of others, without expecting anything in return. This would also act as a protective shield in protecting an individual from getting moved away toward any distractions in situation of crisis (Corner, 2009).

11.5 THE DESIRED ORGANIZATIONAL STRUCTURE AND LEADERS' AGILE ROLE IN INDUSTRY 4.0

There is an extreme requirement of individuals who would possess leadership expertise in order to sustain in the face of constant challenge which are being faced both by the ecosystem and by the society. Thus, it is expected that such energetic or agile leaders would come up and actively engage themselves for protecting both the environment and its inhabitants. We now require a team of leaders who have not been originated from the industrial revolution, rather, they would set examples of leaders who would be ready to adjust, adapt, and reflect (Balser, 2014). There is a need of a leader who would be thoughtful in any action they undertake to perform. They will make use of their empirical skills to understand any difficulty and solve them accordingly (Grandy and Sliwa, 2017). The skill of providing right direction to someone, is the area which is extensively considered in the field of behavioral curriculum related to organization. Dedication for one's place of work or the place from which his livelihood is depended can be considered to be one of the most chief features which every concern expects to see in their employees (Lyndon and Rawat, 2015).

It is only through an efficient leader can we hope to have a better tomorrow. One leader or a group of leaders can motive or inspire others to follow their path towards positivity, thereby helping in the creation of a healthy atmosphere, where sustainable development and growth of all will be ensured. In order to meet the requirement of quality-related reformation activities within an educational institution, the educators seem to take the leading role of making positive changes in the existing teaching-centered activities. For instance, they deliberately come ahead in taking active initiative for making desirable changes in the modules of the syllabus, making arrangements for the personality grooming of the teachers. The educators at the present times are in search of discovering newer technological inputs in the educational sector. They have started incorporating some ingenious ways like improving analytical skills and self-learning abilities. Now, the educators are working hard in order to enable teaching through online means. Over a considerable period of time, it has been understood that making appointment of educators to carry out with their teaching of lessons is not the ultimate thing, which would be restricted within the educator and his student. It is rather a social function, which is spread among all members of the community, and it sequentially develops along a span of years (Berry, 2019).

11.6 CONCLUSION

The leadership qualities in a child must be instilled in him since his early childhood. As parents of the child are considered to be the early educator, therefore, they play a very crucial role in guiding the child in the right path. They are the first person from whom the child will learn or rather try to imitate the expressions or habits they tend to express most often. In the later phase of childhood, the part of an educator becomes important. The things that the child learns from his teacher goes on to have a far reaching effect in his life. The persona of the teacher creates a strong desire in the mind of his pupils that they would also be like him once they grow up. Therefore, possession of a right attitude on the part of the educator becomes essential. It is through their leading abilities in every task they perform, mainly the teaching sessions, the child gradually learns to explore and develop his leading capabilities. From such children, we can expect to have a better society, filled with young, inspiring minds, who deliberately comes ahead in taking responsibilities and fulfilling any task, which will serve to be good for all.

11.7 CHAPTER HIGHLIGHTS

Efficient Leader—Efficient leaders are the reason for a better tomorrow.

Happiness—Leaders can be the source of promoting happiness by maintaining balance at all places of work.

Karma Yoga—The "Karma Yoga" should continuously motivate a leader for working hard at his place of work.

Positive Changes—Leaders are the catalytic agents for allowing positive changes at all places.

Responsibility—The presence of leadership quality in an individual helps a person become more responsible in any work he undertakes to do.

KEYWORDS

- **competitiveness**
- **development**
- **instrumental**
- **leadership**
- **responsible**

REFERENCES

Abulhassn, A. Oil Well Control Training and Safety Leadership Skills of Drilling Personnel. *J. Educ. Thought.* **2016,** *49* (3), 303–320.

Alok, K. "Sāttvika" Leadership: An Indian Model of Positive Leadership. *J. Bus. Ethics* **2017,** *142* (1), 117–8.

Asgari, A.; Mezginejad, S.; Taherpour, F. The Role of Leadership Styles in Organizational Citizenship Behavior through the Mediation of Perceived organizational Support and Job satisfaction. *Innovar: Revista de Ciencias Administrativas y Sociales* **2020,** *30* (75), 87–98.

Balser, T. C. A New Leadership Paradigm? *Bioscience* **2014,** *64* (12), 1065–1066.

Barshefsky, C.; Brock, W. E.; Smith, F. W. Affirming American Leadership: A Call to Action. *Center for Strategic and International Studies (CSIS)* [Online] 2020. www. jstor. org/stable/ resrep26013

Beckerlegge, G. Responding to Conflict: A Test of the Limits of Neo-Vedāntic Social Activism in the Ramakrishna Math and Mission? *Int. J. Hindu Stud.* **2007,** *11* (1), 1–25.

Bendell, J.; Little, R. (Seeking Sustainability Leadership FRGSFCIPD, MBPS. *J. Corp. Citizensh.* **2015,** (60), 13–26.

Berry, B. Teacher leadership. *Phi Delta Kappan* **2019,** *100* (7), 49–55.

Boddy, C. Psychopathic Leadership A Case Study of a Corporate Psychopath CEO. *J. Bus. Ethics* **2017,** *145* (1), 141–156.

Brew, G. COVID-19 and the Oil Price Crash: Twin Crises Impacting Saudi-Iran Relations. *Istituto Affari Internazionali (IAI)* [Online] 2020. www. jstor. org/stable/resrep25274

Brown, B.; Carlucci, R.; Stewart, S. Holding Ourselves Accountable. *Phalanx* **2021,** *54* (2), 36–45.

Bustamante, E. E.; Sawyer, C.; Brown, M. D.; Suman, O. E.; Keith, N. R. The American College of Sports Medicine (ACSM) Leadership and Diversity Training Program (LDTP). *J. Best Pract. Health Prof. Divers.* **2019,** *12* (2), 165–180.

Clayton, O.; Gary, J. Hopps Hope Arrives For Atlanta. *Phylon (1960-),* **2020,** *57* (2), 41–55.

Corke, S. COVID-19 Pandemic Recovery: If the US and Europe Find the Will, Multilateralism Is the Way. *Istituto Affari Internazionali (IAI)* [Online] 2020. www. jstor. org/stable/ resrep27583

Choudhary, A. I.; Akhtar, S. A.; Zaheer, A. Impact of Transformational and Servant Leadership on Organizational Performance: A Comparative Analysis. *J. Bus. Ethics* **2013,** *116* (2), 433–440.

Cheng, M. Y.; Lei Wang, L. The Mediating Effect of Ethical Climate on the Relationship Between Paternalistic Leadership and Team Identification: A Team-Level Analysis in the Chinese Context. *J. Bus. Ethics* **2015,** *129* (3), 639–654.

Corner, P. D. Workplace Spirituality and Business Ethics: Insights from an Eastern Spiritual Tradition. *J. Bus. Ethics* **2009,** *85* (3), 377–389.

Evans, L.; Hicks, C. C.; Cohen, P. J.; Case, P.; Prideaux, M.; Mills, D. J. Understanding Leadership in the Environmental Sciences. *Ecol. Soc.* **2015,** *20* (1), 50.

Fein, E. C.; Tziner, A.; Vasiliu, C.; Felea, M. Considering the Gap between Implicit Leadership Theories and Expectations of Actual Leader Behaviour: A Three-Study Investigation of Leadership Beliefs in Romania. *J. East Eur. Manag. Stud.* **2015,** *20* (1), 68–87.

Gay, H. G. Supporting Women in the Workplace. *Women of Color Magazine* **2020,** *20* (1), 28–31.

Ghose, M. From Vivekananda to Kallat. *Art Institute of Chicago Museum Studies* **2011,** *36* (2), 9–25.
Gourbalova, I. Bulgarian Managers—Specifics in Leadership. *SEER: J. Labour Soc. Affairs East. Eur.* **2015,** *18* (1), 59–81.
Grandy, G.; Sliwa, M. Contemplative Leadership: The Possibilities for the Ethics of Leadership Theory and Practice. *J. Bus. Ethics* **2017,** *143* (3), 423–440.
Huang, K. Minimizing the Social Dilemma. *Investigación Económica* **2021,** *80* (316), 32–55.
Jungwirth, C.; Loren Barth, L. The Entanglement of Intercultural Conflicts and "Bad" Leadership in SMEs. *Die Unternehmung* **2013,** *67* (4), 345–363.
Kim, T. Y.; Liden, R. C.; Kim, S. P.; Lee, D. R. The Interplay Between Follower Core Self-Evaluation and Transformational Leadership: Effects on Employee Outcomes. *J. Bus. Psychol.* **2015,** *30* (2), 345–355.
Lal. B. K. Contemporary Indian Philosophy. In *Motilal Banarsidass Publishers Private Limited*; 2013.
Levine. M. P.; Boaks, J. What Does Ethics Have to do with Leadership? *J. Bus. Ethics* **2014,** *124* (2), 225–242.
Lieten, G. Indian Workers in Multinational Companies. *Econ. Polit. Weekly* **1987,** *22* (18), 810–822.
Lyndon, S.; Rawat, P. S. Effect of Leadership on Organizational Commitment. *Indian J. Indus. Relat.* **2015,** *51* (1), 97–108.
Mcgee, A.; Haworth, P.; Macintyre, L. Leadership Practices to Support Teaching and Learning for English Language Learners. *TESOL Quart.* **2015,** *49* (1), 92–114.
Monaghan, A. Russian Grand Strategy and the COVID crisis. *NATO Defense College* [Online] 2020. www. jstor. org/stable/resrep27746
Mulla, Z. R.; Krishnan, V. R. Karma-Yoga: The Indian Model of Moral Development. *J. Bus. Ethics* **2014,** *123* (2), 339–351.
Pryor. F. L. Industries Behind Bars: An Economic Perspective on the Production of Goods and Services by U. S. Prison Industries. *Rev. Indus. Organ.* **2005,** *27* (1), 1–16.
Quintana, T. A.; Park, S.; Cabrera, Y. A. Assessing the Effects of Leadership Styles on Employees' Outcomes in International Luxury Hotels. *J. Bus. Ethics* **2015,** *129* (2), 469–489.
Rajbhandari, M. How Does Maintenance of Psycho-Socio-Physio Parameters Enrich Educational Leaders? *J. Educat. Thought (JET)* **2015,** *48* (3), 239–254.
Rizzi, W. A.; Reber, G. Leadership Styles: Inertia and changes in the Czech Republic. *J. East Eur. Manag. Stud.* **2013,** *18* (1), 9–35.
Sarwar, C. Future of Ethically Effective Leadership. *J. Bus. Ethics* **2013,** *113* (1), 81–89.
Sutton, J. Market Share Dynamics and the "Persistence of Leadership" Debate. *Am. Econ. Rev.* **2007,** *97* (1), 222–241.

CHAPTER 12

What's Brewing Up? Inventing a New Contemporary Leadership Model in the Era of the Fourth Industrial Revolution and Global Trade Wars

PALVINDER KAUR[1*], KAMALJIT KAUR[2], and GEETA MAHAJAN[3]

[1]*University of Delhi, Delhi, India.*

[2]*PGDAV College, Delhi, India*

[3]*Institute of Information Technology and Management, Janakpuri, Delhi, India*

Corresponding author. E-mail: royalpalvi16@gmail.com

ABSTRACT

Performance and talent of a leader has been changed due to the Fourth Industrial Revolution. Since we are living in an unprecedented time of exceptional change, there is a need to reassess and redefine the role of a leader in a world of advanced and fast-changing technologies, specifically after global trade wars. A trade war always results in uncertainty, insecurity, and smash-up global trade with its contagious effects. These uncontrollable scenarios teach us to make critical decisions aggressively and to collaborate effectively. Thus, the role of humans has changed with advanced and rapidly changing technologies and trade relations in the world, which are uncontrollable and teach us a new thing daily. The aim of the research paper is an attempt to bring out the new model of leadership style that can help the organizations in larger level to survive in the current scenario. A new model on leadership style, namely, "Reinforcing

Agile Leadership for Industry 4.0: An Indispensable Approach for the Digital Era, Tanusree Chakraborty, Tahir Mumtaz Awan, Ashok Natarajan, and Muhammad Kamran (Eds.)

Leadership Style" is proposed based on five attributes of a leader such as crisis leadership, paradoxical leadership, entrepreneurial/intrapreneurial leadership, technological leadership, and emotional quotient leadership. The research is exploratory in nature and based on various articles, research journals, and literature review to design a new concept, and to understand and explore the current trends in international trade. Current paper will be active in formation the practical application and interruption with its future significance.

12.1 INTRODUCTION

We are living in an unprecedented time of exceptional changes. The emergence of the Fourth Industrial Revolution has brought in new fundamentals and changes in every individual, technologies, businesses, and governments. The revolution talks about artificial intelligence, machine learning, data analytics, quantum energy, and other innovative technologies. The rapid move toward digitalization and acceleration of complicated work-related projects forces us to restudy the theories and applications of new ways of leading and organizing, which can help us to suggest possible methods forward. It has changed our mindset about the "performance management" and "talent development."

In the new era, the role of humans has changed with advanced and rapidly changing technologies across world, which are uncontrollable and teach us a new thing daily. Human resources are becoming more efficient and effective by making quick critical decisions with their peers and associates. In the earlier industrial revolutions, human resources were just treated as commodities and their expertise was measured only in the form of money. Now, the human workforce has more liberty with new opportunities as they are the eccentric contributors. Human can choose easily what to produce with designing, supervision of multiple tasks of repetition, and complexity with the help of the advanced technologies. One pertinent fear is always there that it will surge unemployment by those who still believe in earning only on the basis of efficiency but not by their value addition with flexible settings. It is actually the application of unique talent in uncertainties with provided opportunities from utmost support. In this scenario, the leader is not a zenith but is among them and for them. Everyone can influence each other by influencing their leadership skills and can learn from others at the same time. The main work moral principles are teamwork and faith.

The Fourth Industrial Revolution carries with itself the paramount need to change the way we think about both performance and talent of a leader.

We are undergoing a process of a new era in to reassess and redefine the role of leader in a world of advanced and fast-changing technologies due to global trade wars—USA–China trade war or BREXIT from the EU. These are uncontrollable and teach us to make critical decisions fast, and collaborate effectively. A trade war always results in uncertainty, insecurity, and even damage to global trade with its contagious effects. To embrace the uncertainty into opportunity, what is needed is the emergence and development of leaders with the skills by re-infusing the adaptability, trust, and a hope to handle the trade disputes. We require a new theory, a new concept for the Fourth Industrial Revolution era and accordingly require a new leadership model. In unstable situations, leaders must alter their organizations to continue to be appurtenant and aggressive amid prodigious change by guiding their work force by making ready for new possibilities and prosperity, and by engaging them at the center of manufacturing and production. It is a great future investment in the erudition, skills, and mastery to steer the complexities of the time. Unstable changes due to trade disputes or revolutionary technologies, leaders are no longer alone preparing all things and these turbulent forces are already experienced by all organizations. In the future, only the latent qualities can improve efficiency and equity, the quality of life, world's population, and raise income levels. Effective and efficient social institutions play a significant role for supply chain management in the preview of cooperative human capital combination.

The current research paper is an attempt to bring out the new concept of leadership style that can help the organizations at larger level to survive in the current scenario. A new model on leadership style, namely, "Reinforcing leadership style" is proposed on the basis of the following five attributes of a leader:

1. Crisis leadership style;
2. Paradoxical leadership style;
3. Entrepreneurship/Intrapreneurship leadership style;
4. Technological leadership style; and
5. Emotional quotient leadership style.

12.1.1 CRISIS LEADERSHIP STYLE

One of the best qualities in effective crisis leadership is to persuade people to take actions. They inflict hardship in a crisis which makes them stronger like, in today's time, New York Governor Andrew Cuomo. According to Professor Jonah Berger, Wharton School of the University of Pennsylvania,

USA gave emphasis to “we” in place of “I” while responding in crisis. Their directions were clearly communicated to the team for the cumulative benefits as leaders, which was a splendidly emotional encouragement. These leaders know that the crisis will not resolve itself and they have to intervene before the situation worsens.

12.1.2 PARADOXICAL LEADERSHIP STYLE

This attribute of leader deals in holistic thinking, certainty about opinions of open mindedness; risk management; maintaining resonance; making choices with collaboration; rational and logical thinking, importantly, integrating multiple positions, possibilities, and interrelated occurrences; honesty; and honor.

12.1.3 ENTREPRENEURSHIP/INTRAPRENEURSHIP LEADERSHIP STYLE

This includes mergers and acquisitions, greenfield investments, opportunistic, lateral thinking approach for capital investments. There is a growing rate of interest in intrapreneurship development because leaders recognize that they cannot fill this gap only through innovation, they require a strong creativity and entrepreneurial mindset. It is pertinent for a leader to adopt a lateral thinking approach for decision making, sensation, feeling, thinking, and intuition and has to be opportunistic wherein the role of being a strong capital investor in mergers, acquisitions, or brownfield investments and greenfield investments. Organizations have been successful in coming up with novel ideas but need a good leadership style to implement them.

12.1.4 TECHNOLOGICAL LEADERSHIP STYLE

Through the digital aspects of advanced technology such as artificial intelligence, innovation for the organization is incorporated. It is very relevant to inculcate leaders’ innovative approach with the coordinated innovation among industries, academia, and governments. Together, they will build up a strong system approach and take the organization to the next level which will help further to overcome the problems due to the global trade wars. Building through a concept of concurrent engineering, a work methodology is emphasized on the parallelization of tasks.

12.1.5 EMOTIONAL QUOTIENT LEADERSHIP STYLE

Emotional quotient is the capability of a person to pull off, control condition, and acknowledge not only the emotions but also possessing the capabilities to manage others' emotions. Emotional quotient, therefore, measures the cognition to come through one's emotions as well as of others. It means that we must be aware of the simple fact that emotions can have a negative as well as a positive impact on others. Emotional independent leaders should be emotionally separatist and make emotionally intelligent which affect their decisions and actions.

12.2 LITERATURE REVIEW

12.2.1 MAJOR FINDINGS ACROSS LITERATURE ON CRISIS LEADERSHIP

Boin et al. (2010) suggest an effective crisis leadership that recognizes rising dangers, and initiating endeavors to mitigate them and understanding it with its future outcome and when incisive juncture will be passed, re-establish a sense of normality. Klann (2003) also agrees and proposes that a crisis always affects the system and can cause closure and concluding that all stakeholders such as suppliers, customers, and investors lose the confidence that can cause a magnetic inclination in the stock market and bring forward the environmental damage and the psychological anxiety. This suggests that crisis leadership is leading the organization under unfavorable circumstances and it will affect employee's morale, attitudes, and productivity, which will reduce organizational performance even further. This means that an effective crisis leadership comes down to the leader responding to the human needs, emotions, and behaviors caused by the crisis. Heifetz et al. (2009) suggested a crisis leadership with two defined stages. First, in the emergency, a leader's duty is to brace the situation. Second, in the reconciling stage, a leader holds the implicit in the inception of crisis and builds the capability to expand with new dimensions. Nevertheless, whatever the strategy an organization chooses to implement, it has to be effectively communicated to the employees, customers, and other stakeholders. McEntire and Dawson (2007) indicated to design the crisis/emergency management and to act plan of actions and implement the actions which will lessening the casualties, forbid dangers to life and property, protect the environment, and develop and advance harmonization between the relevant organizations during a crisis. Leadership is a process of "cognitive factor" for a group of individuals to execute the public goals (Northouse, 2013; Yukl, 2011). Kellerman (2012) explores the role of leadership to make changes for the developments, and the main role of the management is to delegate and maintain the stability for the organizations (Northouse, 2013).

Shurbagi and Zahari (2012) clarify that each crisis leader has his own way of conducting the government's rules, generalizations, and create a style of management which is effective and set a benchmark of management for the other current and future leaders. Harwati (2013) analyzes the current literature on crisis management within an organizational setup. A crisis is an unanticipated circumstance that threatens the purpose and the evolvement of an organization and it may attack at any point of time. On the other hand, a crisis can also turn out many opportunities for an organization into a reproductive structure. Hence, to carry off a crisis efficaciously, an organization should ascertain the contribution of all stakeholders; therefore, select the leadership traits and strategies accordingly. After re-evaluation of the existing strategies and before implementing the new one, a transformational leadership trait should be proposed for a crisis.

12.2.2 MAJOR FINDINGS ACROSS LITERATURE ON PARADOXICAL LEADERSHIP

According to Sparr (2018), both leaders and their followers experience paradoxical tensions during an organizational change, and they often react very defensively. Therefore, a paradoxical channel is precious to understand individuals' counter actions. There is an affluent discourse in the literature to understand the importance of leaders' sense making about, and managing, contradictions in terms of the organizational change. Contrarily, from the followers' perspective, the leaders influence the followers' sense making regarding the paradoxes, which have largely been neglected so far. To close this disparity, a conceptual model is introduced here, which is based on "Uncertainty Management Theory," highlighting the purpose of the followers' fairness evaluations in paradoxical demands. This theory is based on the own sense-making processes and excites the followers to prosecute in paradoxical sense-making process and suggested a critical extremity state for followers' fairness evaluations and their resultant reactions to paradoxical tensions. This model combines the paradox and sense-making awareness. The model adjudges the paradoxical kind of organizational change and offers a new and special concentration to causing individuals' change reactions positively. Yan Zhang et al. (2014) constructed the model, "Paradoxical Leader Behavior," to manage people. This model studied competing and interrelated behavior and meet the follower demands over time. Out of their two studies, the measurement of paradoxical leader behavior of top management by using five samples from China was analyzed paradoxical leader behavior with the following five dimensions: (1) Combining self-centered attitude with other's same attitude, (2) keeping both distance and closeness in their behavior, (3) handling all subordinates equally, (4) implementing work restrictions with flexibility, and (5) holding decision control with autonomy. In the second study with a field sample of 76 supervisors and 516 subordinates from 6 firms, it probes the prior and outcome to understand paradoxical leader behavior by managing people and explained the extent by which superordinates pursue in holistic thinking and concluded that the combative complexions are positively related to their paradoxical behavior in managing people and are associated with an increased proficiency, captivity, and proactivity among the subordinates.

Lavin (2014) explains that leaders always encounter various management paradoxes in an organization, and try to prepare a balance between power and authorization; efficiency and adaptability; and individuality and political orientation. Waldman and Bowen (2016) elaborates "Conventional Leadership Contingency Theory," which reinforces management paradox and leaders are supposed to make the foremost judgment between choosing authorization and control. Smith and Lewis (2011) advocates this model and concluded that such decisions are favorable only in the short term and leaders should evaluate and accommodate paradoxes to accomplish the long-term performance. De Wit and Meyer (2010) proposed "Strategy Synthesis Approach," which is based on distinguishing the strategy paradox and determines every dimension and step of the following processes of strategic management: (a) Strategic thinking, (b) logic and creativity, (c) strategy formation—deliberateness and emergence, (d) strategic change—revolution and evolution, (e) business level strategy—markets and resources, (f) corporate level strategy—responsiveness and synergy, (g) network level strategy—competition and cooperation, (h) the industry context—compliance and choice, (i) in the organizational context—control and chaos, (j) on the international context—globalization and localization, (k) on the organizational purpose—profitability and responsibility. In predicate and making theories, they faced paradox, everyone tried to find unique and innovative ways of reconciliations of polar. The paradoxes will surround by precariousness and disagreements which needs the solutions.

12.2.3 LEADING FINDINGS ACROSS LITERATURE ON ENTREPRENEURSHIP/INTRAPRENEURSHIP LEADERSHIP STYLE

Entrepreneurial studies have been growing since 1980 (Charan et al., 1980). The significant findings are there by entrepreneurial leadership and related fields research, wherein, 841 youth of age ranging from 14 to 19 were studied through their telephonic interviews. Waistad and Kourilsky (1998) have noticed in their study that the black youth are more courageous and fascinated in protrusile their business. Young people have a conceptualization that prosperous entrepreneurs should give back to society. Boni et al. (2009) and Waistad and Kourilsky (1998) gave the importance of entrepreneurial studies as a component of the education and they tried to bring the importance of innovation for business. Pacione (2015) also detailed that innovation is perfectly indispensable for the perfect role of entrepreneurial leadership whatever be the developing stage of business. Through a case study of family businesses, van der Heyden et al. (2005) gave the value to procedural justice and Shepherd and Haynie, (2009) explained the meaning of family business, choices, and aspiration of struggle and can enhance the public presentation of the family business. Gartner and Liao (2017) investigated 276 emergent entrepreneurs and found that pre-venture planning is associated to enduring the new venture. Emerging entrepreneurs who made a business planning at the initial stage, prevail for a long time compared to those who do not have idea of making planning. Different other studies such as those of Brush, (2002), Galloway et al. (2015), Hamilton (2006), Lechner et al. (2018), Lewis (2015), and Mcgowan et al. (2015) gave importance of gender equality in the venture procedure; on the other hand, Bodolica and Spraggon, (2015) and Dean and Ford (2017) found it is the passion of the leaders irrespective of gender participation who can make entrepreneurship a success through revamping the lives of people, customers, and other competitive entrepreneurs of the society.

Wu et al. (2008) studied 191 combinations of employees and leaders of Chinese firms and creativity and the possible outcomes and the behavior of entrepreneurial leaders is significantly related with the creativity of employees. Various other studies, Gartner (2011), Jayawarna et al. (2011), Yang et al. (2011), and Yang et al. (2015) have exposed the significant value of social, human, and financial capital which influence the funding for new firms, types of funding, right time to get the IPO for new firms, and inclusion of young women as entrepreneurial leaders in business enterprise.

Aygun et al. (2010) and Brown et al. (2010) discussed the control-oriented approach of the business leader where a leader can resign, which is extremely unlikely and any form of intrapreneurship can expand indicating cooperative effort between the business entrepreneur, and these intrapreneurs are potentially molest with problems, issues, and conflicts. Heinonen and Toivonen (2008) studied the collective relationship of entrepreneur and intrapreneur and which can generate an environment in which employees are empowered; without empowerment, the confident intrapreneur may appear to be little more than a problem in the entrepreneur business. Gupta et al. (2004) found assorted functions of leaders for the development of entrepreneurial leadership such as challenge framing, uncertainty absorbing, clearing path, commitment building, and specification limits. Carrier (1997), Rogers (1995) concluded in their study that an entrepreneur should focus on enterprise fulfillment while the intrapreneur, on the basis of personal fulfillment, postulates the opinions of leaders, which directly affects the outcome of an innovation—a powerful way to change agents to affect the diffusion of an innovation. It is substantiated by Prasad (1993) with a view that intrapreneurial employees must be equipped with conflict, politics, tenacity, and imparts considerable strength to convince. Similarly, Pinchot and Pellman (1999) concluded that intrapreneurs must actively look for a leader who is more intrigue to make the innovations happen than to block it to be a successful leader.

12.2.4 MAJOR RESEARCH FINDINGS ACROSS LITERATURE ON TECHNOLOGICAL LEADERSHIP STYLE

The growing role of technology in any organization, especially in e-learning in the domain of education, has increased nowadays and for the need for system thinking and systematic change to manage and lead the rapid change of this complex world. Different studies have analyzed the importance and impact of technology on leadership and how their technical education affects innovative thinking and how technological advancement through joint ventures or alliances also can affect their productivity. Nicholas Clark (2012) built a leadership development model that is based on simple ideas and concepts, which is quite different from complex science. On the other hand, complex leadership development model focuses on the following four key dimensions: Acknowledge the interrelatedness, namely, network externality, mutual leadership, organizational learning and leader's skills, and technological cognition to interpret in complex world better. Umut et al. (2017) conducted research using a relational screening model to investigate the relationship of open leadership with digital behavior and technology adaptability, and understands the self-efficacy in technological leadership positively influences the digital behavior. Kossek (1987) identified the importance of innovations in HRM in organizations as the core of market and then technological leadership becomes essential for the improvement and development of the organizational capability with creative innovative ideas and the adoption of these innovations will become important attributes for the external environmental and social welfares in the future. Chaib et al. (2020) also focused on the relationship of strategic planning and technological development for better organizational performance. Akcil et al. (2017) studied that leader's employee orientation program, will significantly improve the cross-functional team's work climate and possibilities for new innovative learning and development. The outcome considers the behavior of leader than his power, as an important factor to determine the conducive work climate for the successful cross-functional product evolution tasks. Bolden et al. (2009) analyzed and provided two main methods, "developed distribution of leadership" related with top to down and "emergent" related with bottom to top. The "distributed leadership" concentrates more on "leadership" which is most influential through its oratorical value, configure logical representation of individuality, and then influence inherent dynamic power within the organization. Nooteboom et al. (2007) tested the relationship of distance and innovation performance with technology adopted firms and elaborate a "U shaped curve" of the firms and anticipate a positive and high return for those firms which are extremely occupied in revolutionary, exploratory ventures than in exploitative ventures. However, the impact of joint research and development (R&D) from these ventures is unclear. Pedro Mendi et al. (2020) analyzed the PITEC Panel Data of Spanish firms for the period 2005–2013 and found a positive relationship of technology with the previous performance of the organization and it may cause reduction in cost due to linkages among the diverse cooperative partners of firms and research institutions.

12.2.5 MAJOR RESEARCH FINDINGS ACROSS LITERATURE ON EMOTIONAL QUOTIENT LEADERSHIP STYLE

Goleman (1998) represented an emotional intelligence (EI) which is captious for successful leader, several other researchers investigate the relation between EI and leadership. Dulewicz and Higgs (2003) agree with the hypothesis of Goleman. George (2000) explained the importance of moods and emotions for the effective leadership process. Prati et al. (2003) also discussed the competency in EI for effective team presentation. Palmer et al. (2001) found that EI as significantly important for transformational leadership. Higgs and Aitken (2003) furnish evidence to relate EI with leadership and can become a forecaster of better leadership. Rosete and Ciarrochi (2005) also claimed the EI to predict effective leadership. Feyerherm and Rice (2002) understand and manage emotions for better team performance.

12.3 RESEARCH METHODOLOGY

The objective of this paper is to identify, analyze, and create a new leadership style in the era of global trade wars. It focuses on the different leadership attributes such as crisis leadership, paradoxical leadership, entrepreneurial and intrapreneurial leadership, technological leadership and emotional quotient leadership and how these traits are helpful to create a new model of leadership styles. The research is exploratory in nature and various articles, journals, and literature review is studied to design a new concept. The main objective of the research is to understand and explore the current trends in international trade and the role of the leaders in trade wars is very crucial with the attributes in defining the new concept of a leader.

12.4 PROPOSED MODEL OF NEW LEADERSHIP: KPG MODEL OF LEADERSHIP

This new model known as Kamaljit Kaur, Palvinder Kaur, and Geeta Mahajan model (KPG) model has three variables as follows: Antecedent, mediating, and outcome variables. The antecedent variables (crisis leadership management, paradoxical leadership management, and entrepreneurship/intrapreneurship leadership management) play a pivotal role in today's leaders' role in global trade wars. These are the primary or core factors which are responsible in driving the new leadership management style. What further add up to these primary or antecedent variables are the mediating variables, technological leadership management, and emotional quotient leadership management.

These mediating variables act as supporting factors to the antecedent variables to bring in the new leadership management process in this era of fourth industrial revolution and global trade wars. The final outcome variable will be a new leadership style, "reinforcing leadership." The whole cooperation among world leaders and people works together and they share their area of vision regarding trade disputes and how it will reshape economic, social, cultural, and individual lives.

The reinforcing leader are versatile and professional leaders who welcome these disruptions and apprehend their job profile with flexibility and critical rational skills are requisite consequently in the near future. They steer the changes with integrity and changes the business and social environment to the new horizons by providing positive and stewardship.

The unexpected turbulent era desire reinforcing mortal or leader who is emotionally agile and contribute cooperative working places to instruct and mentor, rather than to control and direct out of self-conceit. The fourth industrial revolution needs these anthropoid leadership qualities who can tackle the cooperative intelligence for their organizations and can make the agile teams in dealing with the persistent disturbances of global trade wars.

Since we do not have a crystal ball to forecast in this uncertain world, the reinforcing leader will take hold of the whole situation with its unknown factors, by his skills. Innovative leaders who create new innovative environments, new inventions, tools, and thinking can make organization healthier and future viable. This leader is a game changer and learns to operate in challenging and predictable circumstances.

12.4.1 KPG MODEL OF NEW LEADERSHIP

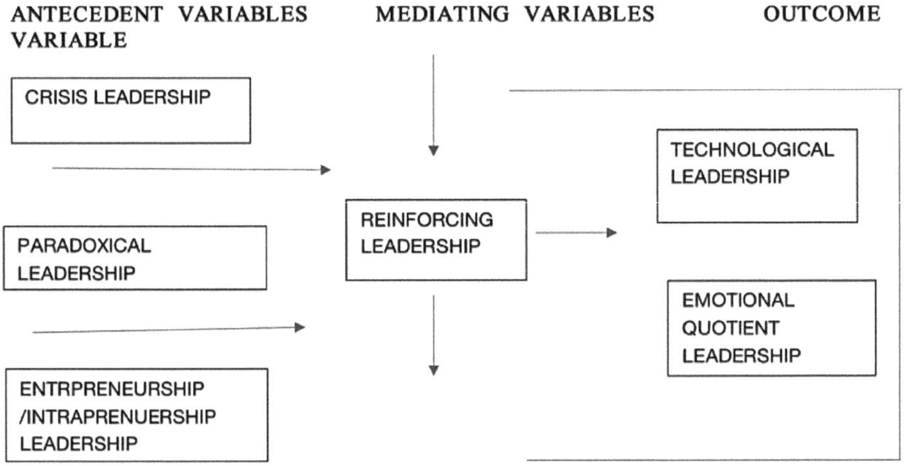

FIGURE 12.1 KPG model (Source: authors' original).

12.4.2 SCOT ANALYSIS OF KPG MODEL

12.4.2.1 CRISIS LEADERSHIP

Strengths

1. Strategic mindset and detail orientation to see a problem;
2. Collaboration and decisiveness that is, listen to others' advice and using the input of those with experience to manage with difficult situation;
3. Building trust by remaining calm in the face of chaos and high-risk environment;
4. Determining the nuances of situation so as to convey a sense of confidence;
5. Delegating the task by gathering everyone and everything together;
6. Calm and focused by not allowing panic to ensue; and
7. Sense of control.

Challenges

1. Chaotic, unanticipated, unprecedented events, where chaos may destroy the organization without urgent and decisive action;
2. Introducing the change in the organization: Change management is the biggest challenge for the crisis manager.

Opportunities

1. Managing the status quo;
2. Ready to adopt to new experiences;
3. Willingness to learn risks;
4. Believe in one thing that all things are possible have solution; and
5. Believe that people and organizations can emerge better off after the crisis than before.

Threats

Leaders free from the feelings of anger, anxiety, guilt, and despair for optimism and hope.

12.4.2.2 PARADOXICAL LEADERSHIP

Strengths

1. Paradoxical leader tends to take the risk and manage it appropriately with reasoning and multiple perspectives;
2. Certainty regarding opinions with open-mindedness;
3. Having feeling of self-assurance with self-improvement;
4. Truthfulness with tact and respect;
5. Fulfilling needs by helping others;
6. Logic with intuition; and
7. Self-motivation for stress management.

Challenges

1. The biggest challenge that the paradoxical leadership face is "The more one learns, the more he realizes how little he knows";
2. Finding several solutions and ways of implementing, and in knowing how to get things to done;
3. Be goal oriented while creating space for the future;
4. Planning to accomplish deadlines with flexibility;
5. Trusting your employees and providing training for their better performance; and
6. Having good vision with achievable objects.

Opportunities

1. Optimism with an anticipation of wrong occurrence;
2. Experiential knowledge of leaders provides a positive organizational culture; paradoxes provide the ultimate opportunity to learn instead as a problem; and
3. The paradoxes leaders invite opposite leaders to find ways to deal with the tension.

Threats

1. It can sometimes lead to the problem of group thinking (a popular term coined by Adolf Hitler wherein each of his commanders were asked about the solution of the problem. The best part was he tends to listen to the solutions but do what he feels he needed to do) and
2. It can lead to complex situation if the strategy of the organization is not clearly defined.

12.4.2.3 ENTREPRENEURSHIP/INTRAPRENEURSHIP LEADERSHIP

Strengths

1. Strong leadership skills for mergers, and acquisitions and green field investments;
2. Creativity and diversity approach in management;
3. Equal passion and innovation for both entrepreneurs and intrapreneurs; and
4. Responding effectively to any situation by strong decision-making skills.

Challenges

1. Entrepreneurship can be etched in words, not by actions. The intent is there without the operational support which make it very difficult for intrapreneurs to be successful. It deals with the hurdles in operating environment puts in their way and
2. Intrapreneurs is often claimed to be entrepreneurial which otherwise they are not. There is hardly any organization that truly embraces entrepreneurship.

Opportunities

1. Prior industry or market experience helps to solve problems greatly and the opportunity to create new innovative solutions for entrepreneurs and intrapreneurs and
2. Process of consistent is the key factor of being a successful organization. Entrepreneurs and intrapreneurs learn from their mistakes without giving up like get back up on the horse and try again.

Threats

1. Every organization is having the experience of new opportunities and threats due to the worldwide technological disturbance, fluctuating economies, demographic changes, and global trades wars and
2. For the current innovations, consequently new changes in the organization does not make entrepreneur and intrapreneur as a captain likewise owning the ship has never made the owner, but the captain.

12.4.2.4 TECHNOLOGICAL LEADERSHIP

Strengths

1. Ebullience is contagious and leaders who are genuinely positive cannot help but motivate others. Inventiveness and creativity help in building positive attitude and
2. Building up the integrated organization and parallelization of task with the help of concurrent engineering techniques.

Challenges

1. Cyber risks have risen in importance and ensuring the security posture of the organization have become the major cause of concern;
2. Lack of coordinated innovation among industries, academia, and governments; and
3. Awful security-by-audit checklist mentality is another drawback as it impacts a lot on the effectiveness of the organization's security controls.

Opportunities

1. Use of artificial intelligence, machine learning, and data algorithms to build up the strong system approach and
2. Companies can use technology planning to take the advantages of economies of scale and economies of scope which help further to make better decisions for strategic planning, more creative and expanded R&D.

Threats

1. Fail to protect the intellectual property rights and
2. Artificial intelligence and quantum sciences hardly help the economies of scale and scope.

12.4.2.5 EMOTIONAL QUOTIENT LEADERSHIP

Strengths

1. The leaders are self-aware and know what others feel, know about their emotions, and know about how their actions will affect others;

2. They regulate themselves effectively and always stays in control;
3. They are self-motivated which encourage them to be consistent to achieve their targets and high benchmarks with better quality of work;
4. Leaders are empathetic and manage successfully their team or organization; and
5. They have strong social skills which make them great communicators.

Challenges

1. Sometimes these leaders do not deal with the poor performers and it yields in mistrust and failure of communication and bitterness among employees;
2. Lack of communication with the poor performers: Leaders do not deal with poor performers so the development of mistrust and poor communication increases the suffocation, frustration, bitter, and confusing relations;
3. Leaders take the things sometimes too personally and think it from their own perspective;
4. The biggest challenge these leaders face is the egoistic behavior, that is, not acknowledging the others for their area of growth;
5. They sometimes fail to take the responsibility at their own end and make people acknowledge the failures. A true leader must act as a scapegoat and take the responsibility; and
6. The biggest challenge leaders face is the act of group thinking. They do not listen, or worse they pretend to.

Opportunities

1. Leaders are emotionally independent and the EI attributes allow them to connect to core emotions and being aware about the outcomes of their actions and
2. They are excellent relationship builders and able to relate behaviors and challenges of EI on workplace performance.

Threats

1. Emotional leaders are narcissist, steely, and actually emotionless with self-possession which could turn out to be a form of social impairment;

2. They will be lower level of innovation and creativity; and
3. Their empathy can sometimes lead to a poor decision making.

12.4.2.6 REINFORCING LEADERSHIP

Strengths

1. *Agility:* Reinforcing leaders bring new thinking and different actions to lead and he can be more agile and quick in this uncertain world and energy to convert ambiguity in an advantage, "small-win" innovations.
2. *Versatility and Flexibility:* Reinforcing leaders helps in being creative and finds solutions to resolve conflicts. They are attentive and active listening skills and learns skillfully from others.
3. *Effectiveness:* Reinforcing leaders can anticipate asymmetrical information which will make their leadership more effective, innovative, and mature.
4. *Adoptability:* Reinforcing leaders are flexible in adopting different situation.
5. *Empathetic:* The inquisitive leaders have questions and questions encourage creativity which enables them to follow courageous routes.
6. *Holistic and Shared Value:* A reinforcing leader has interdisciplinary knowledge with cross-functional experiences to understand and manage holistic digital ecosystems.

Challenges

1. *Knowledge Worker and Human Capital:* Individual knowledge workers' resilience positively influences both their creativity and their performance under pressure.
2. *Transparent and Open Minded:* Managerial positions often go hand in hand with distractions, interruptions, and a high workload.
3. *Sensation:* Often the reinforcing leaders do not take into account the expectations of the potential customers and clients about the product.
4. *Lateral Thinker and Explorer:* Reinforcing leaders should act at right time in this complex world and then can take the advantage of great opportunities.

Opportunities

1. *Competitive Advantage:* Reinforcing leader encourages to come forward to share their creativity and new ideas, and then dedicated the time and money to develop that idea irrespective of the size of the idea.
2. *Latent Thinking:* Innovations are not planned but it can be promoted and the leader provides a conducive environment to flourish the new idea. This idea will be part of the norm.
3. *Responsibility and Inspiration:* Without established moral guidelines, inconsistent and unpredictable decisions are more likely to be made. Keeping our core values in mind will help provide us with direction as well as inspiration for the future.
4. *Accumulator:* Reinforcing leaders understand expects, motivates team members to find adaptable solutions.
5. *Credibility:* Building trust among stakeholders can help in attaining strategic objectives and building favorable environment.

Threats

1. *High Emotional Quotient:* Reinforcing leaders sometimes consider their stakeholders as their enemy who seriously innovate. The threat of taking their places in the organization, becomes prominent in their mind which otherwise not important for the success of the organizations than of the private interest.
2. *Facade:* With instantaneous rate of change of information with shortened knowledge life cycle is a great hurdle for the growth of the organization, even the very famous leaders acknowledge that "there are known unknowns and unknown unknowns."

12.5 IMPLICATIONS OF THE STUDY

The Fourth Industrial Revolution is causing extraordinary changes in ways individuals, societies, and organizations work and function. The revolution made us to re-evaluate different assumptions and practices that have defined our ways of leading the organizations and how to make it more efficient in the near future. This research paper is an attempt to create a baseline model for new age of leadership management process. This model will be of great use for the global professionals. The model will be further used to collect and gather information on the use of the attributes in handling various leadership management in the organizations at large.

12.6 CONCLUSION

The research is on focused attempt to recreate a new model of leadership style base on the five attributes of a leadership. The model will further build up in creating a synchronization of the leaders and society to form a better future wherein empowering by putting the "people first" and prompting with a view that all technologies means are prepared "by people for people." All should be proactive in making and shaping technology and technology disruption.

KEYWORDS

- **Fourth Industrial Revolution**
- **global trade wars**
- **leadership**
- **trade relations**

REFERENCES

Akcil, U.; Aksal, F. A.; Mukhametzyanova, F. S.; Gazi, Z. A. An Examination of Open and Technology Leadership in Managerial Practices of Education System. *Eurasia J. Math. Sci. Technol. Educ.* **2017,** *13* (1), 119–131.

Aygun, M.; Suleyman, I. C.; Kiziloglu, M. In *Intrapreneurship in Small and Medium-Sized Enterprises*, 2nd International Symposium on Sustainable Development, 2010; pp. 217–224.

Bodolica, V.; Spraggon, M. Life on Heels and Making Deals: A Narrative Approach to Female Entrepreneurial Experiences in the UAE. In *Management Decision*; Vol. 53, no. 5; Emerald Publishing Limited, 2015; pp. 984–1004.

Boin, A.; 't Hart P.; Mcconnell A.; Preston T. Leadership Style, Crisis Response and Blame Management : A Case of Hurricane Katrina. In *Public Administration*; Vol. 88, no. 3 MA 02148; Blackwell Publishing: USA, 2010; pp. 706–723.

Bolden, R.; Petrov, G.; Gosling, J. Distributed Leadership in Higher Education: Rhetoric and Reality. *Educ. Manag. Admin. Leadersh.* **2009,** *37* (2), 257–277.

Boni, A.; Weingart, L.; Evenson, S. Innovation in an Academic Setting: Designing and Leading a Business through Market-Focused, Interdisciplinary Teams. *Acad. Manag. Lear. Educ.* **2009,** *8* (3), 407–417.

Brown, C.; Nasarwanji, A.; Catulli, M. In Conflict Over Entrepreneurial. *Intrapreneurial Sensemaking of Business Model Change Initiatives, Conference Paper*; University of Herefordshire: De Haviland Campus, Hatfield, UK, 2010; pp. 1–18.

Bird, B.; Brush, C. A_Gendered_Perspective_On_Organizational.Pdf, 2002.

Carrier, C. Intrapreneurship in Small Businesses: An Exploratory Study. *Entrep. Theory Pract.* **1997,** *21* (1), 5–20.

Chaib Lababidi, H.; Lababidi, R.; Colak, M.; Dayan, M. Contingency Effects of Firm Structure and Environmental Uncertainty on Strategic Planning Process and Firm Performance: Evidence from UAE Enterprises. *Strategic Change* **2020,** *29* (2), 241–252.

Charan, R.; Hofer, C. W.; Mahon, J. F. From Entrepreneurial to Professional Management: A Set of Guidelines. *J. Small Bus. Manage.* **1980,** *18* (1), 1–10.

Clark, N. Model of Complexity Leadership Model. *J. Hum. Resour. Dev. Int.* **2012,** *16* (2), 135–150.

Dean, H.; Ford, J. Discourses of Entrepreneurial Leadership: Exposing Myths and Exploring New Approaches. *Int. Small Bus. J.* **2017,** *35* (2), 178–196.

De Wit, B.; Meyer, R. *Strategy Synthesis: Resolving Strategy Paradoxes to Create Competitive Advantage*; Text and Readings. Cengage Learning EMEA, 2010.

Dulewicz, V.; Higgs, M. Leadership at the Top: The Need for Emotional Intelligence in Organizations. *Int. J. Organ. Anal.* **2003,** *11* (3), 193–210.

Feyerherm, A. E.; Rice, C. L. Emotional Intelligence and Team Performance: The Good, the Bad and the Ugly. *Int. J. Organ. Anal.* **2002,** *10* (4), 343–362.

Galloway, L.; Kapasi, I.; Sang, K. Entrepreneurship, Leadership, and The Value of Feminist Approaches to Understanding Them. *J. Small Bus. Manag.* **2015,** *53* (3), 683–692.

Gartner, W.; Liao, J. The Effects of Perceptions of Risk, Environmental Uncertainty, and Growth Aspirations on New Venture Creation Success *Small Bus. Econ.* **2012,** *39* (3), 703–712.

George, J. M. Emotions and Leadership: The Role of Emotional Intelligence. *Hum. Relat.* **2000,** *53* (8): 1027–1055.

Goleman, D. *What Makes A Leader?* Harvard Business Review, November-December, 1998.

Gupta, V.; Macmillan, I. C.; Surie, G. Entrepreneurial Leadership: Developing and Measuring a Cross-Cultural Construct. *J. Bus. Ventur.* **2004,** *19* (2), 241–260.

Hamilton, E. Whose Story is it Anyway?: Narrative Accounts of the Role of Women in Founding and Establishing Family Businesses. *Int. Small Bus. J.* **2006,** *24* (3), 253–271.

Harwati, L. N. Crisis Management: Determining Specific Strategies and Leadership Style for Effective Outcomes. *Crisis* **2013,** *2* (2), 170–181.

Heifetz R.; Grashow, A.; Linsky, M. Leadership in a (Permanent) Crisis. *Har. Bus. Rev.* **2009,** *87* (7–8), 62-9, 153.

Heinonen, J.; Toivonen, J. Corporate Entrepreneurs or Silent Followers. *Leaders. Organ. Dev. J.* **2008,** *29* (7), 583–599

Higgs, M.; Aitken, P. An Exploration of the Relationship between Emotional Intelligence and Leadership Potential. *J. Manag. Psychol.* **2003,** *18* (8): 814–823.

Kellerman, B. *Cut Off at the Pass: The Limits of Leadership in the 21st Century. (Governance Studies At Brookings)*; The Brookings Institute: Washington, D.C., 2012.

Klann G. *Crisis Leadership: Using Military Lessons, Organisational Experiences and the Power of Influence to Lessen the Impact of Chaos on the People You Lead*; CCL (Centre For Creative Leadership) Press: USA, 2003.

Kossek, E. E. Human Resources Management Innovation. *Hum. Res. Manag.* **1987,** *26* (1), 71–92.

Lechner, C. M.; Sortheix, F. M.; Obschonka, M.; Salmela-Aro, K. What Drives Future Business Leaders? How Work Values and Gender Shape Young Adults' Entrepreneurial and Leadership Aspirations. *J. Vocation. Behav.* **2018,** *107* (March), 57–70.

Lewis, K. V. Enacting Entrepreneurship and Leadership: A Longitudinal Exploration of Gendered Identity Work. *J. Small Bus. Manag.* **2015,** *53* (3), 662–682.

Mcentire, D. A.; Dawson, G. The Intergovernmental Context. In *Emergency Management: Principles and Practice for Local Government*; International City and County Management Association; Washington DC, 2007, pp. 57–70.

Mcgowan, P.; Cooper, S.; Durkin, M.; O'Kane, C. The Influence of Social and Human Capital in Developing Young Women as Entrepreneurial Business Leaders. *J. Small Bus. Manag.* **2015,** *53* (3), 645–661.

Mendi, P.; Moner-Colonques, R.; Sempere-Monerris, J. Cooperation of Innovation and Technology Licensing: Empirical Evidence from Spain. *Technol. Forecast. Soc. Change* **2020,** *154*, 119976.

Nicholls, A. The Legitimacy of Social. *Entrepreneurship Theor. Pract.* **2010,** *44* , 611–634.

Nooteboom, B.; Van Haverbeke, W.; Duysters, G.; Gilsing, V.; Van Den Oord, A. Optimal Cognitive Distance and Absorptive Capacity. *Res. Policy* **2007,** *36* (7), 1016–1034.

Northouse, P. G. *Leadership: Theory and Practice*. Sage:New Delhi; 2013.

Pacione, C. In *The Discipline of Innovation. Proceedings of the Human Factors and Ergonomics Society Annual Meeting*, September, 2015; Vol. 59, No. 1, SAGE Publications: Sage CA: Los Angeles, CA, pp. 303–303.

Palmer, B.; Walls, M.; Burgess, Z.; Stough, C. Emotional Intelligence and Effective Leadership. *Leadersh. Organ. Dev. J.* **2001,** *22* (1), 5–10.

Pinchot, G.; Pellman, R. *Intrapreneuring in Action: A Handbook for Business Innovation*; Berrett-Koehler Publishers Inc, 1999.

Prasad, L. The Etiology of Organizational Politics: Implications for the Intrapreneur. *SAM Adv. Manag. J.* **1993,** *58* (3), 35–41.

Prati, L. M.; Douglas, C.; Ferris, G. R.; Ammeter, A. P.; Buckley, M. R. The Role of Emotional Intelligence in Team Leadership: Reply to the Critique by Antonakis. *Int. J. Organ. Anal.* **2003,** *11* (**4**), 363–369.

Rogers, E. M. *Diffusion of Innovations*, 4th ed.; The Free Press: New York.

Rosete, D.; Ciarrochi, J. Emotional Intelligence and its Relationship to Workplace Performance Outcomes of Leadership Effectiveness. *Leadersh. Organ. Dev. J.* **2005,** *26* (5), 388–399.

Shepherd, D.; Haynie, J. M. Family Business, Identity Conflict, and an Expedited Entrepreneurial Process: Identity Conflict. *Entrep. Theory Pract.* **2009,** (812), 1245–1265.

Shurbagi, A. M.; Zahari, I. B. The Relationship between Organizational Culture and Job Satisfaction in National Oil Corporation of Libya. *Int. J. Hum. Appl. Sci.* **2012,** *1* (3), 88–93.

Smith, W. K.; Lewis, M. W. Toward A Theory of Paradox: A Dynamic Equilibrium Model of Organizing. *Acad. Manag. Rev.* **2011,** *36*, 381–403.

Sparr, J. L. Paradoxes in Organizational Change: The Crucial Role of Leaders' Sensegiving. *J. Change Manag.* **2018,** *18* (2), 162–180.

van Der Heyden, L.; Blondel, C.; Carlock, R. S. Fair Process: Striving for Justice in Family Business. *Fam. Bus. Rev.* **2005,** *18* (1), 1–21.
Waistad, W. B.; Kourilsky, M. L. Entrepreneurial Attitudes and Knowledge of Black Youth. *Entrep. Theory Pract.* **1998,** *23* (2), 5–18.
Waldman, D. A.; Bowen, D. E. Learning to be a Paradox-Savvy Leader. *Acad. Manag. Perspect.* **2016,** *30*, 316–327.
Waugh, W. L.; Tierney, K., Eds. *Emergency Management: Principles and Practice for Local Government*, 2nd ed.; ICMA: Washington DC, 2007; pp. 57–70.
Wu, C.; Mcmullen, J. S.; Neubert, M. J.; Yi, X. The Influence of Leader Regulatory Focus on Employee Creativity. *J. Bus. Ventur.* **2008,** *23* (5), 587–602.
Zhang, Y.; Waldman, D. A.; Han, Y.-L.; Li, X.-B. Paradoxical Leader Behaviours in People Management: Antecedents and Consequences. *Acad. Manag. J.* **2014,** 58 (2), 538–566.
Yang, Q.; Zimmerman, M.; Jiang, C. Characteristics on New Firms' Time to IPO*. *J. Small Bus. Manag.* **2011,** *49* (2), 163–184.
Yang, Z.; Zhou, X.; Zhang, P. Centralization and Innovation Performance in an Emerging Economy: Testing the Moderating Effects. *Asia Pacific J. Manag.* *32* (2), 415–442.
Yukl, G. *Leadership in Organizations*; Dorling Kindersley: New Delhi, India, 2011.

CHAPTER 13

Role of Agile Leadership in the Prevention of Sexual Harassment in the Workplace

ANUBHUTI SAXENA*, ESHA GARG, and ANAND NADAR

Bennett University, Greater Noida, India

**Corresponding author. E-mail: anubhuti.saxena@bennett.edu.in*

ABSTRACT

As shown by the recent #MeToo movement sweeping through social media, 85% of women said that they experienced some form of sexual harassment at workplace (SHW). Sexual harassment is considered as one of the most frightening forms of violence. SHW is regarded as an emotionally bold breach of women's right to equality. A place which encourages sexual harassment creates a very unhealthy and hostile environment for women to work. It discourages women involvement and participation, affecting their social and economic empowerment. In this modern era where we talk about inclusive growth, such incidents prove to be a failure of justice and humanity at offices. Sexual harassment is often ignored in India, despite the #MeToo movement exposing many public figures in various domains such as politics, business, entertainment, and beyond. Several victims shared their stories and traumatic experiences which exposed the harsh truth about working as a woman in these domains. To make women feel secure and discourage such practices, the legislature formulated the Sexual Harassment of Women at Workplace (Prevention, Prohibition, and Redressal) Act 2013, also called POSH Act, 2013. This act was formulated in 2013, but so far, there has been

Agile Leadership for Industry 4.0: An Indispensable Approach for the Digital Era, Tanusree Chakraborty, Tahir Mumtaz Awan, Ashok Natarajan, and Muhammad Kamran (Eds.)

no reduction in the occurrence of sexual harassment incidents, in fact, the graph is rising upward. It is the need of the hour that the organizations must make sure to develop a culture which allows women to grow. Whenever we talk of changing culture, the most powerful and easiest way to do it is to make changes in the leadership. Therefore, this chapter discusses about how the leadership can bring about changes in the mindset of people and make more responsible workplaces so that the occurrences of sexual harassment could be controlled. The purpose of this study is to bring forth the causes of sexual harassment by exploring the data available, emphasizing practices adopted by the modern organizations to avoid SHW, highlighting what kind of leadership style should be adopted to curb sexual harassment and supply suggestions to address the issue. It also intends to give an insightful understanding of the important legal framework of the POSH, Act 2013.

13.1 INTRODUCTION

"A murder can dismantle the physical frame of the sufferer, while sexual harassment can demean and degrade the soul of an unprotected woman"

Women victimization by men has always been an area of focus for academicians since a long time (Brownmiller, 1975; Koss and Heslet, 1992; Watts and Zimmerman, 2002). The society has brought into the limelight several cases of violence against women; by and large, the most typical and severe kinds of oppression done on women include forced prostitution, sexual coercion, rape, sexual abuse by both an intimate partner as well as non-intimate partners, and ill treatment of labor (Crowell and Burgess, 1996; Watts and Zimmerman, 2002). Our constitution talks about equality between the sexes, Article 14 of the Constitution of India talks about equality of opportunity and status for all citizens. Thus, a safe workplace is a legal right of every working woman. Articles 14, 15, and 21 of the Indian Constitution clearly express that every individual lives a life devoid of any kind of discrimination. This can only be achieved if we ensure that women are safe at work. The emphasis should not only be on prevention but also on penalties issued out of such charges.

We talk of women empowerment and making women independent in all levels of society. It is often said that women are the key to the nation's overall development. However, how much effort has been taken to overcome the violence against women? The Sexual Harassment of Women at Workplace

(Prevention, Prohibition, and Redressal) Act was instituted in 2013 to make sure women are safe at work. It focuses on building work environments that would respect women in terms of both equality and opportunities. At this juncture, the key question that arises is "Has the implementation of the provisions given in act been done in the most appropriate way by organizations or not?" If we have such stringent rules and regulations, then why the statistics show the graph of cases still soaring high? However, sexual harassment in the workplace does not solely mean male-to-female harassment. Extensive research and numerous reports suggest that male-to-female harassment is the most common type (O'Leary-Kelly et al., 2009). Sexual harassment is detrimental to an individual's mental well-being if they had any history of depression or stress-related issues (Buchanan and Fitzgerald, 2008; Langhout et al., 2005; Willness et al., 2007)

Sexually oriented violence has been an extremely sensitive issue for many due to which it is often underreported (UNFPA, 2007; Watts and Zimmerman, 2002). Regardless of the statistics that is put forward, many women experience fierceness and must live with brutal impacts every day in life (AIUK, 2008; Watts and Zimmerman, 2002; World Health Organization, 2008). Sexual harassment has been regarded as a stressor at work and is considered as a function of many factors influencing organization settings such as the perception of employees regarding how tolerant the organization is with respect to the incidents of sexual harassment, that is, organizational tolerance of sexual harassment (OTSH), how open people are to talk about it at workplace, and the men-to-women ratio in the immediate setting (Lee, 2018). Higher OTSH is found when people perceive that the perpetrators are not punished appropriately and the complaints are often ignored (Hulin et al., 1996). Many organizations do not take sexual harassment complaints seriously and offer fewer penalties too for the immoral acts (Hulin et al., 1996). Organizational tolerance of sexual harassment gets encouraged due to ineffective leadership styles and an ignorant attitude which disregards the policies related to controlling sexual harassment (Fitzgerald et al., 1997). Contemporary literature on sexual harassment falls short on leadership responsibilities as well as organizational approach for controlling sexual harassment. Some research in literature has explored and articulated the ways through which organizations can control the occurrence of sexual harassment by training and development of the employees and enforcing policy aimed at curbing sexual harassment (e.g., Goldberg, 2007; Perry et al., 1998; Reese and Lindenberg, 2004).

This study aims to contribute to the literature addressing sexual harassment. This study analyzes the numerous factors leading to sexual harassment at work. There have been several antecedents and repercussions of violence against women that have been documented in the past research (Lapierre et al., 2005; Willness et al., 2007). Even though leaders can significantly inhibit sexual harassment cases at the workplace, virtually, this area remains unexplored. The study also sheds light on the legal framework related to sexual harassment in the Indian context.

13.2 PROBLEM STATEMENT

Although SHW has become a serious matter in organizations worldwide, women still hesitate to come forward and report the matter to the concerned authorities, maybe out of fear of losing professional standing and a good name in the work place. The Sexual Harassment of Women at Workplace (Prevention, Prohibition, and Redressal) act was instituted in 2013 to make sure women feel safe at work. It also focuses on building a workplace that exemplifies the importance and need of the right to equal opportunity and dignity. It has been many years since the act has been enacted, however, the number of sexual assault cases has not reduced. This puts forward the question, "Why even after having such strong legislation, our women are unsafe at the workplace?" Women workers often experience sexual harassment but due to a lack of awareness of their rights, they do not put it forward. These demand increasing awareness of the act among both employees and employers. The data about women's participation in the organized sector is scarce. A substantial part of the women workforce is not even captured in official data accounts. A total of 93% of women workers are employed in the unorganized sector, where they are unprotected by laws and often exploited. The reason for these harrowing statistics is the ignorance of privileged people and the inefficiency of the government. It is extremely important to prioritize women safety. Not only does sexual harassment tarnish India at the highest level but also remains as a gross human rights violation. The first step is to make people more aware about what is sexual harassment and the steps that could be taken to address it. We propose that ethical leadership can play supreme role in influencing employees and making them more aware of the provisions and policies related to sexual harassment so that it does not act as a stigma for women anymore.

13.3 LITERATURE REVIEW

13.3.1 CONDITION OF WOMEN EMPLOYEES DURING INDUSTRY 4.0

Fitzgerald in his research in 1993 explained sexual harassment as an unwelcome behavior in the workplace that is sexual and sexist in origin. It is a complicated concept that includes behaviors which create intimidating and humiliating environment. Since 1980, the sexual harassment has been classified into five categories, namely, *gender harassment* (sexist comments or behavior), *seductive behavior* (inappropriate sexual advances), *sexual bribery* (sexual favors for some reward), *sexual coercion* (sexual advances with a treat of punishment), and *sexual imposition* (assault). Sexual harassment can take place in public places such as schools, universities (Chuang and Kleiner, 1999; Leach and Sitaram, 2007), public transport (Gekoski et al., 2015; Graham, 2015), restaurants (Macmillan et al., 2000), shopping malls (Lenton et al., 1999), or workplaces (Mushtaq et al., 2015). This research restraints the focus to SHW only.

Even if one has no direct interest in learning about SHW, it is significantly important to gain knowledge about it, to have a better understanding of the gender inequality, especially at professional workspaces (Cassino and Besen-Cassino, 2019). If a woman fears to get into a department wherein she fears getting sexually harassed, it creates occupational segregation at workplace, which is also known as gender in equality. To support action against such behaviors, in the recent past, the #MeToo movement came into highlight as a game changer. The movement #Metoo was initiated by Tarana Burke in 2006 in the USA, with the main motive to aid the low-income women of color and all those who are the survivors of harassment. The main motto of this movement was to find path and rediscover new opportunities and independence for women (Das, 2019). It is a painful job for the victims of sexual harassment to deal with the trauma and abuse (Das, 2019) physically, mentally, and emotionally. In addition to that, many women fear on the impact it will make on their career and the repercussions it will have, but this social campaign gave strength to victims in need (Das, 2019).

A huge development came into action when after a decade later in 2017 #metoo was re-popularized by Alyssa Milano, an American actress, who shared her story accusing the famous film producer Harvey Weinstein (Das, 2019). The film producer was accused of misconduct to over 30 actresses in Hollywood and was fired from The Weinstein Company which he cofounded himself (Zambas, 2017). Alyssa Milano encouraged women to share their stories with #MeToo tag. Following her, many other prominent female personalities around the world joined in, making sure that the issue is everywhere heard and given attention to (Das, 2019). India was not ignorant to it. Many famous actresses and women from various fields came out on social media and this time with no fears and hesitation. They demanded and insisted for strict actions against the sexual abuse and misconduct which was happening with a constant pace (Das, 2019). After a huge commotion, actions were taken against the abusers and there was an enormous impact on the career of these famous actors, politicians, position-holders; to name a few, Nana Patekar, painter Jatin Das, journalist Vinod Dua, and many other prominent figures (Das, 2019). The movement empowered and heightened awareness among a lot of women of all ages to ventilate their lives (Gupta et al., 2019). Despite the huge strides made #MeToo movement, sexual harassment has not declined; on the contrary, there has been an increase in the number of sexual harassment cases around the world (Times, 2020). Due to the extremely highlighted and popular case of Harvey Weinstein in the Western world, where he was accused of misconduct to 70 women, it eventually led to a revolution and many women started to come out with the exposure of harassers (Kathayat, 2018). One such case in India was that of Miss Tanushree Dutta, former Miss India, who accused Mr. Nana Patekar, Indian actor, for misbehaving with the actress. This became the head start for the women of India to come up and expose more such harassment stories (Kathayat, 2018).

The question that arises now is "What was the need for it?" "What made women from all corners of the world come up and speak about sexual harassment with such intensity and why that now corporations are acting against it more seriously than ever?" The questions need to be answered. The sexual harassment has a lot to do with the repetitive portrayal of the suppressed status of the victim (Leskinen and Cortina, 2014). The research on the sexual harassment displays a comprehended classification of the sexual harassing conduct inclusive of gender harassment (sexual malice that is verbal and physical sexual behaviors), unwanted sexual attention (unwelcoming and offensive interest of sexual and inappropriate nature), and sexual coercion (verbal and physical behaviors not sexual but are gender targeted) (Fitzgerald, 2018). According to the legal and the psychological aspects of sexual harassment, it deals with behavioral and sex-based perspectives and is often an objective phenomenon (Chawla et al., 2020). According to Equal Employment Opportunity Commission (EEOC) (Cassino and Besen-Cassino, 2019), sexual harassment is unwelcome advantages, favors, and physical and verbal misconduct of sexual nature.

The privilege of being a commoner should not be used to suppress a common woman; privilege plays an important part in determining the opportunities and skills, they also play a significant role in carving young minds. Privilege is a sensitive matter (Gruber, 1998), which is why gender-based perception has a lot to do with male privilege (Times, 2020). According to Chawla et al., 2020, the behavior that seems to devalue, demean, and insult an individual based on individual's gender will eventually start to underestimate and sabotage the victim's personality. "Sexual harassment" issue is between the people in power and the people who do not have the power or are easily targetable by the authoritative people mostly women who are taken advantage of. In general cases, the existence of the authority is within the hands of the male-dominating positions who at times tend to misuse the power they have, be it in academia, firms, or the big corporations (Somasundran et al., 2015). Sexual discrimination at workplace gives shape to sexual harassment and tend to make it more unwelcoming for women (Cole et al., 2020; Karami et al., 2020). Women have little awareness and courage to come up and act against such issues, considering the potential perpetrators are quite close to the victims such as their head in the company, spouse, co-worker and like that (Gupta et al., 2019). Harassment may lead to a long-term adverse impact on the economic growth of the country as women may quit their jobs due the reason and the consequences of harassment (Cassino and Besen-Cassino, 2019). "Sexual harassment is a genuine issue that needs to be addressed at all workplaces urgently and sensitively. Women are valuable human resources. Their contribution in all spheres of life can never be belittled, whether at the home and hearth or away from it, in more impersonal office spaces. In either sphere, they are entitled to a congenial and dignified environment to live their life fully and attain their full potentiality," a bench of Justices Rajiv Sahai End law (Times, 2020). To attain their potential, women are very much entitled to have a congenial and dignified workplace and should have zero tolerance against the sexual harassment issues, says the High Court of Delhi (Times, 2020). Sexual harassment is seen as the goal-driven act which is assumed to have a high amount of success rate and a smaller number of penalties. The harasser usually goes through a considerable amount of unethical decision-making process and plans his/her actions in accordance with his motive level (Chawla et al., 2020). Women have a little in say for such issues mainly because of the belief that it will sabotage the image of many corporates. This is exactly why the research done on the topic is also considerably less until today and after the #MeToo movement (Chawla et al., 2020).

13.4 IS SHW A CHALLENGE DURING INDUSTRY 4.0?

Sexual harassment has no halt to it in any form; but with the new factors such as feminism, media, and politics, the topic has changed its form, the voices are recognized, and the media portrayal is more upfront and talked about (Amir Karami, 2021). Women feel that if they talk about their harassment experiences, they will be treated differently and when they raise their voice, they will be judged and segregated from work. This will raise just not raise questions on the company but also its authority and its leadership styles (Klassen, 2021). Therefore, it is a necessity to make sure leadership is in the favor of the right people in the organization. To understand that the females are in risk is vital. At the same time, to determine how vulnerable they are and what are the factors that hamper their safety at work seem to be thoroughly investigated (Hatch-Maillette and Scalora, 2002). Male workforce cannot just get away with the strong redressed systems in the organizations that support them, from the vexatious complaints taking them away from being guilty. Cooperate industry is slowly shattering in terms of gender equality as the organizations are becoming more revengeful and defensive. They show their power by retaliating against the abused and at times deliberately hamper the growth of the individual, leading to a decrease in the number of reported cases and huge turnover rates (Cassino and Besen-Cassino, 2019). There should be a smooth system to support women in the organization that addresses their problem and issues, no matter how small it is. The companies should make sure that they erase bias and unfair treatment from their systems as fast as they can, says the High Court of Delhi (Times, 2020). According to a survey conducted by Amir Karami, 90% of the harassers are males and the kind of the harassments included gender harassment, unwanted sexual attention, and sexual coercion. Modifying the culture of the organizations and making sure that the women at work feel inclusive in all aspects, is the utmost duty of the management and human resource (HR) professionals (Cole et al., 2020). It is due to such landmark cases of negative professionalism that the victimized women are expected to follow a very formalized system of filing complaints against sexual assault. The system within organizations is usually biased and narrow-minded, favoring the wrong doers more (Gupta et al., 2019).

This eventually results in huge female turnover, low performance, and hamper psychological wellness causing anxiety, depression, trauma, etc. (Karami et al., 2020). The informal sector is one sector that rarely informs or files complaints against the sexual harassment issues (Reuters, 2020). In fact, the sector is not even covered under the POSH Act, 2013. It has been observed that the victims often step away from complaining as they do not want to put their job and hard-earned position in danger (Karami et al., 2020). However, thanks to the rise of social media and omnipresence of internet, some of the issues got dragged into limelight and made people more aware about it (Karami et al., 2020). It is illegal to harass a person verbally and even making comments to a woman in general, which can/does change into more intense situations if timely corrective measures are not taken as it may lead to unpleasant exchange between the parties, which is where leadership comes in handy (Cole et al., 2020). Under the category of the modesty of women, section 509 of Indian Penal Code (IPC), the National Crime Records Bureau (NCRB) started collecting data regarding SHW. Since 2014, they have collected 57 cases in 2014; 119 cases in 2015; and 142 cases in 2016 under "insult to the modesty of woman" category.

The number of complaints recorded under SHW during the last 3 years is as follows:

The National Commission for Women (NCW) has started maintaining the data under the specific category of "sexual harassment at workplace" and the data below are such cases which were registered during January 2019:

The ministry of women and child development has initiated an online complaint management system known as sexual harassment electronic box (SHe-box). This was made for women for registering complaints regarding sexual harassment including all government and private employees. Till today, 423 complaints have been recorded from which 114 have been disposed of.

13.5 SEXUAL HARASSMENT AT WORKPLACE AND THE ROLE OF LEADERSHIP DURING INDUSTRY 4.0

Sexual harassment was once conceived solely as a sex-related trouble: Forceful sex-related breakthroughs that derive from natural sensations of libido or romance. The study has shown that by far the most typical manifestation of sexual harassment is sexual harassment, which has contempt at its core; this conduct aims to put individuals down as well as humiliate them, not just sex (Lilia et al., 2021). The *Sexual Harassment of Women* report calls out and highlights sexual harassment as a crucial culture and climate matter. It signifies the importance of institutional leaders to set reduction and prevention of sexual harassment as an area of utmost priority. It is up to the organizational leaders whether the organization will have an inclusive culture with a zero-tolerance policy against such acts of harassment or will promote. The organizational climate determines the happening of sexual harassment. If the attitude toward prevention of sexual harassment is not severe or strict, it is taken for granted. Through such culture it is often contemplated that those who get into sexual harassment acts will be protected, on the other hand those who report will be overlooked and even penalized. If such inappropriate acts are allowed at workplace, it may create a very unhealthy culture. Therefore, leadership interventions are imperative and should be made mandatory to create and maintain a healthy environment leading to zero tolerance (Sadler et al., 2018). The USA law regards sexual harassment as an act of employment discrimination and consequently the organizations are held liable for sexually harassing behaviors, unless they prove that they took hold of all rational measures to prevent the occurrence of such cases or that they took prompt steps to correct things as they became evident (McDonald, 2012).

Some instances of behavior that constitute sexual harassment at the work environment: (Government of India, Ministry of Women and Child Development, 2015)

1. Making sexually suggestive remarks or references.
2. Severe or duplicated offensive statements, such as teasing pertaining to an individual's body or looks.
3. Offending remarks or jokes.
4. Unsuitable questions, pointers, or remarks regarding an individual's sex life.
5. Displaying a variety of other sexist insulting images, pictures, MMS, SMS, WhatsApp, or e-mails.
6. Intimidation, threats, and blackmail around sexual favors.
7. Hazards, scare tactics, or retaliation against a staff member who speaks up regarding unwelcome behavior with sex-related overtones.
8. Undesirable social invites, with sexual overtones generally understood as flirting.
9. Unwelcome sexual advancements which may or may not be gone along with by threats or promises, implied or specific.

Some examples of behavior that might show underlying workplace sexual harassment and also merit inquiry:

1. Criticizing, insulting, and condemning an employee or reprimanding in public.
2. Exclusion from team activities or tasks without a legitimate factor.
3. Declarations damaging an individual's track record or job.
4. Eliminating locations of responsibility, unjustifiably.
5. Regularly overruling authority without just reason.
6. Condemning an individual regularly for errors without cause.
7. Continuously selecting a worker by designating her with demeaning words as well as giving her jobs that are not component of her routine duties.
8. Disrespecting or humiliating a person. duplicated efforts to isolate an individual or leave out.
9. Systematically hindering normal job conditions, messing up areas or instruments of work.
10. Humiliating a person in front of associates, participating in smear campaigns.
11. Arbitrarily taking disciplinary activity versus a staff member.
12. Managing the individual by holding back sources (time, training, budget plan as well as autonomy) necessary to do well.

Literature also supports frequency of occurrence of women rape cases as significantly associated with leader behaviors; for example, permitting or initiating sexually demeaning remarks, *quid pro quo* behaviors etc.) (Sadler et al., 2003). Many researchers found that the leaders not only guide their immediate subordinates; they also build the organizational culture, leading to a trickle-down effect to other individuals in the organization. The extent of the sexual harassment depends on the leader's tolerance to the sexual discriminatory and offensive behaviors (Gruber, 1998; King et al., 2012). Regarding what is appropriate interpersonal behavior within the organization, confusion and ambiguity can exist when leaders pay little or no attention to their employees (Robinson and Bennett, 1995). This creates a high chance of considering sexually harassing behaviors as acceptable and not barred (Pryor et al., 1993). If sexual harassment seems permissive, it will lead to fostering inadequate behaviors in the organization (Fitzgerald et al., 1997; Miner-Rubino and Cortina, 2004). Doty and Gelineau (2008) stated, "By passing be or disregarding a violation of a unit's command climate, you are establishing a new command climate." Due to lack of monitoring and overseeing by the passive management the chances of being observed and caught decrease. This certainly strengthens the motives of the potential abusers resulting in the increase of number of cases of sexual harassment.

One major area of focus for the leadership should be establishing an equal representation of men and women in the workplace and creating an agile relationship between the genders (Dhatt et al., 2017). Small behaviors that are left unchallenged can enable an environment whereby minor offenses turn into more serious offenses. Leaders have enormous power in removing sexual violence from their organizations. The link we discover between the leaders' conduct and the reality of sexual harassment is consistent with past research findings (Buchanan et al. 2014; Firestone and Harris 2003; Pryor et al. 1993). We build on to this background by putting forth that the leaders play a particularly vital role in shaping the control over SHW.

13.6 ANALYSIS OF SEXUAL HARASSMENT ACT, 2013

Gender-based discrimination prevention is the basic right of a human that is very well addressed in our preamble and our constitution. It is noteworthy that for a considerable amount of time women were not protected by the constitution of India. Due to globalization, women have shown a growing interest in taking up the role of the bread winners of the houses. They are coming forward to bring radical changes in their status and are rapidly uplifting their living standards (Considerations, 2020). The workplace sexual harassment in India was recognized by the Supreme Court of India following the landmark judgement of Vishaka V. State of Rajasthan where in the law-making body framed revolutionary guidelines and certain directions to enact the appropriate law for combating workplace sexual harassment. Later, the guidelines laid by the court were known as the Vishaka Guidelines as this was the pinnacle of long fight for protection against injustice for the rape victim (Madhu Sivaram Muttathil, 2020). As a result of growing importance of this issue, 354A was added to the IPC by making amendment to the Criminal Law Act, 2013, which lays down the list of acts which constitute the offence of sexual harassment. Prior to this, there were no laws related to it in the IPC. It has always been extremely important to recognize and act against the issue of SHW, keeping in view the importance of basic human rights (Madhu Sivaram Muttathil, 2020). In 2013, the Indian Ministry of Women and Child Development came forth with the first legislation specifically addressing the issue of sexual harassment of women at workplace act, POSH Act, 2013 (Considerations, 2020).

Women will be able to use the system more comfortably when it is supportive with right legal process and that's how gender equality would be promoted (Klassen, 2021). Keeping litigation at the bay, the ability to look at the situation is immensely important, and comprehensive understandings of the POSH Act have been humbly accepted by the Indian courts. The POSH Act was a great step taken by Indian Government to prevent the abuse against the women. Although it has made today's working women more aware of their rights and ways to protect themselves legally, even after its enactment, a significant increase in the number of cases of SHW has been reported (Saumya, 2020). Being a new law in India, this act provides clarity not only for the employers and the IC members but also for the public at large. Although this act came into force in the year 2013, still many companies lack internal committees. Those having ICC usually do not conduct meetings on regular basis and often do not support women in organization. Low awareness in most women in organizations about the complaint mechanism is mandated by the law therefore, there is an immense need to take pro-active actions to prevent and effectively redress the complaints of sexual harassment. The respective HR and in-house legal teams must make sure to make sure to give prompt response to the complaints. The major flaws are in the enactment and capitulation of these laws in the workplace is that the employers usually do not go for doing the systematic analysis of the work environment, thus end up saving the abusers. There is an urgent need to give importance awareness and a strict implementation of the rules and action plan to reduce the abuse and harassment against women at workplace.

13.7 ROLE OF LEADERSHIP IN CONTROLLING SHW

Sexual harassment is sexual conduct that is unwelcome or unwanted behavior or an action that makes a person feel offended, insulted, and humiliated. Sexual harassment in the workplace happens more frequently than we think. According to CNBC, 20% of the American population in the workplace have been sexually harassed (Jones and Wade, 2020). Under the international law of equality, all employers should have safeguards prohibiting sexual assault and harassment (Jones and Wade, 2020). Sexual harassment in the working environment can be prevented by the organization's leaders implementing consistent and effective sexual harassment policies, training both employees and their supervisors on sexual harassment and how to deal with it and adopting a trusted and confidential reporting system. The organization's leadership should also ensure that it takes the sexual harassment allegations seriously.

Organization leadership should train and educate employees about sexual harassment. Better-managed firms show a great quality of management in dealing with issues about sexual harassment and any sort of misconduct, and they play a significant role in hiring more women in the positions they deserve and filling the gap in between the genders for the growth of the organization (Renee Adams, 2021). Attorney Sachi Barreiro advises organizations to train and educate their employees about sexual harassment, prevent sexual harassment, and handle sexual harassment in the workplace (Quick and McFadyen, 2017). Managers should also be educated regularly on how to deal with sexual complaints. With training and education, employees and supervisors are continually reminded of the policies of sexual harassment. The American Federation of State (2021) notes that it is essential to encourage employees to speak up when they experience or see sexual harassment in the workplace. If everybody speaks up in the workplace, sexual harassment cases will decrease. Forbes insists that each organization should "build a zero-tolerance harassing-related language into the performance objectives of each manager" (McDonald and Charlesworth, 2016). Managers should address all complaints with seriousness and ensure zero tolerance for employees.

Leaders in an organization are responsible for developing a trustworthy and reliable reporting system. A confidential and reliable reporting system is essential for every organization that values its employees' views and complaints. With a well-structured and organized reporting system, employees can report their sexual harassment issues without fear (Fitzgerald and Cortina, 2018). The reporting system should also be efficient and effective with a zero-tolerance policy. Organizations should have a department of human resources dealing with employees' complaints about sexual harassment (Hersch, 2015). The human resources managers' role is to spread awareness of the confidential and secure system of reporting.

Leadership is essential in implementing consistent and effective sexual harassment policies. Sexual harassment policy should include the interpretation of sexual harassment, investigation of sexual allegations, discipline, and punishment (Quick and McFadyen, 2017). A transparent and precise sexual harassment policy encourages workers and managers to recognize it, leaving no space for misinterpretation. It allows all employees, regardless of their position in the organization, to be mindful of the consequences. In addition, the organizational leadership should encourage the employees to sign that they understand and agree to the sexual harassment policies and the consequences if violated (Hersch, 2015). Employees' signatures add additional protection and a pledge to the organization not to carry out sexual harassment behaviors and actions. The signatures also help employees be more responsible for their decisions. Once an appropriate sexual harassment policy has been established, organizations must educate their managers and workers to cope with various sexual harassment situations.

Sexual harassment causes the humiliation of employees in the workplace. It makes employees feel uncomfortable at work and can lead to prosecutions. Sometimes managers and employees are not aware that they are sexually harassing their colleagues. Organizational leadership should, therefore, develop and implement proper guidelines to guide their employees. Organizational leadership can effectively prevent sexual harassment by implementing sexual harassment policies, training, and educating managers and employees, and developing a reliable and trusted reporting system. We propose that if leaders adopt "ethical leadership style," sexual harassment could be prevented to a great extent.

13.8 AGILE LEADERSHIP AND SHW

Every leader is obligated to ensure that basic rights are respected and acknowledged at the workplace. Agile leaders ensure that the workplace is a safe space for women. In a large organization it is important to institute a mechanism which tackles SHW head on. Sexual harassment at workplace could be controlled by employing a framework of transparency, accountability, and action (TAA). A transparent medium for communication will facilitate the identification and reporting of inappropriate exchanges and incidents between employees. Additionally, transparency tackles the issue of power dynamics between the employees as the appropriate rules and ethics will be applicable to every employee equally. The safest and most desired workplaces in the world are the most transparent organizations. There should be zero-tolerance to sexual assaults as well as high focus on anti-discrimination policies should be encouraged. Accountability is vital for controlling SHW as it establishes the consequences of inappropriate conduct. Without accountability, responsibility is dangerous and is often neglected. If the leadership of the organization ensures that the employees are accountable for their actions, the employees will uphold their responsibilities. To tackle

SHW and uproot this moral blot from our organizations, it is crucial for the leadership of the organizations to establish a policy of zero tolerance toward inappropriate, heinous, and vulgar behavior.

Action against SHW should not be conceptualized as a post-incident phenomenon. Action should be undertaken on day one by the leadership by analyzing and predicting the numerous possible incidents that could occur as well as the vast number of triggers that cause these incidents. The best cure is prevention. Ineffective precautions and guidelines for SHW is an ugly truth of our organizations. This can be rectified by instilling good organizational values and ensuring that all the loopholes of the system are corrected. A strong, clear, and unified action policy of the leadership against SHW bolsters confidence among the employees and testifies for the integrity and safety of the workplace. Sometimes, the eyewitness even after knowing the details of the matter hesitate to come in the light, they fear retaliation against them from the organization and its people with authority, which is extremely unjust (Klassen, 2021). A right leadership can provide the best solution with justice to the victim. According to Jay Starkman, CEO of Engage PEO, a professional employer organization, training for the employees is essential at least once a year. Training sessions should touch on how sexual harassment is reported and the ramifications of the same. The entire organization should attend these sessions, including the CEO and other executives. This measure would counterbalance the power dynamics in the organization and set the precedent that every single individual is accountable and responsible.

The leadership should also publicly declare their stance against SHW. Action against SHW should be taken swiftly when evidence is overwhelming. Policies for protection to the accuser against retaliation after reporting harassment should be made. The procedure for reporting sexual harassment should be articulated and communicated to every employee. Confidentiality, wherever required by the employees, must be assured. A strong commitment to a safe workplace could be shown by allocating resources for anti-SHW measures. Women are subjected to sexual harassment significantly more than men. According to the Washington Post, 54% of women reported unwanted sexual advances at the workplace. Furthermore, what is shocking is that 95% of the men go unpunished, as reported by the women who reported them. The #MeToo movement proved that sexual harassment would no longer be tolerated, and the perpetrators would be held accountable. Social media was leveraged to expose the ugly truth of oppression and hostile work environment. Employers must also train their IC members on the complexities of the POSH Act and ensure that they understand the investigation procedures.

While employers are trying their best to educate the employees of their respective departments about the POSH Act, the employers should also make a note to follow the rule and guidelines religiously and secure the department from any reputational risk.

TABLE 13.1 Number of Complaints Recorded Under SHW During the Last 3 Years.

Year	**2016**	**2017**	**2018**
Number of cases	539	570	965

TABLE 13.2 No. of Registered SHW Cases.

Category	**Registered cases**
Number of cases	29

Source: National Commission for Women

KEYWORDS

- **sexual harassment in the workplace**
- **#MeToo movement**
- **leadership**
- **POSH Act, 2013**
- **legal framework**

REFERENCES

Adams, R. B.; Akyol, A. C. ; Grosjean, P. A. Corporate Gender Culture. *UNSW Business School Research Paper Forthcoming*, SSRN [Online] 2021. https://ssrn.com/abstract=3880650 or http://dx.doi.org/10.2139/ssrn.3880650 (accessed July 4, 2021).

Adams-Roy, J. ; Barling, J. Predicting the Decision to Confront or Report Sexual Harassment. *J. Org. Behav*. **1998,** *19* (4), 329–336.

AFSCME at Work [Online] 2021. https://afscmeatwork.org/ (accessed Jan 16, 2021).

Arora, K. Sexual Harassment at Work: Understand the Rules of The PoSH Act In 6 Minutes. *Ungender* [Online] 2020. https://www.ungender.in/sexual-harassment-at-work-understand-the-rules-of-the-posh-act-in-6-minutes/ (accessed Sept 22, 2020).

Auroshree. Decoding POSH: Analysis of India's Regime Against Sexual Harassment. *The SCC Online Blog* [Online] 2018: https://www.scconline.com/blog/post/2018/12/29/

decodong-posh-analysis-of-indias-regime-against-sexual-harassment/ (accessed Dec 29, 2018).

Brownlie, I. ; Goodwin-Gill, G. S., Eds .; *Brownlie's Documents on Human Rights* ; Oxford University Press, 2010.

Brownmiller, S. *Against Our Will: Men, Women, and Rape*. Simon and Schuster: New York ; 1975.

Buchanan, N. T.; Fitzgerald, L. F. Effects of Racial and Sexual Harassment on Work and the Psychological Well-being of African American Women. *J. Occup. Health Psychol.* **2008,** *13* (2), 137.

Buchanan, N. T .; Settles, I. H .; Hall, A. T . ; O'Connor, R. C. A Review of Organizational Strategies for Reducing Sexual Harassment: Insights from the US Military. *J. Soc. Issues* **2014,** *70* (4), 687–702.

Burgess, A. W .; Dowdell, E. B . ; Prentky, R. A. Sexual Abuse of Nursing Home Residents. *J . Psychosoc. Nurs . Mental Health Serv.* **2000,** *38* (6), 10–18.

Cassino, D . ; Besen-Cassino, Y. Race, Threat, and Workplace Sexual Harassment: The Dynamics of Harassment in the United States, 1997–2016. *Gender Work Org.* **2019,** *26* (9), 1221–1240.

Chawla, N .; Gabriel, A. S .; O'Leary Kelly, A . ; Rosen, C. C. From #MeToo to #TimesUp: Identifying Next Steps in Sexual Harassment Research in the Organizational Sciences. *J . Manag.* **2021,** *47* (3), 551–566.

Cole, C. M .; Burrell, D. N . ; Springs, D. Exploring Organizational Development Intervention Around Sexual Harassment in Technical Firms. *Int. J . Hyperconnect . Inter . Things* **2020,** *4* (1), 29–42.

Considerations, H. R. *Prevention of Sexual Harassment at the Workplace (POSH). December,* 2020.

Das, N. R. Me Too Stories India: Everything You Need to Know About The #MeToo Movement in India. *POPXO* [Online] 2019: https://www.popxo.com/trending/stories-of-sexual-harassment-581351/ (accessed Aug 23, 2019).

DeSouza, E. R .; Wesselmann, E. D . ; Ispas, D. Workplace Discrimination against Sexual Minorities: Subtle and not-so-subtle. *Canad. J . Adm. Sci.* **2017a,** *34* (2), 121–132.

DeSouza, E. R .; Wesselmann, E. D . ; Ispas, D. Workplace Discrimination against Sexual Minorities: Subtle and not-so-subtle. *Can. J . Adm. Sci.* **2017b,** *34* (2), 121–132.

Dhatt, R .; Kickbusch, I . ; Thompson, K. Act Now: A Call to Action for Gender Equality in Global Health. In *the Lancet* , Vol. 389, Issue 10069 ; Lancet Publishing Group, 2017; p. 602.

Dhatt, R .; Theobald, S .; Buzuzi, S .; Ros, B .; Vong, S .; Muraya, K .; … Jackson, C. The Role of Women's Leadership and Gender Equity in Leadership and Health System Strengthening. *Global Health Epidemiol . Genom.* **2017,** *2*, e8.

England, D. *The Essential Guide to Handling Workplace Harassment & Discrimination* . Nolo: Berkeley, Calif, 2009.

Firestone, J. M . ; Harris, R. J. Perceptions of Effectiveness of Responses to Sexual Harassment in the US Military, 1988 and 1995. *Gender, Work Org.* **2003,** *10* (1), 42–64.

Fitzgerald, L. F. Still the Last Great Open Secret: Sexual Harassment as Systemic Trauma, 2017.

Fitzgerald, L. F. Sexual Harassment in Work Organizations: A View from the 21st Century. In *APA Handbooks in Psychology®. APA Handbook of the Psychology of Women: Perspectives on Women's Private and Public Lives* ; 2018, pp. 215–234.

Fitzgerald, L. F . ; Cortina, L. M. Sexual Harassment in Work Organizations: A View from the 21st Century. In *APA Handbooks in Psychology®. APA Handbook of the Psychology of Women: Perspectives on Women's Private and Public Lives*; Travis, C. B., White, J. W.; Rutherford, A.; Williams, W. S.; Cook, S. L.; Wyche, K. F., Eds.; American Psychological Association, 2018; pp. 215–234.

Goldberg, C. B. The Impact of Training and Conflict Avoidance on Responses to Sexual Harassment. *Psychol . Women Q.* **2007,** *31* (1), 62–72.

Gruber, J. E. The Impact of Male Work Environments and Organizational Policies on Women's Experiences of Sexual Harassment. *Gender Soc.* **1998,** *12* (3), 301–320.

Gupta, R .; Gupta, A . ; Nehra, D. Going Forward with # MeToo Movement: Towards a Safer Work Environment. *J. Psychosex. Health* **2019,** *1* (2), 174–179.

Hatch-Maillette, M. A . ; Scalora, M. J. Gender, Sexual Harassment, Workplace Violence, and Risk Assessment: Convergence Around Psychiatric Staff's Perceptions of Personal Safety. *Aggress . Violent Behav.* **2002,** *7* (3), 271–291.

Hersch, J. Sexual Harassment in the Workplace. *IZA World of Labor,* 2015.

Hulin, C. L .; Fitzgerald, L. F . ; Drasgow, F. *Organizational Influences on Sexual Harassment* ; Sage Publications, Inc., 1996.

Jones, T . ; Wade, E. E. Me Too: Race, Gender, and Ending Workplace Sexual Harassment. *Duke J. Gender L. Poly* **2020,** *27*, 203.

Karami, A .; White, C. N .; Ford, K .; Swan, S . ; Yildiz Spinel, M. Unwanted Advances in Higher Education: Uncovering Sexual Harassment Experiences in Academia with Text Mining. *Inform. Proc . Manag.* **2020,** *57* (2), 102167.

Kathayat, D. #Me Too Movement —Sexual Harassment of Women at Workplace. *Legal Bites* [Online] 2018 , https://www.legalbites.in/me-too-movement-sexual-harassment/ (accessed Oct 11, 2018).

King, K. M .; Nguyen, H. V .; Kosterman, R .; Bailey, J. A . ; Hawkins, J. D. Co-occurrence of Sexual Risk Behaviors and Substance Use Across Emerging Adulthood: Evidence for State-And Trait-Level Associations. *Addiction* **2012,** *107* (7), 1288–1296.

Klassen, S. Challenges in Bringing Gender Equity into the Workplace: Addressing Common Concerns Women Have When Deciding to Hold Employers Accountable for Gender Discrimination. *J . Race Gender Ethn.* **2021,** *10* (1), 5.

Koss, M. P . ; Heslet, L. Somatic Consequences of Violence Against Women. *Arch . Family Med.* **1992 ,** *1* (1), 53–59.

Kunal Talgeri, E. B. Why India is far Behind the US in Addressing Sexual Harassment at the Workplace. *The Economic Times* [Online] 2017 . https://economictimes.indiatimes.com/news/politics-and-nation/why-india-is-far-behind-the-us-in-redressing-sexual-harassment/articleshow/61312681.cms (accessed Oct 29, 2017).

Langhout, R. D .; Bergman, M. E .; Cortina, L. M .; Fitzgerald, L. F .; Drasgow, F . ; Williams, J. H. Sexual Harassment Severity: Assessing Situational and Personal Determinants and Outcomes 1. *J . Appl. Soc. Psychol.* **2005,** *35* (5), 975–1007.

Lapierre, L. M .; Spector, P. E . ; Leck, J. D. Sexual Versus Nonsexual Workplace Aggression and Victims' Overall Job Satisfaction: A Meta-Analysis. *J . Occup. Health Psychol.* **2005,** *10* (2), 155.

Lee, J. Passive Leadership and Sexual Harassment. *Pers. Rev*. 2018.

Leskinen, E. A . ; Cortina, L. M. Dimensions of Disrespect: Mapping and Measuring Gender Harassment in Organizations. *Psychol . Women Q.* **2014,** *38* (1), 107–123.

Letke, A. 7 Steps to Prevent Sexual Harassment in the Workplace. *Integrity* [Online] 2014 . https://integrityhr.com/hrblog/7-steps-to-prevent-sexual-harassment-in-the-workplace/#:~:text=%207%20Steps%20To%20Prevent%20Sexual%20Harassment%20In,together%20and%20cutting%20loose%20for%20a...%20More (accessed July 18, 2014).

Cortina, L. M.; Areguin, M. A. Putting People Down and Pushing Them Out: Sexual Harassment in the Workplace. *Annu. Rev . Organ. Psychol . Organ. Behav .* **2021,** *8*.

Madhu Sivaram Muttathil, A. G. India's POSH Act — A Snapshot. *ACC docket* [Online] 2020 . https://www.accdocket.com/indias-posh-act-snapshot (Dec 14, 2020).

Marg, S. *Ministry of Women 8*. Child Development, 2015.

McDonald, P. Workplace Sexual Harassment 30 Years on: A Review of the Literature. *Int. J . Manag. Revi.* **2012,** *14* (1), 1–17.

McDonald, P . ; Charlesworth, S. Workplace Sexual Harassment at the Margins. *Work Employ. Soc.* **2016,** *30* (1), 118–134.

Miner-Rubino, K . ; Cortina, L. M. Working in a Context of Hostility Toward Women: Implications for Employees' Well-Being. *J . Occup. Health Psychol.* 2004, *9* (2), 107.

O'Leary-Kelly, A. M .; Bowes-Sperry, L .; Bates, C. A . ; Lean, E. R. Sexual Harassment at Work: A Decade (Plus) of Progress. *J . Manag .* **2009,** *35* (3), 503–536.

Perry, E. L .; Kulik, C. T . ; Schmidtke, J. M. Individual Differences in the Effectiveness of Sexual Harassment Awareness Training 1. *J . Appl. Soc. Psychol.* **1998,** *28* (8), 698–723.

Pina, A .; Gannon, T. A . ; Saunders, B. An Overview of the Literature on Sexual Harassment: Perpetrator, Theory, and Treatment Issues. *Aggress . Violent Behav.* **2009,** *14* (2), 126–138.

Pryor, J. B .; LaVite, C. M . ; Stoller, L. M. A Social Psychological Analysis of Sexual Harassment: The Person/Situation Interaction. *J . Vocat. Behav.* **1993,** *42* (1), 68–83.

Putting People Down and Pushing Them Out: Sexual Harassment in the

Quick, J. C . ; McFadyen, M. Sexual Harassment: Have we Made any Progress? *J . Occup. Health Psychol.* **2017,** *22* (3), 286.

Reese, L. A . ; Lindenberg, K. E. Employee Satisfaction with Sexual Harassment Policies: The Training Connection. *Public Pers. Manag.* **2004,** *33* (1), 99–119.

Reuters. #MeToo, 2 Years on: Indian Women in Informal Sector Rarely Report Workplace Harassment Over Fear of Losing Job. *Economic Times* [Online] 2020 . https://economictimes.indiatimes.com/magazines/panache/2-years-after-metoo-movement-poor-indian-women-fear-losing-their-jobs-over-workplace-harassment-complaintsormetoo-2-years-on-indian-women-in-informal-sector-rarely-report-workplace-harassment-over-fea (accessed Oct 16, 2020).

Robinson, S. L . ; Bennett, R. J. A Typology of Deviant Workplace Behaviors: A Multidimensional Scaling Study. *Acad . Manag. J.* **1995,** *38* (2), 555–572.

Sadler, A. G .; Booth, B. M .; Cook, B. L . ; Doebbeling, B. N. Factors Associated with Women's Risk of Rape in the Military Environment. *Am. J . Indus. Med.* **2003,** *43* (3), 262–273.

Sadler, A. G .; Lindsay, D. R .; Hunter, S. T . ; Day, D. V. The Impact of Leadership on Sexual Harassment and Sexual Assault in the Military. *Mil. Psychol.* **2018,** *30* (3), 252–263.

Saumya. Sexual Harassment at Workplace: Has the Situation Changed? *JUDICATEME* [Online] 2020 . https://judicateme.com/sexual-harassment-at-workplace-has-the-situation-changed/ (July 28, 2020).

Schneider, K. T .; Swan, S . ; Fitzgerald, L. F. Job-Related and Psychological Effects of Sexual Harassment in the Workplace: Empirical Evidence from Two Organizations. *J . Appl. Psychol.* **1997,** *82* (3), 401.

Somasundran, V .; Sudan, P . ; Mohanty, L. *Handbook on Sexual Harassment of Women at Workplace. November* 1–58. 2015. https://bopter.gov.in/assets/pdf/handbook-on-sexual-harassment-of-women-at-workplace2.pdf

Terrell Carver, V. M. *Politics of Sexuality* ; Taylor and Francis Group, 1998.

Times, T. E. Women are Valuable Human Resource, There Should be Zero Tolerance Towards Gender Insensitivity: HC [Online] 2020. https://economictimes.indiatimes.com/news/politics-and-nation/women-are-valuable-human-resource-there-should-be-zero-tolerance-towards-gender-insensitivity-hc/articleshow/79783637.cms (Dec 17 2020).

Watts, C . ; Zimmerman, C. Violence Against Women: Global Scope and Magnitude. *Lancet* **2002,** *359* (9313), 1232–1237.

Wikipedia. Vishaka Guidelines. Wikipedia [Online] 2020 . https://en.wikipedia.org/wiki/Vishaka_Guidelines#Background (Mar 16, 2020).

Willness, C. R .; Steel, P . ; Lee, K. A Meta-Analysis of the Antecedents and Consequences of Workplace Sexual Harassment. *Pers. Psychol.* **2007,** *60* (1), 127–162.

Word Health Organization. WHO (2008). Multi-Country Study on Women's Health and Domestic Violence against Women: Initial Results on Prevalence, Health Outcomes and Women's Responses [Online]. September 2008. http://www. who.int/gender/violence/who_multicountry_study/en/index.html

World Health Organization. *Strengthening country office capacity to support sexual and reproductive health in the new aid environment: report of a technical consultation meeting: wrap-up assessment of the 2008-2011 UNFPA-WHO collaborative project*, Glion : Switzerland, 21–23 March 2011 (No. WHO/RHR/11.29). World Health Organization, 2011.

Zambas, J. 10 High-Profile Sexual Harassment Cases in the Workplace. *Career Addict* [Online] 2017. https://www.careeraddict.com/work-harassment-stories#:~:text=%2010%20HighProfile%20Sexual%20Harassment%20Cases%20in%20the,TV%20host%20Bill%20O%E2%80%99Reilly.%20He%20has...%20More (Oct 20, 2017).

Zambas, J. How to Prevent Sexual Harassment in the Workplace. *Career Addict* [Online] 2018 . https://www.careeraddict.com/prevent-sexual-harassment-work (July 2, 2018).

CHAPTER 14

Agile Personality: Machiavellianism and Positive Self-Impression Management Behavior Via Interactive Role of Situational Mechanism

HAFSAH ZAHUR

National University of Modern Languages (Management Sciences), Islamabad 440000, Pakistan
E-mail: hafsazahoor6585@gmail.com

ABSTRACT

There is a need to develop the understanding of the functional and agile aspects of Machiavellianism personality with respect to their in-role behavioral tendency at workplace. Therefore, the current study is not only focusing on the direct association of Machiavellianism trait and behavioral outcome, but it also investigates the situational mechanism for predicting such relationships. This study extends the TAT (trait activation theory) by investigating the direct association of Machiavellianism personality with impression management and the moderating role of situational mechanism, that is, ambiguous work roles in predicting the Machiavellian personality and in-role behavioral outcome of impression management by using the field sample of ($N = 420$) employees working in higher education institutes. The results revealed that role ambiguity is a significant situational mechanism to activate the positive agile attributes of Machiavellian personality trait for predicting self-presentable behavior. It is also proved that Machiavellianism personality has a significant positive association with managing their soft impressions that can lead to better organizational functioning.

Agile Leadership for Industry 4.0: An Indispensable Approach for the Digital Era, Tanusree Chakraborty, Tahir Mumtaz Awan, Ashok Natarajan, and Muhammad Kamran (Eds.)

14.1 INTRODUCTION

Personality traits are perceived according to how people behave. The theoretical perspective advocates that personality is the underlying outlook that is present in every individual until it is activated and shows in the behavioral display of an individual. The significance of trait and situation exists when situational-based mechanisms allow the manifestation of these traits (Tett and Burnett, 2003). The appearance of one's personality traits through behavioral display must be investigated in the particular situational context/mechanism. Such investigations are beneficial for organizations to help them in recognizing the importance of situational factors for trait relevant behavioral display (Mischel and Shoda, 1995).

In interpersonal relations, it has been observed that individuals who possess Machiavellianism personality traits may indulge in strategic in-role behaviors (such as positive impression management) for maintaining interpersonal relations (Jonason and Schmitt, 2012; Jones and Mueller, 2021); it has also been observed that such individuals are adaptive to impression management behaviors (IM) to gain as many resources and benefits as possible from those they interact with and in doing so, they have less uncertainty to exploit situations for achieving the objectives (O'Boyle et al., 2012; Hilbig and Zettler, 2015; Jonason et al., 2012; Walters, 2021).

Trait Activation Theory emphasizes "on the inter-individual uniqueness's" as a significant factor in Machiavellian personality tendency to behave distinctively in accordance with the workplace situational demand/mechanism (Tett and Guterman, 2000: 398) that means the dormant prospective present in the individuals is voiced only in reaction to trait-based cues that provide room for the expression of such traits. A plethora of literature is available which reflects that trait-based situational mechanism moderates the relationship in between Machiavellian personality and its consequent behavior (Farh et al., 2012; De Hoogh et al., 2005; Genau et al., 2021).

This study focuses on task and social-based situational cues within the work place. Task-relevant cues are routine-based job responsibilities (such as role ambiguity). Social level workplace cues arise from the expectations of individual personalities to act in a socially approved manner (Tett and Burnett, 2003). Machiavellianism personality seems to be significantly influenced by situational mechanisms. The evolutionary point holds that our society is more prone toward getting used to the negative personality traits (i.e., Machiavellianism in this study) (Penke et al., 2007), and that "no personality trait is unanimously adaptive or maladaptive." The presence of

Machiavellianistic attributes in individual employees is reflective of the fact that this trait has adaptability in certain situations (Paulhus and Williams, 2002) which demand the display of positive self-image (through the positive impression management tactics). For the display of positive self-impression, the job should be based on weak situational factors (such as role ambiguity).

The significant role of the study is to investigate the missing link, which is the logical consideration of Trait Activation Theory (TAT) in the agile role of Machiavellianism personality and the positive self-impression management (IM) relationship. Activation of desirable traits is the basis to recognize the positive attributes of Mach's at energetic work environment by investigating the situational indicators. The relationship between Machiavellianism personality and assertive impression management tactics provided with weak job role situation (such as role ambiguity) is yet to be explored in an agile workplace. The curvilinear effect between the Machiavellianism agility and situational factors within workplace needs to be explored, as linear models are not appropriate to study the complex association in between dark personalities such as Machiavellianism and in-role behavioral outcome, that is, carrying positive self-impressions. The multiplicative models provide the new theoretical outlook. The objective of this study is to enhance the personality-impression management literature that includes the positive and flexible association in between Mach personality and soft taxonomy of IM (impression management). Second is by examining the multiplicative effects of the workplace situational mechanism, that is, role ambiguity and Machiavellianism personality to activate their functional in-role behavioral expression (i.e., positive self-impression management).

14.2 LITERATURE REVIEW AND THEORETICAL FRAMEWORK

Machiavellianism is regarded as an antisocial behavior that avoids social interaction and one of its inherent traits is inadaptability. However, the recent researches focus on its adaptability factor and have revealed its positive aspects that include popular characteristics of in-role behavior such as impression management (IM) skills. Belschak et al. (2013) observe that Mach's behave in a highly cooperative manner when they have self-benefits to achieve. Impression management is the willingness of an individual to exert effort for managing the impressions that others have toward them (Leary and Kowalski, 1990; Goffman, 1959; Walters, 2021). Machiavellianism is a wide concept that is studied in the context of both intelligence and personality. Machiavellianism trait means the same as manipulation, and those individuals who strongly possess this trait are known for utilizing strategic behaviors (such as impression management) that serve their purpose.

Jones and Pittman (1982) developed the soft taxonomy of impression management behavior that includes self-promotion (highlighting the accomplishments to be seen as a capable employee), ingratiation (use the flattery, giving away the favors and support the opinions of peers to be perceived as likable), exemplification (accomplish the tasks by going extra mile to be known as a devoted worker by the superiors and subordinates). Machiavellian personality is skilled in managing their positive impressions; they can use in-role behavioral tactics (such as IM) effectively in different work situations for achieving constructive consequences. Evidence has proven that soft impression management behaviors have a more profound impact than hard impression management tactics (such as intimidation) when individuals present themselves either to a superior or a colleague (Wayne and Liden, 1995).

Rauthmann (2011) observed that Machiavellianism reflects assertive in-role behavior by creating a positive self-impression for fitting in the organization and to avoid criticism. Machiavellians enhance their self-image by being motivated to avoid disappointing others. Machiavellianism is also having a tendency to control others (Sherry et al., 2006; Jones and Paulhus, 2009; Christie and Geis, 1970; Genau et al., 2021) and appear noticeable for the self-beneficial gains (Wilson et al., 1996). Therefore, it may well be interrelated with the use of soft Impression management behavior.

A Machiavellianism personality trait is somehow convinced that to engage in the organization-sponsored in-role behavior (i.e., IM) is in their best interest to appear a supportive and dedicated employee. For example, the in-role behavior of "ingratiation" and giving and taking favors may be helpful in creating workplace friendships. These friendships, in the long run, may be used and exploited to fulfill the work obligations. The demonstration of the in-role IM behavior by Mach's leaves the exploitation undetectable by the coworkers because they consider it doing a favor to a friend (Farmer et al., 1997).

As suggested by some of the researchers that the functional attributes of Machiavellianism personality activate at the strategic level for behaving in a socially acceptable manner. It is hard to detect the successful manipulation of situations by Mach's, as they make use of socially influencing behavior (such as positive Impression management) (Jonason and Webster, 2012). It has been suggested by Jonason et al. (2012) that positive persuasion appears to be strongly dependent on Machiavellianism personality. Machiavellianism trait is linked more closely to a "self-presentation and self-protective" orientation that is typically demonstrated by compliance with the standards and avoiding disapproval. The self-presentable behavior covers a wide variety of human behaviors that form the basis for social functioning of Mach's (Schlenker, 1980; Jones and Pittman, 1982; Sadler et al., 2010; Leary, 1995) Machiavellianism is such a personality attribute that characterizes an effective tendency to uphold their resources and public privileges with the strategic display of impression management behaviors; therefore, a Machiavellianistic personality is logical, rational and pre-emptive in nature" (Christie and Geis, 1970; Jones and Paulhus, 2009). Gu et al. (2017) propose that the service-oriented autonomous positions possessed by the Mach's prompt them to have a self-beneficial in-role behavioral display (i.e., assertive impression management).

- ***H1: Machiavellianism personality is positively associated with impression management behavior.***

Trait activation theory (TAT) has been proposed to investigate the multiplicative effects of personality and work-related situational mechanism (i.e., Machiavellianism * role ambiguity) to predict in-role behavior (such as impression management) (Lievens et al., 2006; Tett et al., 2021). Role ambiguity is the lack of job-related information which includes performance expectations and job duties (Rizzo et al., 1970). When the job roles are not strictly defined then there is much freedom for Machiavellian personality to enhance their job-related roles according to their personality traits (such as power, entitlement, and self-serving instinct) (Morgeson et al., 2005), such ambiguity in job roles provides opportunity for Mach's to display their positive image especially when they try to amend their malevolent impressions into positive self-image for achieving the self-set broader roles.

Machiavellian personalities reflect positive traits in the job settings where they can set the benchmark for their self-prescribed roles and have the freedom to deal with the problems related to role interpretation (Abdel-Halim, 1978; Leonelli et al., 2021). In such role ambiguous situations Mach's feel highly motivated and exercise better choices while making decisions and personal judgments particularly when it comes to the projection of their own role, which can be the triggering force for them to refine their self-impression.

Machiavellianism does not necessarily directly relate to the outcomes associated with behavior. The situational context seems to play a significant role in the relationship between Mach's and the outcomes. Mach's are low on conscientiousness and bloom in unstructured roles, where rules are not explicitly communicated, such a job relevant situational mechanism can shape up the positive impressions of Mach's (Jones and Paulhus, 2009; Becker and O'Hair, 2007). The evidence holds that the Mach's behave intimidatingly in highly controlled job roles (O'Connor and Morrison, 2001). A weak situational mechanism (such as ambiguous roles) builds psychological pressure on an individual to be involved in positive in-role behavior and refrain from undesirable behavior display (Beaty et al., 2001). The undertaken study investigates the weak situational mechanism of ambiguous role expectations to activate the functional attributes of Machiavellianism for self-presentable behavioral expression.

- ***H2*: *Role ambiguity moderate the relationship in between Machiavellianism and IM (impression management), the relationship is stronger in the presence of high role ambiguity and vice versa.***

14.3 RESEARCH MODEL

This section of the chapter discusses the methodology adopted in the study. It highlights the details of the sample, data collection methods and tools, and the measures of the study.

14.4 METHOD

14.4.1 POPULATION

The research population is comprised of the service sector that includes banks and telecommunication companies (Fig. 14.1). Machiavellianism personality performs extraordinarily in jobs that involve influencing others (such as in higher educational institutions) due to their ability to effectively employ situations for achieving self-presentable behaviors (i.e., soft impression management behavior) (Van der Linden et al., 2017).

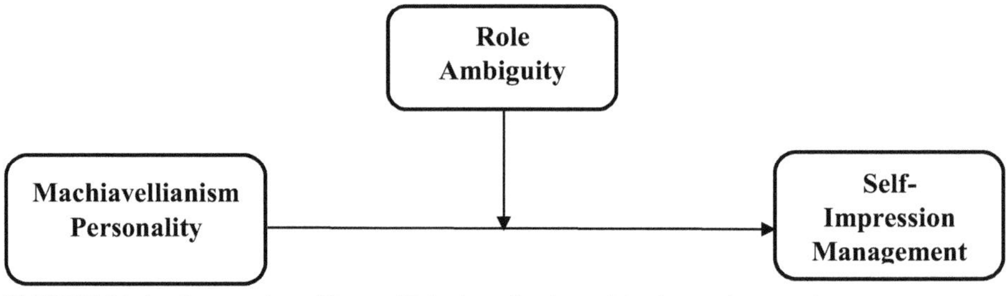

FIGURE 14.1 Interactive effects of Mach and role ambiguity on impression management.

14.4.2 *SAMPLE SIZE AND PROCEDURE*

A time-lagged procedure was adopted for data collection. According to Kumar et al. (2013) causal studies based on behavioral phenomenon shall be analyzed efficiently by using the time-lagged data. At Time 1 the independent variable (Machiavellianism) and moderating variable (role ambiguity) responses were collected. Outcome variable data (i.e., soft impressions management behavior) was collected at Time 2 with the time interval of 30 days. This study overcomes the common method bias by collecting the data at two points in time (Podsakoff et al., 2012). 420 useable structured questionnaires were received out of a total of 500 distributed questionnaires. 50 questionnaires were not returned and 30 were having incomplete responses. 84% response rate was achieved which consist of 37% females and 63% males. The mean value of respondents age was 26.4 years (SD=14.3).

To perform the CFA, structural equation modeling (SEM) software was used. Confirmatory factor analysis was used to attain the conditions of model fit indices, factor loadings, both convergent and discriminant validities and composite reliability of measures. SPSS v21 was used to analyze the data. For finding the association in between the variables, the Pearson Correlation test was applied and direct effects were analyzed through regression analysis. Hayes (2013) "PROCESS" method in SPSS was used to investigate the moderation effects. Interaction plots and slope test were also conducted as per suggestion of Aiken and West (1991). Questionnaire items' internal consistency was also measured by Cronbach alpha (α) value.

14.4.3 *MEASURES*

14.4.3.1 *MACHIAVELLIANISM*

To measure the Machiavellianism personality trait, Jones and Paulhus (2014) nine-items scale was used. Likert scale based on five options ranging from 1 to 5 (strongly disagree to strongly agree) was used for collecting the responses. The sample statement is "Do not involve in direct conflict with peers as they can be useful in future."

14.4.3.2 *ROLE AMBIGUITY*

Rizzo et al. (1970) six-items/questions scale was adapted to quantify the level of role ambiguity. Five-point Likert scale was undertaken for collecting the responses. The defined range was 1 = definitely not true to 5 = definitely true. All the items/questions were reverse scored; therefore, high values reflect a high level of role ambiguity and vice versa. The sample statement is "I know my responsibilities very well."

14.4.3.3 IMPRESSION MANAGEMENT

Bolino and Turnley (1999) 12-item-based scale was used to collect the responses. The scale is based on Jones and Pittman (1982) soft self-impression management taxonomy which includes four items of exemplification, four items of ingratiation, and four items of self-promotion. Impression management (IM) is taken as an overall construct on a theoretical basis in this study, which is associated with the prior study of Brouer et al. (2015). Responses are rated on a five-option Likert scale based on 1=not once behave this way to 5=constantly behave this way. The example questions are "Let other realize that you are very valuable to the organization" and "Reach at work early to show dedication."

14.5 RESULTS INTERPRETATION

Table 14.1 shows the standard deviation and mean values; Pearson correlation tells the direction and strength of the relationship in between variables (Cohen et al., 2014). Machiavellianism shows a positive association with impression management ($r = .25$, p value$<.01$); the results are according to the previous study of Hogue et al. (2013). IM is negatively related with gender ($r = -.09$, $p < .05$) and age ($r = -.10$, $p < .05$). Machiavellianism has a significant relationship with age ($r = .09$, $p < .05$). Role ambiguity has a significant positive relationship with Machiavellianism ($r = .16$, $p < .01$) and age ($r = .09$, $p < .05$), prior studies also support the positive relationship of role ambiguity and Machiavellianism (Hollen, 1983).

TABLE 14.1 Descriptive analysis, Pearson Correlation, and Internal Consistency.

	Mean	SD	Gender	Age	IM	Mach	RA
Gender	–	–	–				
Age	–	–	–.05	–			
IM (T2)	3.02	.69	–.09*	.10*	**(.88)**		
Mach(T1)	3.48	.82	–.07	.09*	.25**	**(.83)**	
RA (T1)	3.91	.76	.05	.09*	–.01	.16**	**(.85)**

Note: Total sample $N = 420$. *Shows the correlation significance at less than 0.05 level. **Reflects correlation significance at less than 0.01 level. IM(T2), impression management time 2; Mach (T1), Machiavellianism time 1; RA(T1), role ambiguity time 1.

14.5.1 CONFIRMATORY FACTOR ANALYSIS (CFA) OF COMPLETE MODEL

CFA test was applied to analyze the interdependence of all the variables in the theoretical model. Relationship in between the variables and all its latent constructs along with the p value or significance value has been achieved by the CFA test. Maximum Likelihood method was performed in CFA for analyzing the factor loadings and validities of the undertaken factors. = 2.15 (threshold range is in between 1 and 5), RMSEA (absolute fit measures) = .05 and SRMR = .05 (range should be less than .09), parsimony model fit indices AGFI = .87 (cut criteria is greater than .08), CFI (incremental fit index) = .92 that meets the adequate criteria of greater than .90 (Gaskin and Lim, 2016). The results are shown in Table 14.2.

TABLE 14.2 Model Fit Indices of Complete Model.

Complete model CFA	Cmin	Df	Cmin/df	CFI	SRMR	RMSEA	AGFI
	631.6	293	2.15	.92	.05	.05	.87

AVE is the measure of convergent validity against the acceptable value of greater than .5 and discriminant validity of the constructs is measured with MSV which should be less than AVE (Malhotra et al., 2012). Composite reliability (CR) is also analyzed as recommended by Hair et al. (2010). The factor loadings of Machiavellianism items range from .53 to .75, item loadings of role ambiguity range from .68 to .77, and item loadings of impression management taken as single latent factor range from .65 to .79. The validity (MSV and AVE), (CR) composite reliability, and internal consistency (Cronbach alpha) values are given in Table 14.3.

TABLE 14.3 Reliabilities, Construct Validities, and Factor Analysis.

Constructs	Alpha (α)	Ave	Msv	CR	Factor loadings
Mach (T1)	.82	.51	.21	.86	.53–.75
IM (T2)	.88	.56	.20	.86	.65–.79
RA (T1)	.85	.50	.11	.85	.68–.77

Note: $N = 420$.

IM (T2), impression management time 2; Mach (T1), Machiavellianism time 1; RA (T1), role ambiguity time 1.

14.5.2 REGRESSION ANALYSIS

Hayes (2013) PROCESS method has been used to investigate the direct effects. Only significant demographic variables should be controlled as suggested by Becker (1998). Therefore, age has been controlled for impression management ($\beta = -.13$, $p = .00$). The proposed Hypothesis 1 depicts the positive significant association of Mach and IM behavior. The result shows the positive relationship of Machiavellianism with impression management ($\beta = .26$, $p = .00$). 26% variance is explained by the Machiavellianism personality trait in impression management which is the outcome variable. The stated results are similar to the previous studies (Corral and Calvete, 2000; Rauthmann, 2011). Hence H1 is supported as shown in Table 14.4.

TABLE 14.4 Regression Results of the Direct Effects.

	R	ΔR^2	B	S.E.	t	p
Step 1						
Age	.10	.00	–.13	.02	–2.80	.00
Step 2						
Mach → IM	.28	.07	.26	.05	5.66	.00

Note: $N = 420$, age is used as control (covariate). Standardized regression coefficients. IM, impression management; Mach, Machiavellianism; S.E., standard error.

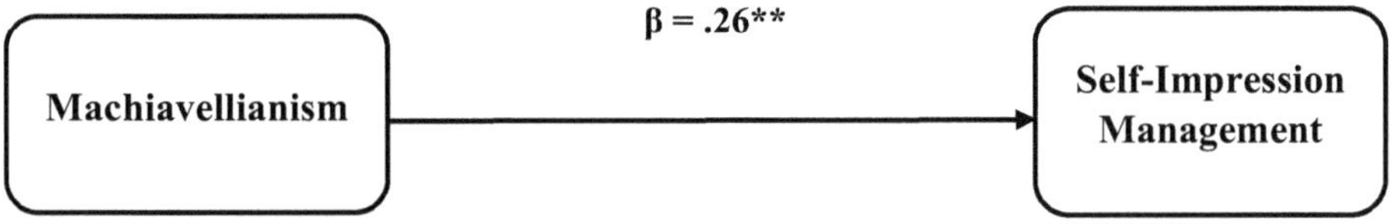

FIGURE 14.2 Direct effects.

14.5.3 MODERATION ANALYSIS

Hypothesis 2 predicted that the role ambiguity (RA) moderates the relation in between Mach personality and impression management (IM), the relation is stronger if the work roles are not strictly defined and vice versa. The latest Macros PROCESS technique has been used for moderation analysis. It reports the R^2 change along with the slope test values. The multiplicative effects of Mach and ambiguous roles are significant ($\beta = .27$, p value $< .01$, $\Delta R^2 = .02$, p value $< .01$) as reflected in Table 14.5. In Table 14.6 the slope test shows the significant slope at high ($\beta = .38$, p value $< .001$) and at the average ($\beta = .19$, p value $< .05$) value of role ambiguity and insignificant at low ($\beta = .00$, $p > .05$) value of role ambiguity (RA).

TABLE 14.5 Moderation Analysis (95% Confidence Level).

	β value	Boot std error	LLCI	ULCI
Step 1				
Mach	0.19*	.09	.00	.38
RA	−0.02	.08	−.19	.13
Step 2				
Mach × RA	0.27**	.11	.05	.49
ΔR^2 (interaction term)		**.0222****		

*p < .05. **p < .01. *** p < .001. Boot std error, bootstrap standard error; LLCI, lower-level confidence interval; ULCI, upper-level confidence level.

TABLE 14.6 Slope Test.

Moderator (role ambiguity)	Effect	Boot std error	LCI	UCI
−.75	.00	.13	−.27	.25
.00	.19*	.09	.00	.38
+.75	.38***	.11	.15	.61

*p < .05. **p < .01. ***p < .001. Boot std error, bootstrap standard error; LLCI, 95% lower confidence interval; ULCI, 95% upper confidence interval.

The interaction plots presented in Figure 14.3 show that the significant positive interactive effects on impression management are stronger when roles are ambiguously defined as compared to lack of ambiguity which proved insignificant in this study. Hence H2 is fully supported.

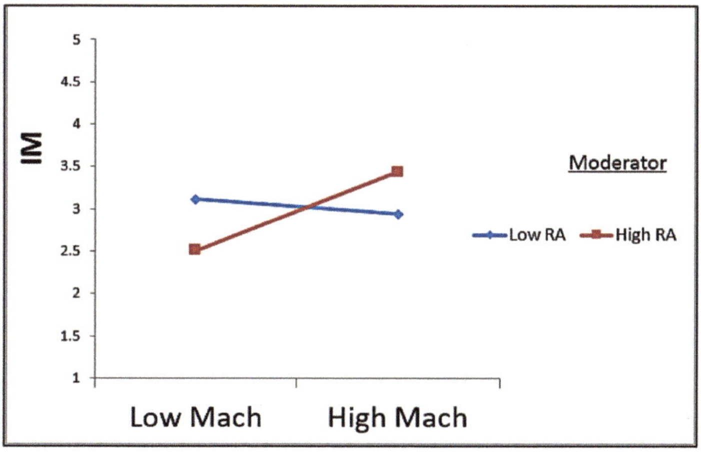

FIGURE14.3 Interactive plots of Mach and RA (role ambiguity) on impression management (IM).

14.6 CONCLUSION

This study has investigated the strategic side of agile Machiavellian personality and its association with the positive impression management behavior that includes the attributes of exemplification, ingratiation, and self-promotion in the presence of a job-based situational mechanism under the Trait Activation Theory. The proposed model proved that even though the Machiavellian personality is quoted as a dark personality in literature, but the demands of the social context are helpful in suppressing their negative predispositions for achieving positive impressions. The interactive effects of ambiguous job roles in the relationship of Machiavellianism personality and impression management have been proved significant. When the role ambiguity is high (means roles are not strictly defined) within the jobs that require frequent social interaction, the Machiavellianism and IM relation would be stronger. Therefore, role ambiguity is proved to be a significant situational mechanism for activating Machiavellian traits to carry positive self-impressions under the practical premise of Trait activation theory.

14.7 DISCUSSION AND RECOMMENDATIONS

Machiavellian personality utilizes self-impression management quite often because they perceive such in-role behaviors as a social mechanism, which is useful for triggering their achievement-oriented attributes to constructive outcomes. Rauthmann (2011) studied that Machiavellianism is mainly associated with the positive presentation of self, such social orientation helps them to fit with the demands of the agile organizational environment and to evade criticism. Mach's are able to manage their soft impression management taxonomies in different interpersonal/social situations for effective accomplishment of their objectives. Moreover, according to Sherry et al. (2006) Machiavellianism personalities avoid to disappoint others that is why they develop their aptitude to exhibit socially acceptable self-promotable behaviors (such as positive self-impression management).

According to CEO of Lean Kanban: ***"The core of Agile is recognizing that we need to get to and maintain an Agile personality and positive impressions. If I have an organization with an Agile personality and really rock-solid Self-impressions of employees, Agile processes will evolve out of that. If you have the Agile Personality with positive self-impressions and an awesome connection with your customers and are solving their problems, things will evolve in the right way. You won't even realize you're being Agile. It's just good business."***

Mach's count on others for fulfilling their self-seeking interests and in doing so they must retain the sociability in their relationship. Certainly, a dark trait of Machiavellianism is related to the projection of an appealing self-impression that others hold of them for building strategic alliances (Jonason et al., 2012; Rauthmann, 2012). The persons who possess the attributes of Mach's can manipulate the social situations to fulfil their self-serving desires (Paulhus and Williams, 2002). In line with the Paulhus and Williams argument, this study proves that Machiavellian personality does make use of their inter-individual behavioral propensity to control their encounters with others in different social situations by managing their soft impressions. The causal relationship between Mach and in-role IM behavior is significantly vindicated through role ambiguity (i.e., job relevant situational mechanism) under the disguise of TAT (trait activation theory). To succeed in being agile, employees carrying positive self-impressions need to extend and excel their competencies and accompanying behaviors that enable them to adjust and navigate the unknown situations in the business world.

14.7.1 THEORETICAL INFERENCES

The undertaken study contributes in the personality and behavioral literature by hypothetically proving the positive relation in between Machiavellianism personality and soft side of impressions management behavior. The findings are in-line with the prior study of Hogue et al. (2013). Second, the results reveal that the situational mechanism of role ambiguity activates the socially acceptable attributes of Machiavellianism which leads to positive association with soft impression management behavior. The role ambiguity does prove itself a significant situational trigger to activate the responsible side of Machiavellianism (when they set extraordinary job roles) which helps them to manage the socially acceptable impressions. The study proved the theoretical significance of Trait Activation Theory as a triggering force to activate the functional attributes of Machiavellianism in the collectivist culture. In a collectivist culture, the social context demands that individuals behave in a presentable way by displaying positive self-impressions, especially when their job roles are loosely defined.

14.7.2 PRACTICAL IMPLICATIONS

The emphasis of the undertaken study is on the strategic and positive side of Mach personality by investigating the situational mechanisms that can prevent the Machiavellian personality from practicing their negative instincts in the service-oriented corporations where such personality trait is directly involved in frequent social interactions. In a collectivist culture such as ours, to carry positive self-impression is the social and occupational demand of today's agile business world, which is the activating force behind Machiavellian personality to behave in their best profile for achieving occupational-focused objectives. The employees of service-based firms indulge in frequent interactions with others so they are likely to set their own roles. As Machiavellian personalities are very ambitious so they go for challenging roles, which can only be achieved if they will carry the impression of being a capable, accomplished, achievement-oriented and as a likable employee. This study gives a good insight into the top management for channelizing the strategic side of agile personality, that is, Machiavellianism by using the situational mechanisms as a driving force to accomplish the business targets through them.

14.7.3 LIMITATIONS AND FUTURE DIRECTIONS

The study limitations include the self-reported responses of Machiavellian personality. The peer reports of dark personality might give a different insight of such traits and can minimize the self-report bias. Since the Machiavellian personality has some of the attributes similar to a psychopath, so there is also a need to investigate the display of psychopathic personality in-role behaviors and to explore the other work and environment-related situational mechanisms that may activate the positive aspects of such dark personalities. In the future, the impression management may be investigated as a dimensional construct with respect to dark personality traits instead of addressing it as an overall construct.

KEYWORDS

- **agile Machiavellianism personality**
- **self-impression management**
- **role ambiguity**
- **trait activation theory**
- **service organizations**
- **strategic agile leadership**

REFERENCES

Abdel-Halim, A. A. Employee Effective Responses to Organizational Stress: Moderating Effects of Job Characteristics. *Pers. Psychol.* **1978,** *31*(3), 561–579.

Aiken, L. S.; West, S. G.; Reno, R. R. *Multiple Regression: Testing and Interpreting Interactions*; Sage: USA, 1991.

Beaty Jr, J. C.; Cleveland, J. N.; Murphy, K. R. The Relation Between Personality and Contextual Performance in "Strong" Versus "Weak" Situations. *Hum. Perform.* **2001,** *14*(2), 125–148.

Becker, J. A.; Dan O'Hair, H. Machiavellians' Motives in Organizational Citizenship Behavior. *J. Appl. Commun. Res.* **2007,** *35*(3), 246–267.

Becker, T. E. Integrity in Organizations: Beyond Honesty and Conscientiousness. *Acad. Manage. Rev.* **1998,** *23*(1), 154–161.

Bolino, M. C.; Turnley, W. H. Measuring Impression Management in Organizations: A Scale Development Based on the Jones and Pittman Taxonomy. *Organ. Res. Methods* **1999,** *2*(2), 187–206.

Bragozzi, R. P.; Verbeke, W. J.; Diet Vorst, R. C.; Belschak, F. D.; van den Berg, W. E.; Reddick, W. J. Theory of Mind and Empathic Explanations of Machiavellianism: A Neuroscience Perspective. *J. Manage.* **2013,** *39*(7), 1760–1798.

Christie, R.; Geis, F. Scale Construction. In *Studies in Machiavellianism*; Academic Press: USA, 1970; pp 10–34.

Cohen, P.; West, S. G.; Aiken, L. S. *Applied Multiple Regression/Correlation Analysis for the Behavioral Sciences*; Psychology Press, 2014.

Corral, S.; Calvete, E. Machiavellianism: Dimensionality of the Mach IV and its Relation to Self-Monitoring in a Spanish Sample. *Spanish J. Psychol.* **2000,** *3*, 3–13.

De Hoogh, A. H.; Den Hartog, D. N.; Koopman, P. L. Linking the Big Five-Factors of Personality to Charismatic and Transactional Leadership; Perceived Dynamic Work Environment as a Moderator. *J. Organ. Behav.* **2005,** *26*(7), 839–865.

Farh, C. I.; Seo, M. G.; Tesluk, P. E. Emotional Intelligence, Teamwork Effectiveness, and Job Performance: The Moderating Role of Job Context. *J. Appl. Psychol.* **2012,** *97*(4), 890–900.

Farmer, S. M.; Maslyn, J. M.; Fedor, D. B.; Goodman, J. S. Putting upward influence strategies in context. *J. Org. Behav.* **1997,** *18*(1), 17–42.

Gaskin, J.; Lim, J. Model Fit Measures. *Gaskination's StatWiki*, 2016.

Genau, H. A.; Blickle, G.; Schütte, N.; Meurs, J. A. *J. Pers. Psychol.* 2021.

Goffman, E. (1959). The Presentation of Self in. *Butler, Bodies that Matter*.

Gu, G. X.; Takaffoli, M.; Buehler, M. J. Hierarchically Enhanced Impact Resistance of Bio-inspired Composites. *Adv. Mater.* **2017,** *29*(28), 1–7.

Hair Jr, J. F.; Black, W. C.; Babin, B. J.; Anderson, R. E.; Tatham, R. L. SEM: An Introduction. In *Multivariate Data Analysis: A Global Perspective*; Pearson Education: UK, 2010; pp 629–686.

Hayes, A. F. In *Introduction to Mediation, Moderation, and Conditional Process Analysis: A Regression-Based Approach*; Guilford Press: New York, 2013.

Hilbig, B. E.; Zettler, I. When the Cat's Away, Some Mice Will Play: A Basic Trait Account of Dishonest Behavior. *J. Res. Pers.* **2015,** *57*, 72–88.

Hogue, M.; Levashina, J.; Hang, H. Will I Fake it? The Interplay of Gender, Machiavellianism, and Self-Monitoring on Strategies for Honesty in Job Interviews. *J. Bus. Ethics* **2013,** *117*(2), 399–411.

Hollon, C. J. Machiavellianism and Managerial Work Attitudes and Perceptions. *Psychol. Rep.* **1983,** *52*(2), 432–434.

Jonason, P. K.; Schmitt, D. P. What Have You Done for me Lately? Friendship-Selection in the Shadow of the Dark Triad Traits. *Evol. Psychol.* **2012,** *10*(3), 400–421.

Jonason, P. K.; Slomski, S.; Partyka, J. The Dark Triad at Work: How Toxic Employees Get Their Way. *Pers. Individ. Differ.* **2012,** *52*, 449–453.

Jonason, P. K.; Webster, G. D. A Protean Approach to Social Influence: Dark Triad Personalities and Social Influence Tactics. *Pers. Individ. Differ.* **2012,** *52*(4), 521–526.

Jones, D. N.; Mueller, S. M. Is Machiavellianism Dead or Dormant? The Perils of Researching a Secretive Construct. *J. Bus. Ethics* **2021,** 1–15.

Jones, D. N.; Paulhus, D. L. Introducing the Short Dark Triad (SD3) A Brief Measure of Dark Personality Traits. *Assessment* **2014,** *21*(1), 28–41.

Jones, D.; Paulhus, D. Machiavellianism. In *Handbook of Individual Differences in Social Behavior*; Horowitz, L., and Strack, S., Eds.; Guilford Press: New York, NY, 2009; pp 93–108.

Jones, E. E.; Pittman, T. S. Toward a General Theory of Strategic Self-Presentation. *Psychol. Perspect. Self* **1982,** *1*(1), 231–262.

Kumar, M.; Talib, S. A.; Ramayah, T. In *Business Research Methods*; Oxford Fajar/Oxford University Press: Oxford; 2013.

Leary, M. R. In *Self-presentation: Impression Management and Interpersonal Behavior*; Brown and Benchmark Publishers, 1995.

Leary, M. R.; Kowalski, R. M. Impression Management: A Literature Review and a Two-Component Model. *Psychol. Bull.* **1990,** *107*(1), 34–47.

Leonelli, S.; Di Pietro, F.; Masciarelli, F. Narcissism, Machiavellianism and Psychopathy: How Do Displayed Entrepreneurs' Personality Dark Traits Influence Crowdfunding Success? In *The Entrepreneurial Behavior: Unveiling the Cognitive and Emotional Aspect of Entrepreneurship*; Emerald Publishing Limited: Bingley; 2020.

Lievens, F.; Chasteen, C. S.; Day, E. A.; Christiansen, N. D. Large-Scale Investigation of the Role of Trait Activation Theory for the Understanding Assessment Center Convergent and Discriminant Validity. *J. Appl. Psychol.* **2006,** *91*(2), 247–258.

Malhotra, N. K.; Mukhopadhyay, S.; Liu, X.; Dash, S. One, Few or Many? An Integrated Framework for Identifying the Items in Measurement Scales. *Int. J. Market Res.* **2012,** *54*(6), 835–862.

Mischel, W.; Shoda, Y. A Cognitive-Affective System Theory of Personality: Reconceptualizing Situations, Dispositions, Dynamics, and Invariance in Personality Structure. *Psychol. Rev.* **1995,** *102*(2), 246.

Morgeson, F. P.; Delaney-Klinger, K.; Hemingway, M. A. The Importance of Job Autonomy, Cognitive Ability, and Job-Related Skill for Predicting Role Breadth and Job Performance. *J. Appl. Psychol.* **2005,** *90*(2), 399–406.

O'Boyle Jr, E. H.; Forsyth, D. R.; Banks, G. C.; McDaniel, M. A. A Meta-Analysis of the Dark Triad and Work Behavior: A Social Exchange Perspective. *J. Appl. Psychol.* **2012,** *97*, 557.

O'connor, W. E.; Morrison, T. G. A Comparison of Situational and Dispositional Predictors of Perceptions of Organizational Politics. *J. Psychol.* **2001,** *135*(3), 301–312.

Paulhus, D. L.; Williams, K. M. The Dark Triad of Personality: Narcissism, Machiavellianism, and Psychopathy. *J. Res. Person.* **2002,** *36*(6), 556–563.

Penke, L.; Denissen, J. J.; Miller, G. F. The Evolutionary Genetics of Personality. *Eur. J. Person.* **2007,** *21*(5), 549–587.

Podsakoff, P. M.; MacKenzie, S. B.; Podsakoff, N. P. Sources of Method Bias in Social Science Research and Recommendations on How to Control it. *Ann. Rev. Psychol.* **2012,** *63*, 539–569.

Rauthmann, J. F. Acquisitive or Protective Self-Presentation of Dark Personalities? Associations Among the Dark Triad and Self-Monitoring. *Pers. Individ. Differ.* **2011,** *51*(4), 502–508.

Rauthmann, J. F. The Dark Triad and Interpersonal Perception: Similarities and Differences in the Social Consequences of Narcissism, Machiavellianism, and Psychopathy. *Soc. Psychol. Personal. Sci.* **2012,** *3*(4), 487–496.

Rizzo, J. R.; House, R. J.; Lirtzman, S. I. Role Conflict and Ambiguity in Complex Organizations. *Adm. Sci. Q* **1970,** 150–163.

Sadler, M. E.; Hunger, J. M.; Miller, C. J. Personality and Impression Management: Mapping the Multidimensional Personality Questionnaire onto 12 Self-Presentation Tactics. *Pers. Individ. Differ.* **2010,** *48*(5), 623–628.

Schlenker, B. R. *Impression management*; Brooks/Cole Publishing Company: Monterey, CA.

Sherry, S. B.; Hewitt, P. L.; Besser, A.; Flett, G. L.; Klein, C. Machiavellianism, Trait Perfectionism, and Perfectionistic Self-Presentation. *Pers. Individ. Differ.* **2006,** *40*(4), 829–839.

Tett, R. P.; Burnett, D. D. A Personality Trait-Based Interactionist Model of Job Performance. *J. Appl. Psychol.* **2003,** *88*(3), 500.

Tett, R. P.; Guterman, H. A. Situation Trait Relevance, Trait Expression, and Cross-Situational Consistency: Testing a Principle of Trait Activation. *J. Res. Pers.* **2000,** *34*(4), 397–423.

Tett, R. P.; Toich, M. J.; Ozkum, S. B. Trait Activation Theory: A Review of the Literature and Applications to Five Lines of Personality Dynamics Research. *Ann. Rev. Organ. Psychol. Organ. Beh.* **2021,** *8*, 199–233.

Van der Linden, M. *Transnational Labor History: Explorations*; Routledge: UK, 2017.

Walters, R. L. The Predictive Power of Machiavellianism, Emotional Manipulation, Agreeableness, and Emotional Intelligence on Counterproductive Work Behaviors. Doctoral Dissertation, Wright State University, 2017.

Wayne, S. J.; Liden, R. C. Effects of Impression Management on Performance Ratings: A Longitudinal Study. *Acad. Manag. J.* **1995,** *38*(1), 232–260.

Wilson, D. S.; Near, D.; Miller, R. R. Machiavellianism: A Synthesis of Evolutionary and Psychological Literature. *Psychol. Bull.* **1996,** *119*(2), 285–299.

CHAPTER 15

Amul: An Agile Brand and Its Leadership: "Road to Self-reliance"

SMITA SHUKLA, JAWED KALANGADE, AND RADHA IYER*

Alkesh Dinesh Mody Institute for Financial and Management Studies, University of Mumbai, Mumbai, India.

**Corresponding author. E-mail: radha.mohan258@gmail.com*

ABSTRACT

The world today is seeing an era of adopt or perish. That is what marks the businesses to be agile. The present chapter focuses on business agility and the practices that Amul practices toward agility. In the era of industry 4.0 businesses are striding toward much more agility and that is evident in business practices around the world. Amul is one such company. The chapter discusses how the various practices of Amul make it more agile and help it to run in the business with excellence. The chapter relies on the interviews taken from the top leaders of Amul. They state in their discussion, how Amul has been leading toward excellence by following the path of agility and adopting the best practices of Industry 4.0.

15.1 INTRODUCTION

In today's agile business world, it is very important to be very dynamic, open, and engage in continuous learning so that the company is able to grow and achieve its goals. The Agile Business approach believes in agile resources that are extensive and are available globally and that it strives to reach out

Agile Leadership for Industry 4.0: An Indispensable Approach for the Digital Era, Tanusree Chakraborty, Tahir Mumtaz Awan, Ashok Natarajan, and Muhammad Kamran (Eds.)

to its customer with a new approach and gives a new experience (Loss and Crave, 2011). Through digitalization as well, businesses are turning agile and adopting newer ways of doing business (Perkin and Abraham, 2021). Today, most of the organizations are racing to become agile; business agility has become a desirable trait for businesses. Though the positive outcomes of agility are extensively recognized by business executives, those considering a business-wide agile transformation are doubtful about the prospective, capabilities, and the outcomes that agility accompanies. Still on an overall understanding, business agility has been gaining momentum and that led to the business processes, practices rules, and even leadership undergo a change. To thrive in this agile business world, organizations need leaders who handle the workforce differently. However all businesses and organizations are not agile in nature. One such company that is known for its agility is Amul.

15.1.1 AMUL: THE BRAND

Amul as a brand has a unique way of dealing with its stakeholders. It takes everything differently, be it politics, entertainment, sports, Industry 4.0, COVID-19 pandemic, etc., turning each situation into a billboard to tell stories on their butter/jam/milk product capturing the attention of the customers. So, it always keeps thriving in the mind of customers and the business world, what is going to be the next move of Amul? We have all seen the image of joker character from the Batman movie into Amul's milk boy. This is an everlasting story in the mind of audience. This attitude of Amul states that it does not require big market budgets, huge campaigns, and what is required is genuine interest to interact with customers, tell them a story, and keep them engaged in a conversation even if common commodity like butter is being sold. Amul has used the strategy of storytelling to convey its messages and that is how we remember the stories of Amul so well and Amul products seem to be so very of our own. That is where Amul evokes nostalgia among customers, satisfaction among customers and also brings a sense of pride. Story telling for brand building is one of the best strategies to build long-lasting impressions in customer's minds (Kim et al., 2016; Lund et al., 2018; Peho, 2020).

15.1.2 BUSINESS AGILITY AND ITS IMPORTANCE

Hugos (2009) in his book, "Business Agility: Sustainable Prosperity in a Relentlessly Complex World," defined Business Agility as, "the ability of an organisation to:

a. Adapt quickly to market changes both internally and externally.
b. Respond rapidly and flexibly to market/customer demands.
c. Adapt and lead change in productive and cost-effective ways without compromising quality.
d. Adaptability, balance, and flexibility are the three characteristics that are essential to the long-term business agility.

When a business becomes agile, its value systems become differently noticed by the outside world. The value system of the business gets into the DNA of an organization through its people and their leaders. Since these changed values are generally nurtured by the founders of the organization. In a world where change is the only constant and the leadership position is temporary, only being agile creates competitive advantage for companies/businesses. In every sphere of business internet and information technology has brought about changes (Chakraborty and Nag, 2018, Chakraborty and Saha, 2018).

It is common knowledge that business intelligence is a crucial factor to the success of the business. IoT provides too much information that change faster than the information systems. Business agility gives companies an ability to deliver fast and respond quickly. It offers a great experience to organize quickly and learn rapidly.....business agility gives one of the tools to pivot and remain necessary to customers and changing market needs" (Smet, 2015). Craig et al. (2013) have identified factors influencing agility and he has grouped them into three categories and they are: (1) marketing, (2) organization, and (3) IT technologies. It has been found that Cloud Computing and BI are two important factors which influence the agility of the business. The research further states that the top three IT technologies that draw massive investment in 2015 are: BI systems and advanced analytics, cloud computing and mobile technologies. The study also shows that Gartner Group considered business intelligence as "the nexus of forces." and is expected to change the way the society looks at businesses and expected to create new and fresh business models. This platform is expected to either change or modify the way the businesses interact with their customers, their collaboration with employees and partners. It will also undergo a change, thereby improving business agility. Other factors that create an impressive difference in business agility are agile leadership, agile communication, learning mindset, flexibility, readiness to change, etc.

15.1.3 AGILE LEADER AND INDUSTRY 4.0

An agile leader is defined as, "a person who not only fosters agile ways of thinking and working but also creates an environment where teams can do their best work, and champions agile throughout their organisation" (Hayward, 2021). Agile Leaders center on major three things: (1) they make and look after customs in which there is scope of experimentation and learning, (2) they work together with teams, and (3) they make an organizational structure that strengthens and rewards the previous two behaviors. In current trend of automation and data exchange, Industry 4.0 is producing technologies. It consists of cyber-physical systems, the internet of things, and cloud computing. "Smart Factory" is created by Industry 4.0.

15.2 AMUL'S CASE

In the context of agile leadership, we bring to light the case of Amul. Amul has adopted a lot of practices that make the brand agile. In the following sections, we discuss certain practices of Amul that make it stand out as an agile business.

15.2.1 AMUL ONLINE APP

The most important achievement in Amul's journey is its online app. Morning walks to the nearest milk booths to buy milk will be past. This is due to the fact that dairy majors like Amul plan to reach its customers directly through mobile applications that the company wishes to launch. We all are aware of the fact that major dairies like Mother Dairy and Amul are marketing their products through e-grocers like Big Basket, Grofers, and Big Bazaar to attract the customers living in urban areas. This decision will impact consumer engagement with the brand, Amul, and will surely increase sales and make a name in the field of milk products for its agile operations. Why Amul made such a decision is to make all the products available to the customer which a customer may not get at the nearby store. The mobile application will help the customer to order a big assortment. The Managing Director of GCMMF, R S Sodhi, in an interview said that the mobile application will be first tested in Ahmedabad market and then will be launched across other markets (Goh, 2020, October 21).

15.2.2 LOCATE AMUL

Amul uses another App, Locate Amul, which connects sales executives with distributors and enables them to trace information of distribution at Amul parlours. There are eight thousand Amul parlours located throughout India. Locate Amul App is useful for internal sales executives and distributors only. However, it has the potential to be used for hyper-local delivery in the future. Sodhi (Goh, 2020, October 21) said that Amul will make Locate Amul for customers through which they can order the company's products and this is expected to happen in the next phase.

15.3 AMUL AND ITS INNOVATIVE PRODUCTS

Amul uses the digital push to launch new categories. The company has recently diversified its business by producing beverages, chocolates, sweets, and even bakery items in addition to its leading products like milk, cheese, and butter. Abneesh Roy, Executive Director of Edelweiss Financial Services states that "To survive in the crowded consumer market, Amul needs to constantly innovate." He further states that innovation by Amul can be witnessed through their innovative products such as AmulTru, Roti Softner, and Camel Milk. A combination of milk and fruit is AmulTru in a bottle for INR 10. Kishore Jhala, GCMMF Chief Operating Officer, states that the concept of producing Roti Softner originates from the traditional idea of adding milk to rotis so that it remains fresh and soft for longer time and thus can be packed for journeys, the practices of grandmothers at individual houses (Goh, 2020, October 21).

When consumers suspected that palmolein oil is used to produce biscuits, Amul decided to launch its own butter cookies in which the customers can actually taste the butter. Sodhi stated that the customers were ready to pay a dear price for the cookies as it tasted better (Goh, 2020, October 21). Amul also started producing fresh bread for the same reason as butter cookies. For the production of butter cookies, Amul has decided to set up plant one each in east, south, and north though it does not wish to go as a big national player. Amul is ready to produce any product right from sweets to fresh bread to butter cookies as milk is the core product of all these.

15.3.1 AMUL'S CHOCOLATES

In 1973, Amul launched its chocolates. The farmers of Cocoa from Kerala and Karnataka approached Dr. Verghese Kurien for fair prices. That was cue to Cadbury whose market share was 66%. To this effect Sodhi added that he fixed in on dark chocolate (Goh, G., 2020, October 21). Amul claims to be the leader with 65% share. The examples provided show how agile is Amul as a brand and how its leaders have contributed to the same.

15.3.2 AMUL AND INFIBEAM

Amul engaged Infibeam to develop the application. Infibeam is an Ahmedabad-based firm. Infibeam creates the last-mile connectivity. If a customer places an order through this App, Infibeam arranges for the product, meaning it picks up the product from the distributor and delivers the same directly to the customer. Sodhi further stated that Amul's share of online sales is less than 1% of its turnover (Goh, 2020, October 21). The company has not tasted success with modern trade as it contributes approximately 7% of its overall sales. Senior official (does not wish to be disclosed) in the cooperation pointed out that in current times several major retails that were not profitable have closed down several outlets. Therefore, modern retail has seen ups and downs. In nineties Amul had employed different modes of reaching out to customers. Amul App that is available in the play store can be downloaded. This is Amul Online App. This will deliver available hygienic Amul products quickly, meaning within stipulated time period making Dr. Verghese Kurien's vision true.

15.3.3 UNPARALLELED SLOT SYSTEM AND SPEEDY DELIVERY

Amul Online App has unparalleled and remarkable idea to select an appropriate time slot to place the order by the customers. This solves the problem of delay. This App consists of same-day delivery option permitting the customer to select the products that he/she needs immediately. Delivery of the products on the same day within a stipulated time as desired by the customers will be of great help to the customers. As such, it is to be noted that whoever joins the Amul team immediately connects with Amul's values. For instance, integrity was something that Dr. Kurien used to emphasize upon. Now, at Amul all employees are known as a person of integrity. The same is true to other important values like "excellence" or "efficiency."

15.3.4 EASY PAYMENT OPTION AND RETURN POLICY

Amul Online App has several payment options including conventional and unconventional methods for the customers. Amul Online in addition to cash on delivery option offers other options such debit card/credit card and Net Banking, RTGS, etc., understanding the comfort of the customers. Customers can order all products as there are no delivery charges. Customers incur only minimal costs. Amul accepts payment through debit card and net banking from all the leading banks. However payment through credit card is possible if credit cards are Visa, Master, or Amex cards. The sensitive details of the customers are kept confidential as the online transactions are processed through secured and trusted payment gateway. The products that are sold on Amul Online are checked and verified before delivery. Amul Online also provides easy return policy but under certain conditions. The complaints made by customers are immediately solved if registered.

15.3.5 IBM AND AMUL

Another agile movement of Amul is the signing of an MoU with IBM having taken consent from both the parties. As per the MoU, IBM would operate and manage Information Technology sector of GCMMF for a given period of time, i.e., 10 years. It further stated that IBM's foremost setup would be a technology medium based on Enterprise Resource Planning (ERP) along with a data center supported by a Disaster Recovery Centre. The project would start by the implementation of ERP application and other essential systems which would increase Amul's operational efficiency. Mr. Sandip Patel, MD of Global Business Services in IBM India and South Asia, stated that this enterprise will aid Amul in strengthening customer focus, improving supply chain, thereby achieving stronger growth (Babu, 2020, September 25).

The General Manager, Lingaraju Sawkar of Global Technology Services of IBM India/South Asia, said in an interview that IBM had created a private cloud computing infrastructure with a disaster recovery infrastructure which reinforces spontaneously for GCMMF (Chandrasekaran and Baruah, 2019, December 8). He further added that Amul should fill Disaster Recovery (DR) on pre-programmed basis. The systems run seamlessly between production and DR. To carry out the smooth functioning, IBM said that it will monitor and manage the processes 24/7 using remote technology which are available at their command centers. Amul has approved a deal with IBM and this deal will transform the Information Technology (IT) environment of Amul and the milk unions. Further, it will support their growth plans. Amul is also incorporating new technological mediums created by IBM with farmers and retailers. GCMMF Ltd. Is amalgamating all distribution retails: every shop down to every item that is being sold. This 10-year outsourcing deal had cost INR 80 crores for Amul. The deal also emphasizes that IBM would manage and operate Information Technology environment for the next 10 years through SAP and ERP on IBM technology platform. The strategy between Amul and IBM will involve five business transformation initiatives laying ground for growth plans. The initiatives for Information Technology were identified and designed. This initiative will enhance the clarity and translucency by creating a value chain and as a result improved operational efficiency. Four solutions were chosen: SAP for Information Technology infrastructure and applications; Enterprise Resource Planning (ERP) solution and distributor management were chosen for a Village Co-operative Society; business analytics solution was chosen for progressive planning capabilities; information technology infrastructure to provide an integrated platform for better clarity for the entire distribution chain. The technology medium would assure better BCDR (Business Continuity and Disaster Recovery) solutions. The critical objects that were identified include scalability and flexibility. A plan was customised and designed. This comprehensive plan would upgrade to reliable, scalable, and flexible transaction processing systems supporting growth. GCMMF began its ERP infrastructure journey on IBM Power Systems 10 years back. Now, it has been metamorphosized to POWER9 servers supporting growth vision and expanding operations across different cooperative societies. Thus, IBM built the state-of-the-art technology medium supported by disaster recovery center. This medium would not only describe the policy for security but also maintain and manage the entire technology and application.

15.3.6 AWAZSETU – A SPOKEN WEB

Amul-Awaz-Setu – a spoken web service was launched by Amul on January 1, 2012. Amul is the first diary to come up with such a service. This service would connect hundred thousands of farmers living in approximately 1000 villages. This spoken web is designed by IBM. 370 farmers already use this.

14.4 HOW AMUL REACTED TO MARKET DEMAND EARLIER

Digital was not new to Amul. Around 1998 to 1999 Amul reached out to its customers through electronic mails. A customer would send an electronic mail ordering for products (a minimum value of Rs. 200 or so). The mails were exchanged with other branches of Amul which would then organize delivery of the products via its distributor. This did not see much traction as the internet at that time was slow.

15.4.1 HOW AMUL SURVIVED COVID-19 PANDEMIC

On March 24, 2020 Indian Government announced lockdown for the entire county due to Coronavirus (COVID-19) Pandemic. GCMMF Ltd., one of the largest FMCG Companies in India, was in a major fix. During this period the company witnessed an increase in the demand for dairy products and a decrease for frozen products. The dairy's supply chain was under pressure. The delivery of the frozen products lying idle, the dairy giant optimized the usage of resources and infrastructure through the establishment of the frozen one for delivering dairy products overnight.

Sodhi mentions that it was very critical for Amul because it is a vital service (Goh, 2020, October 21). On March 17, 2020 the company thoroughly examined various facets of their performances including but not limited to precautions, social distancing, sanitisation, invoicing, warehousing. It found out that a robust Information Technology platform is the backbone to carry out their functions smoothly. Expecting disorder lines the dairy giant piled up the products in their seventy-seven warehouses. It transported greater than what it had. During lockdown but through their IT infrastructure, Amul was able to track what was happening at every point right from their supply chain of 36 lakhs farmers, eighteen thousand seven hundred (18,700) societies, five thousand (5000) milk tankers going to two hundred (200) chilling stations, making 750 Stock Keeping Units (SKU) then going to sixty-two (62) branches and ten thousand (10,000) distributors and ten lakhs (one million) retailers. The IT infrastructure enabled the company to identify where milk tankers were terminated, from the village where milk was uncollected and locations where milk vans could not go, etc., RS Sodhi added that technology was very critical for them.

Goh, October 21, 2020 states that according to Sodhi food is a vital source for a healthy living. There were benefits of being multiple products, multiple locations, and multiple channel company. With approximately 84 plants in Mumbai, only 4–5 plants were affected by the COVID-19 pandemic while other plants worked overtime.

Amul says that it realized many things during the pandemic and the important ones are stated below:

- People buy more and more branded products.
- Relationships with customers are important.
- Bonding with customers strengthened.

During COVID-19 pandemic, Sodhi said that demand for food did not decline instead saw a sharp increase (Goh, 2020, October 21). Amul decided not to lag in production, supply, and transport of products. It procured 17% to 18% (approx. 40–50 lakh liters) more milk from Gujarat at an additional cost of Rs. 18,000 crores.

GCMMF Ltd. renewed its contract with IBM for another 10 years. The purpose is to consolidate all its members on a single business activity (enterprise) platform and to authorize visibility of its business logistics management, all while supporting commercial development. The first MoU was signed in 2009. The business transformation was covered through the infrastructure and application management services that were implemented in the form of Enterprise Resource Planning and fourteen member unions. Through technology planning and strategy study IBM developed Information Technology strategy roadmap for the dairy giant.

Amul's ERP solutions solved one of the biggest problems in the supply chain solutions, stated Kamal Singhani, the Managing Partner of Global Business Services for IBM India and South Asia. This was one of the objectives of GCMMF Ltd. He further claimed that it not only takes care of financial balancing but also is in charge of the stock supply for the business activities. The technology medium was created around the planning environment ensuring that Amul sustains and maintains the pace. In addition to the above ones, GCMMF Ltd. has established other sophisticated Information Technology solutions such as purchase of milk, testing of milk, management of warehouse, and even Points of Sales solutions.

15.5 AMUL'S SUPPLY CHAIN

Amul's supply chain is not only complex but also multilayered and multidimensional. The supply chain does not start from a plant or subsidiary supplier. It starts from the origin production of milk from the cattle and the suppliers of milk. It needs a superior agile, highly responsive, and high performance platform. This platform would be able to give commanding position and actual time clarity to their complex 24-h supply chain management. BCDR (Business Continuity and Disaster Recovery) solution for the products provided by Amul which are highly perishable was a must. The solution had to get the processes better and quick with minimum interlude. The channels of distribution should always be operative for optimum benefits of the customers. Managing Director, B M Vyas of GCMMF Ltd., states that Amul has crafted IT architecture from the base, that is, Village Dairy Cooperative Societies by establishing technological mediums such as AMCS (Automatic Milk Collection Systems). The company has also installed various innovative business processes distinctive to India and illustrious to the globe as "Amul's Pattern."

1. He commented that functioning together of SAP, Amul, and IBM will nurture the Information Technology processes. Through this the entire dairy sector of India will benefit innovative technology mediums.
2. The project will start by administering SAP ERP (Enterprise Resource Planning) application in addition to the essential modules and systems that will enable Amul to increase its operational efficiencies.
3. As a result, the dairy giant will improve its logistics management, solidify customer focus, and attain multifold growth which is the objective of GCMMF Ltd.

15.6 RESULTS

- GCMMF Ltd. has accomplished decennial growth through the adoption of IBM data centre.
- This has helped GCMMF improve data accuracy of Amul products. Further, it has helped the company to achieve a seamless integration with the trading system.
- IBM's ability to think has given uninterrupted mobile-based applications and automation abidance for managing applications for the dairy giant.
- The dairy giant has gained a visibility in the logistics and inventory process of daily supply of milk and milk products though this adoption.

Kamal Singhani explained the following on enquiry about the renewal of the contract with IBM:

a. The contract should help Amul achieve smart logistics management that is more adaptable and invigorating.
b. The contract should also help to introduce lots of Artificial Intelligence and Machine Learning in their logistics management which will save costs and enhance the potential to achieve growth in multi-dimensions in future years.

15.7 FUTURE PLANS OF GCMMF LTD.

- GCMMF Ltd. expects 15% to 20% increase in turnover.
- Amul is planning to introduce new products like Haldi (turmeric) milk, tulsi (herb) milk, and adarak (ginger) milk as they are perceived as immunity boosters.
- Amul has decided to spend double on its advertisements.
- It plans to acquire Punjab Dairy for INR 21.2 crores.
- Amul has three predictions:
  - People will understand the importance of relationship, communication, and speed.
  - Traveling will reduce and people will move toward online purchasing.
  - Marketing companies will keep exclusive media budget.
- Amul plans to go for IPO soon.

15.8 CONCLUSION

It has been seen that as a case Amul has always been agile and adopted Industry 4.0. It also prepared itself to face and survive the pandemic by taking bold steps such as fast decision-making. Amul always survived and sustained due to its continuous and mindful efforts. Collaboration with Infibeam and IBM on Business Intelligence System and setting up of spoken web that is designed by IBM will surely help Amul to get connected with millions of farmers easily. Spoken Web that was launched in January 2012 will be fully operational and is expected to help one lakh members who will get connected directly to the dairy union. The steps that Amul has taken so far stand out as living examples of agility in business.

KEYWORDS

- **Industry 4.0**
- **agility**
- **agile leadership**
- **contract between Amul and IBM**

REFERENCES

Babu, A. The World Has Entered into an Extreme Digitisation MODE: IBM India MD. *The Week* [Online], Sept 25, 2020. https://www.theweek.in/news/biz-tech/2020/09/25/the-world-has-entered-into-an-extreme-digitisation-mode-ibm-india-md.html

Chakraborty, T.; Nag, K. Sports Entrepreneurship: Ontology Based Review Through Digital Marketing. *Global J. Enterprise Inf. Syst.* **2018,** *10*(3), 57–72.

Chakraborty, T.; Saha, R. Digital Innovations: Breakthrough Opportunities to Build Novel Business Models. *Bus. Perspect. Aims Scope*, **2018,** *1*, 46.

Chandrasekaran, A.; Baruah, A. IBM India Says Bullish on Tech Ecosystem. *Mint* [Online], Dec 8, 2019. https://www.livemint.com/industry/infotech/ibm-india-says-bullish-on-tech-ecosystem-11575829295847.html

Craig le Clair, B. J.; Cullen, A.; Mines, C.; Keenan, J. *The 10 Dimensions of Business Agility. Enabling Bottom-Up Decisions in a World of Rapid Change*; Forrester: USA, 2013.

Gartner Executive Programs' Worldwide Survey, Business Intelligence, Mobile and Cloud Top the Technology Priority List for CIOs. [Online], Dec, 2014 http://www.gartner.com/newsroom/id/1897514, 2010-2015

Global Dairy Trade. [Online], 2018. Amul. https://www.globaldairytrade.info/en/gdt-events/gdt-events-sellers-andproducts/amul/amul.

Goh, G. Amul India's Dr. R S Sodhi: Once We Take a Position, We Keep to It: Warc. [Online], (2020, October 21). WARC AnAscential Company. https://www.warc.com/newsandopinion/opinion/amul-indias-dr-r-s-sodhi-once-we-take-a-position-we-keep-to-it

Hayward, S. *The Agile Leader: How to Create An Agile Business in the Digital Age*; Kogan Page Publishers: London, 2021.

Hugos, H. *Business Agility: Sustainable Prosperity in a Relentlessly Complex World*; Wiley Publishers: New York, 2009.

Kim, J. E.; Lloyd, S.; Cervellon, M. C. Narrative-Transportation Storylines in Luxury Brand Advertising: Motivating Consumer Engagement. *J. Bus. Res.* **2016,** *69*(1), 304–313.

Loss, L.; Crave, S. Agile Business Models: An Approach to Support Collaborative Networks. *Prod. Plan. Control* **2011,** *22*(5–6), 571–580.

Lund, N. F.; Cohen, S. A.; Scarles, C. The Power of Social Media Storytelling in Destination Branding. *J. Dest. Mark. Manage.* **2018,** *8*, 271–280.

Peho, C. Storytelling as a Means of Brand Building-Building Authenticity and Emotions through Brand Stories, 2020.

Perkin, N.; Abraham, P. *Building the Agile Business Through Digital Transformation*; Kogan Page Publishers: London, 2021.

Smet, A. D. *The Keys to Organizational Agility.* [Online], Feb 16, 2018. McKinsey & Company. https://www.mckinsey.com/business-functions/organization/our-insights/the-keys-to-organizational-agility.

CHAPTER 16

HR Challenges in the Post-Covid-19 World: Role of Agile Leadership towards Organizational Effectiveness

L. GANDHI

SDM Institute for Management Development (SDMIMD), Mysore, Karnataka, India
E-mail: gandhi@sdmimd.ac.in

ABSTRACT

The pandemic struck people like a bolt of lightning, destroying everything in its path. Organizational HR departments have been at the forefront of the global pandemic response, providing workers with as much data as possible. During the lockout, they also attempted to maintain staff morale and efficiency by setting new job objectives and offering environmental and mutual rewards. To comply with modern operating procedures, they must plan and make modifications to systems, frameworks, and systems. The formulation and implementation of these changes include all stakeholders in the organization, necessitating collaboration and communication. Their obligations, however, are far from straightforward, as the new standard of functioning demands them to answer a series of questions before they can begin preparing. The author of this paper attempted to present the HR issues that have occurred because of COVID 19. This article is divided into four sections: the introduction, section two, which gives the opinions of HR specialists, section three, which elaborates on HR difficulties post-COVID, and section four, which includes arguments and conclusions.

Agile Leadership for Industry 4.0: An Indispensable Approach for the Digital Era, Tanusree Chakraborty, Tahir Mumtaz Awan, Ashok Natarajan, and Muhammad Kamran (Eds.)

16.1 INTRODUCTION TO THE HR SCENARIO AND CHALLENGES IN INDUSTRY 4.0

Managing people is never out of the question in the ever-changing, dynamic corporate world. Because the changes and difficulties are huge and rapid, there is always a need for new equipment and approaches. Technological advancements are a boon to the creative use of numerous methodologies and aid in the production of better results. In a competitive market context, the best utilization of human capital is required to meet the organization's finishing line goal. The gap that organizations must navigate from 'Good' to 'Great' is largely determined by their grasp of organizational needs and their capacity to achieve long-term business success. With most workers needing to work remotely, HR teams are handling many issues, for example, giving a detailed attendance model. They must develop new guidelines, if needed. Working concepts such as part-time, full-time, and remote will demand modified employment criteria. New methods will be necessary to track the productivity of distant staff. It is probable that the performance management system will need to be entirely revamped, with new goals, metrics, and incentives. To allow employees to work remotely, revisions to policies and guidelines should include a strategy for resource availability and consumption.

16.2 THE VUCA WORLD AND CURRENT ORGANIZATIONS

We flourish in today's corporate world, which is defined by two acronyms: VUCA and SMAC. Employees have different communities, mindsets, and levels of trust in the value system. In essence, corporate organizations do not have complete control over all the factors that influence their operations. SMAC, which appropriately organizes the convergence of diverse technologies, drives the industry and its priorities. The acronym VUCA was coined by the Army College of the USA to describe the perilous situations in Afghanistan and Iraq in the 1990s. The concept was repeatedly embraced by business executives to chronicle the dramatically developing world company models. Aside from changing business models, consumer tastes are changing, as are the options accessible to them, and pricing wars are adding gasoline to the fire. It used to be enough for managers to draw up plans to cope with uncertainty, but in today's business world, uncertainty must be constantly resolved, which necessitates the leaders' involvement.

In emerging countries, evidence suggests that the number of Gen-Y and Gen-Z workers in organizations is increasing significantly (Gaidhani et al.,

2019), India has many working persons under the age of 32, with over 50% of them under the age of 32. India will surpass China as the globe's biggest country with a young working-age population by 2030 (Ritchie 2019).

When it comes to using organizational technologies, Gen-Y outperforms Gen-X. Furthermore, the organization's boundaries do not often limit this generation and desire a significant lot of independence in the workplace. SMAC creates an environment that allows the HR department to enhance its tasks and increase employee performance. In making workers feel great working in the company, all the four SMAC factors are uniformly relevant.

16.3 EXPERT VIEWS ON HR CHALLENGES IN THE VUCA WORLD

GenY's personnel managers face various challenges when selecting the right applicants, developing talent awareness and skills, motivating success, and using employees as a company's resource. Manufacturing companies should demonstrate their capabilities in human resource activities, which is a win-win situation for all key participants in the organization.

Raj (2020) claims that the pandemic has plunged our world into uncharted territory. Experts believe that Covid19 may seriously affect the Indian IT industry by US$180 billion. The bright side is that this may only be a temporary problem for the IT business. Even though IT investment would initially drop in 2020, market prospects for IT companies, particularly product companies, will progressively grow. Remote work will accelerate the adoption of automation and resource sharing, as well as the shift to cloud computing. To manage this future collectively, it is critical to sustain employee passion and enthusiasm. During these times, the HR department's role is to guarantee that the team has the necessary skills and participation, while also bracing for some difficult decisions.

Raghavendra Kadkol, HR at Infosys feels that mental wellbeing tops the list of post COVID HR challenges as the company is thinking of wellbeing for employees who are working from home. Also, relevant communications to keep employees updated about the organization and address concerns are another major challenge mentioned by Kadkol. Training had to be reimagined given that our trainees are in various parts of the nation and most importantly strengthening cybersecurity.

Baker (2020) emphasizes that 32% of companies replace full-time employees with temporary workers to save costs. Gartner Distinguished Vice President Brian Kropp (Brian Kropp) said that It is critical for company executives to recognize that significant changes are occurring in the way individuals operate and conduct business. Hiring effective and responsive managers can ensure that their company stands out from the competition.

Garg and Pawar (2020) believe that COVID19 has taken the lead and accelerated one of the biggest workplace changes in our lives. Organizations are revisiting their mission, innovating swiftly, and bringing themselves closer to the community in these challenging times. For CEOs, developing deep, meaningful, and personal ties with customers and coworkers becomes even more critical. Companies that now invest in building trust, emotional attachment, and friendliness should ensure the long-term and moral loyalty of their employees. As we moved from the office to the home, our way of life and the way we continue to do business have undergone a fundamental change. We've seen corporations develop quickly to deal with the future's uncertainties, thanks to smart technology, artificial intelligence, and mobile supercomputing. It has become increasingly necessary in this day to create new ways to communicate with employees through virtual media rather than physical or team-based approaches to anticipate the demand for emotional support.

The 3R method is described in a Deloitte paper on HR options for recovering from COVID-19. A typical crisis has three stages: response, in which the firm reacts to the present situation and maintains continuity; recovery, in which the company learns and grows stronger; and flourishing, in which the organization plans.

In an interview conducted by the author, one respondent, Sreejith, Country GM, India and South Asia, Horizontal Talent feels that "the real challenge begins now. While the outside world is opening and keeping the discipline and productivity of individual. People in customer facing roles will face challenges in adapting to a world where you traditionally built trust through meeting and sharing a coffee or meal. Nearly 100% of the business deals are made on trust. Training and developing - managers and learning departments will find it increasingly challenging to drive coaching and training. Most of it needs collaboration and free flow of ideas and trials. Social capital- there is an important social capital that drives locality when people work together in offices. Companies will have this daunting task of figuring out growing this social capital without having them work together."

16.4 GOOGLE TRENDS AND CURRENT AGILE MANAGEMENT SCENARIO

This article uses Google Trends analysis and results to show interest over time. Use keyword search on Google to analyze them.

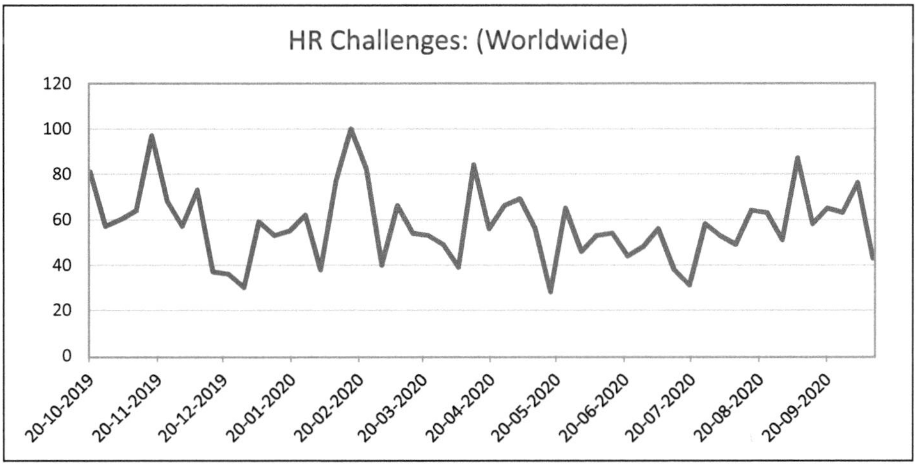

CHART 16.1: HR Challenges post COVID and the need for agile leadership (Worldwide).
Source: https://trends.google.com/trends/explore?q=HR%20Challenges.

The image above shows the number of search queries using the keyword HR Challenges on Google. As can be clearly seen from the above figure, the popularity of the HR Challenges keyword worldwide reached its peak in February 2020.

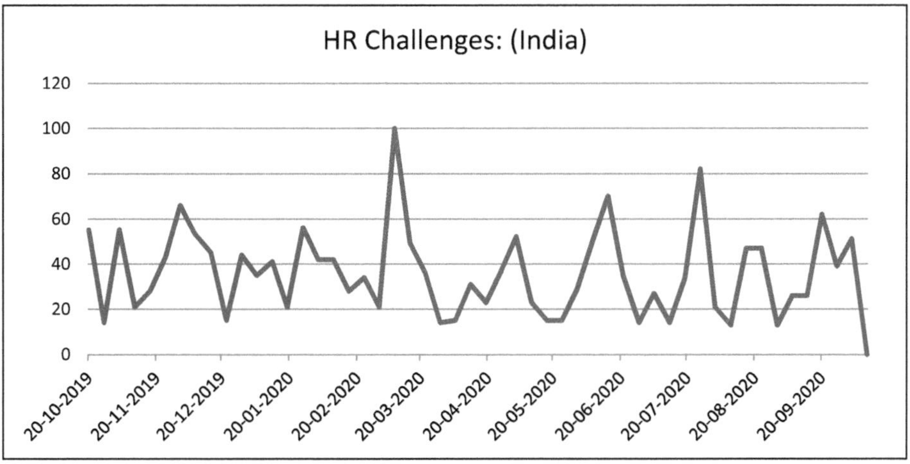

CHART 16.2. HR Challenges post COVID and the need for agile leadership (India).
Source: https://trends.google.com/trends/explore?geo=IN&q=HR%20Challenges.

The above chart depicts the number of searches made using the key word HR Challenges on Google. It is clear from the above chart that the key word HR challenges reached the peak popularity at the country level during Feb-Mar 2020.

16.5 HR CHALLENGES POST COVID AND THE NEED FOR AGILE LEADERSHIP

In the post-pandemic environment, employees must adjust their ways of working, and the entire human resources department should shift to a more people-oriented approach (Ahmed et al., 2020). Become obsolete (Caligiuri, 2020). Restructuring policies to reflect the new normal is critical for HR. To keep the business culture stimulating – for both existing and new employees. In the coming years, Hr practitioners will encounter a slew of challenges and advancements that will redefine their responsibilities.

FIGURE 16.1 Post COVID HR Challenges.
Source: Author Compilation.

16.5.1 AGILE LEADERSHIP IN MANAGING CRISIS

Companies are building up contingency plans for any future unanticipated problems after surviving the lockdown. Whether it occurs or not, HR teams need to persuade top management to put aside a larger amount of revenue for crisis management. Most businesses lacked remote operating capabilities when it came to crisis response. Employees in data-driven industries including consulting, technology, and finance will be reminded of the additional responsibilities that come with getting remote access to sensitive information. Building trust between employees and organizations will be a primary goal for today's HR (Sull, 2020).

16.5.2 AGILE LEADERSHIP IN COMMUNICATION SYSTEM

It is challenging to find a communication tool that fits the organization's culture. HR departments are racing against the clock to install a networking solution that is applicable for their staff as telecommuting becomes progressively pervasive. In uncertain times like these, when governments modify their policies daily, top leadership must set aside some time each day to address potential employee issues. That is what it takes to make everyone feel like they are a part of the organization. Regular mass mailers based on HR managers' general experience should be avoided, as they might easily be misunderstood as spam.

16.5.3 AGILE LEADERSHIP IN EMPLOYEE WELLNESS

The most critical challenge that HR managers face in the post-COVID era is this. On the shop floor, many people are trained how to understand their coworkers' psychology. So, in a remote working situation, how can one identify an employee's mental state? The forced boarding of various businesses across industries revealed that most people felt unable to distinguish between family time, personal time, and work duties. They were able to focus on one aspect while putting the other on hold due to the physical separation of business and home space. They were able to manage their time effectively, with little overlap in responsibilities. HR managers will have to consider how to get ideas back to standard, or their employees' mental health would suffer severely. A better employee engagement programme should be able to keep staff working during this period.

16.5.4 AGILE LEADERSHIP IN EMPLOYEE ENGAGEMENT

Companies can make a remarkable distinction in forming a more active workforce if they focus on the right things. Organizations play a critical role in establishing an engaged workforce and keeping their people productive. The difference between the extrinsic benefits required by employers for competitive advantage and the ability of employers to elicit this effort from a large section of their staff should be understood by the employer. Employee engagement is a factor that drives results in success, according to the world's best-performing firms. Employees that are engaged are more productive, according to Gallup study (Gallup, 2021). They are more successful, more client-focused, safer, and less inclined to abandon ship. The most successful companies understand that a strategy for increasing employee engagement tied to the accomplishment of business objectives will improve them win the battle (Chakraborty et al., 2020).

16.5.5 AGILE LEADERSHIP IN QUALITY OF WORK-LIFE

Maintaining a high quality of life has become a new scale in an array of industries. The COVID-19 problem has acted as a catalyst for this constructive reform to take hold. HR managers must assess whether job tasks have become obsolete since the lockdown or are on the verge of being shut down and attempt to laterally align workers with those roles. Reinventing training programmes to reflect the changing environment would make them more successful and, as a result, more cost-effective. Changes in employee leave regulations will ensure that employees are not affected by 'pandemic-related' health emergencies. HR managers must create an employment experience that not only challenges but also encourages their employees to overcome their fears of uncertainty and fully commit to personal and professional development.

16.5.6 ROLE OF ARTIFICIAL INTELLIGENCE IN AGILE BUSINESS

Artificial intelligence is being employed more in the HR job in the post-pandemic era. AI can help recruiters find potential applicants among gigabytes of required data, helping companies to not only meet their diversity targets during the hiring process, but also retain top talent and engage with employees through smart apps. According to Baker (2020), 48% of personnel to be expected work remotely for at least part of the period following COVID-19. According to a rising number of business mentors, the remote work environment is here to stay, and for companies trying to decrease real estate expenditures to compensate for losses incurred during the lockout. This will entail increased digitalization and technology integration, as well as the shrinkage and transformation of physical workplaces into virtual, digitally enabled environments. One of the primary advantages of this method is that managers will be able to choose the best qualified candidate after getting a bigger number of job applications from around the country because recruitment will no longer be based on location.

16.5.7 REBIRTH OF THE GIG ECONOMY

Employers can more readily choose their people when they hire temporary labor. According to a Gartner survey, "32% of companies are replacing full-time employees with contract workers to save money. HR managers who want to fill vacancies with contract workers, on the other hand, may run into problems with the pertinence of employment laws to workers. Employee productivity has long been a top priority for businesses, but in the post-COVID-19 era, resilience will take primacy. At the time of hiring, recruiters would look at a job candidate's technical dexterity and technical teaming abilities, keeping in mind the rising remote work environment and the risk of infection".

16.5.8 LEARNING AND DEVELOPMENT IN AGILE BUSINESS

Choose which online technologies will be used to give on-the-job learning to supervisors, trainers, and instructors. It focuses specifically on the concept of remote communication. These programmes can help people manage stress, cultivate a positive mindset, and make progress when working from home. In the VUCA environment, anticipating, interpreting, adapting, and delivering in new conditions is all part of the job. The Internet revolution and the sub-prime financial crisis in the United States in 2008, for example, were major events that shook the old business eco-system and changed how corporations saw their personnel. Adversity fosters talent enhancement, this is where VUCA and SMAC collide. The way businesses are conducted has changed dramatically because of digital technologies (Chakraborty and Saha, 2018). In VUCA and SMAC times, companies should recognize the fact that researching and adapting to operate the organization for the long term or as a credible alternative to closure benefits only the leaders.

16.5.9 FLEXIBLE PLANS AND NONLINEAR HR

Plans, especially the long-term ones, should not be extremely linear because this makes implementation challenging for management and staff. Because depressions can create new growth chances and pave the way for new prospects, strategic or long-term plans must be adaptive. As a result, the organization must be reminded to use the potential to redefine itself and progress in its business difficulties. Some C-suite administrators will need to get more for less in this fast-paced climate. Companies will need people who heed prudently, think on hovering applicable matters, have no longing to command but naturally co-create therapies, and get immense satisfaction by seeing the active and passive beneficiaries of their pains when they expect leaders who are the source and precursor of energy, job commitment, sense of belonging, and motivation. VUCA and SMAC are here to stay, keeping managers and employees on their toes and ensuring that the firm runs smoothly.

16.5.10 GENERATIONAL CHALLENGES DURING INDUSTRY 4.0 AND POST COVID WORLD

With terms like VUCA and SMAC circulating, firms must remember that employees are not just GenY or Z, but also multi-generational. Companies must have a strategic strategy in place to move information from Baby Boomers, who are slated to withdraw in the coming years, to GenY or Z, who do not have time constraints. GenY does not believe in waiting long for promotions, raises, or direction responsibilities, according to an expert quote. Despite their eagerness to hire the young, businesses have realized that they will need crucial skills that can only be taught by the more experienced. Businesses today must allow employees to voice their concerns while remaining trustworthy.

16.5.11 CREATING A KNOWLEDGE SHARING CULTURE FOR SUSTAINABILITY

It is all about enhancing the firm's degree of expertise. To develop a knowledge sharing cultural organization, it is critical to enable individuals to work together more efficiently, cooperate, and share—eventually to make organizational knowledge more productive. However, there are a few matters that employees must maintain in mind: it's about sharing information and specifics, not simply statistics. The goal of knowledge sharing is to help a corporation achieve its objectives. The proper solution is to assist individuals in seeing for themselves why sharing information is beneficial to them.

16.6 CONCLUDING REMARKS AND SCOPE FOR THE FUTURE RESEARCH

As HR professionals, it is critical to keep our eyes and ears on the strategy table to collaborate on changes, quickly learn to navigate the growing talent market, and close the talent gap within our organizations. Building an organizational structure to handle these changes will also be vital, which brings me to the second critical pillar of organizational effectiveness. There are exciting times ahead, but they will be fraught with challenges. HR managers have an opportunity to shape the destiny of their organizations currently. Organizations in the present context would require a positive organizational culture (Chakraborty and Ganguly, 2019), will need positive approaches to understand employees by appreciative questioning (Chakraborty and Mishra, 2019; Chakraborty et al., 2019). Organizations are more agile after a disaster, so the role of human resource professionals in managing aftereffects of a crisis are very crucial (Chakraborty and Saha, 2017).

Individuals, policy, and practises will be the three primary foundations on which they should build a new framework for HR greatness. It's past time for HR professionals to take a chance. Managers must overcome the mentality of clinging to old procedures and plans to deal with the hazards of VUCA and SMAC. As a result, to meet the varying market times induced by VUCA and SMAC, more agile strategies, and techniques, as well as firewalls, must be built. Apart from that, to succeed in the VUCA and SMAC business sectors, organizations and HR managers must be curious, imaginative, and original.

KEYWORDS

- **HR challenges post-COVID**
- **agile leadership**
- **staff morale**
- **incentives**
- **generational challenge**

REFERENCES

Ahmed, T.; Shahid Khan, M.; Thitivesa, D.; Siraphatthada, Y.; Phumdara, T. Impact of Employee's Engagement and Knowledge Sharing on Organizational Performance: Study of HR Challenges in COVID-19 Pandemic. *Hum. Syst. Manag.* **2020**, (Preprint), 1–15.

Baker, M. 9 Future of Work Trends Post-COVID-19. 2020 https://www.gartner.com/smarterwithgartner/9-future-of-work-trends-post-covid-19/.

Caligiuri, P.; De Cieri, H.; Minbaeva, D.; Verbeke, A.; Zimmermann, A. International HRM Insights for Navigating the COVID-19 Pandemic: Implications for Future Research and Practice. 2020.

Chakraborty, T.; Ganguly, M. Crafting Engaged Employees through Positive Work Environment: Perspectives of Employee Engagement. In *Management Techniques for Employee Engagement in Contemporary Organizations,* IGI Global, 2019; pp 180–198.

Chakraborty, T.; Mishra, N. Appreciative Inquiry: Unleashing a Positive Revolution of Organizational Change and Development. *Int. J. Econ. Commerce Bus. Manag.* **2019**, *6*(2), 32–37.

Chakraborty, T.; Ganguly, M.; Natarajan, A. Predicting Entrepreneurial Satisfaction: The Role of Non-Financial Incentive Factors and Quality of Life among Women Digital Entrepreneurs. *J. Glob. Bus.* **2019**, *12*(3), 328–355.

Chakraborty, T.; Saha, R. Proactive and Reactive Role of Human Resource Professionals in an Aftershock Management. *Train. Dev. J.* **2017**, *8*(1), 1–11.

Chakraborty, T.; Saha, R. Digital Innovations: Breakthrough Opportunities to Build Novel Business Models. *Bus. Perspec.–Aims Scope* **2018**, *1*, 46–54.

Chakraborty, T.; Gohain, D.; Saha, R. What Comes in the Way of Engagement? Moderation Analysis of Stress on Women Marketing Executives' Work Life Balance. *Int. J. Hum. Resour. Dev. Manag.* **2020**, *20*(3-4), 349–368.

Gaidhani, S.; Arora, L.; Sharma, B. K. Understanding the Attitude of Generation Z towards Workplace. *Int. J. Manag. Technol. Eng.* **2019**, *9*(1), 2804–2812.

Garg, M.; Pawar R. *Future of Human Resources (HR) Post Covid-19*, KPMG, 2020. https://home.kpmg/in/en/blogs/home/posts/2020/05/re-thinking-human-resources-post-covid-19.html.

Gallup. *How to Improve Employee Engagement in the Workplace*. Gallup.com., 2021 https://www.gallup.com/workplace/285674/improve-employee-engagement-workplace.aspx.

Ritchie, H. *India will Soon Overtake China to Become the Most Populous Country in the World.* Our World in Data, 2019. https://ourworldindata.org/india-will-soon-overtake-china-to-become-the-most-populous-country-in-the-world.

Raj, S. J. HR to Play a Bigger Role During and After Covid-19. **2020**. https://hr.economictimes.indiatimes.com/news/hrtech/hr-to-play-a-bigger-role-during-and-after-covid-19/75740727.

Sull, D.; Sull, C.; Bersin, J. Five Ways Leaders Can Support Remote Work. *MIT Sloan Manag. Rev.* **2020**, *61*(4), 1–10.

WEBSITES:

HR Challenges n.d., Google Trends. https://trends.google.com/trends/explore?q=HR%20Challenges

HR Challenges n.d., Google Trends. https://trends.google.com/trends/explore?geo=IN&q=HR%20Challenges

CHAPTER 17

Agile Entrepreneurship: Probing into Women Entrepreneurial Innovation in Rural India

TANUSREE CHAKRABORTY[1] and MADHURIMA GANGULY[2*]

[1]*Administrative Staff College of India, Hyderabad, India*

[2]*Heritage Business School, Kolkata, India*

[*]*Corresponding author. E-mail: mganguly10@gmail.com*

ABSTRACT

The need for innovation toward sustainable entrepreneurial development is a prerequisite across developing countries, especially in India, where development of women entrepreneurship is still not at its optimum level, more so in the rural areas. Identifying a dearth of research examining the determinants of sustainable innovative rural entrepreneurship, and agile entrepreneurship, the purpose of this chapter is to determine the aspects enabling sustainability of the Indian rural women entrepreneurs. Data was collected in June 2019 for conducting the current research. The researchers met with the rural women entrepreneurs in person, conducted focused group discussions, and subsequently developed a questionnaire. Data was collected from the subjects on a person-to-person basis. A total of 355 filled-in responses were collected from women rural entrepreneurs. The data collection through interview was followed by a semi-structured interview as well. The study reveals the existence of a relation between psychological and environmental factors leading to the success of entrepreneurial innovation. The findings also bring out the fact that innovation has a moderating role in the sustenance and success of

Agile Leadership for Industry 4.0: An Indispensable Approach for the Digital Era, Tanusree Chakraborty, Tahir Mumtaz Awan, Ashok Natarajan, and Muhammad Kamran (Eds.)

women-run rural business. The study is significant as its findings can be used in supporting the governmental policies and schemes aimed at the improvement of women rural entrepreneurs. The environments in the family, society, and institutions are yet not sufficiently encouraging for the rural women to take up entrepreneurship as a career. Last, the paper discusses its shortcomings as well as prospective research possibilities in the future.

17.1 INTRODUCTION

Among the various other widely discussed issues on entrepreneurship, central are entrepreneurial innovation and innovative entrepreneurship. As businesses become more and more agile and the world of entrepreneurship sees agility, it is important that entrepreneurial innovation takes a robust stride. Entrepreneurial Innovation is the versatile course of action that an entrepreneur engages in to try to introduce new items and technology to the commercial market that are inventive and possibly disruptive, while the innovative entrepreneur is the individual who can undertake to produce innovative products responding to the shifting needs of the market. Start-ups are also seeing a huge scale of innovation these days (Tsolakidis, 2020). In today's market, the confidence of young entrepreneurs is mainly due to their enthusiastic acceptance of technology. Entrepreneurs are expanding novel technologies like social media, mobile computing, analytics, and the cloud to generate innovation in business and ever more disrupting existing value chains. India is keeping ahead in prudent invention which constitutes 18 percent of the global inhabitants, majority of which are deprived and belong to the agrarian society. Rural innovation in the fields of agriculture, domestic utilities, micro-enterprises, and importantly, energy gadgets has soared high. Rural women may improve their individual potential and decision-making standpoint in the household and society as a whole by developing their entrepreneurial skills (Sharma, 2013). Kulkarni (2010) investigated the barriers to women's business growth and identified that maximum women entrepreneurs could not progress forward on account of economic problems arising out lack of finance and low exposure to modern technology. Also it is true that many entrepreneurs fail because they fail to learn from the past courses of their action (Amore et al., 2021).

Rural entrepreneurships are businesses defined as occupied in activities dependent on the natural and physical capitals and are based on resources from the rural environment as the major source of income and they employ local labor (McElwee and Smith, 2014).

In India, the rise of women entrepreneurs and the way they influence the economy are well known, but there are instances of their being held back (Korreck, 2019). The attitude at the policy level to support and develop women entrepreneurship is visible and encouraging women to participate in entrepreneurial activities. However, the society and culture still operate on a framework of stereotypes and disgrace when it comes to a woman stepping out into the big world of business. The efforts to develop women need to be validated and supported (Chakraborty et al., 2018). The need for sustainable entrepreneurial innovation is clearly visible across developing countries, especially in India where development of women entrepreneurship is still not at its optimum level, more so in the rural areas. Besides an impressive collection of research on the entrepreneur, her personality traits and characteristics and her impact on the success of the small business have not been done much with respect to service firms (Glancey and Pettigrew, 1998).

"India lives in its villages" – This statement by Mahatma Gandhi holds true even today, from the social, economic, and the political perspective. Thus, empowering the rural sector by creating means of livelihood needs to be an agenda in the growth of any developing country, more so in the case of India (Gaddefors and Anderson, 2019). Rural entrepreneurship broadly refers to industrialization of rural sector. A regional entrepreneur is a valuable resource for a country's economic growth. She is the one who starts industrial and entrepreneurial ventures in rural areas and carries out entrepreneurial activities there. Rural entrepreneurship can be used as a key mechanism to alleviate poverty, curb migration to cities, discourage economic discrimination (Sá et al., 2019), eradicate ignorance, and reduce unemployment. However, industrialization cannot start or be sustained without the presence of innovative entrepreneurship. Innovation research has copious literature on technology-driven innovations and entrepreneurship. But, research literature on an Indian rural innovator and entrepreneur is not sufficient (Patel and Chavda, 2013; Kumbhar, 2013).

Identifying a dearth of research examining the determinants of sustainable innovative rural entrepreneurship, the goal of the current research is to recognize the factors enabling sustainability of the Indian rural women entrepreneurs and understanding the role of innovation toward entrepreneurial success.

17.2 CONCEPTUAL MODEL AND HYPOTHESES DEVELOPMENT

17.2.1 LEADERSHIP AND ENTREPRENEURIAL SUCCESS

There have been studies that relate to leadership and entrepreneurial success. It has been stated by early studies that leadership qualities of the entrepreneur lead to entrepreneurial success (Chatterjee et al., 2019; Harper-Anderson, 2018). These studies had not covered rural entrepreneurship or rural innovation.

Studies have also explored the challenges of women entrepreneurship (Basit et al., 2020). So, it becomes pertinent to study the relationship between leadership and entrepreneurial success among rural women. There we hypothesize that:

H1: There is a positive relationship between Leadership and Entrepreneurial Success among Rural Women.

17.2.2 PROACTIVITY AND ENTREPRENEURIAL SUCCESS

Various personality factors have been studied with respect to entrepreneurial success (Hachana at al., 2018. Zhou et al., 2019). Gender of the entrepreneurs has been studied in this respect as well. In general, studies have attempted to examine personality factors and entrepreneurial passion with proactivity as a variable (Li et al., 2020). But specifically women entrepreneurs' proactively as a personality factor is to be tested yet. So, we hypothesize that:

H2: There is a positive relationship between Proactivity and Entrepreneurial Success among Rural Women.

17.2.3 CREATIVITY AND ENTREPRENEURIAL SUCCESS

Creativity has been found in studies to be associated with entrepreneurship (Ko and Butler, 2007). Creativity and entrepreneurship has positive relations (Ndofirepi et al., 2018); creativity and entrepreneurial intentions have positive relations (Kusmintart et al., 2017); and leadership has positive relations with entrepreneurial success (Newman et al., 2018). Entrepreneurs are by nature creative and innovative in presenting new products and services to their customers. Women have been shown to be more creative than men (Hora et al., 2021), in general. This leads to an interesting question as to whether women entrepreneurs in rural environment possess similar characteristics that lead to entrepreneurial success.

Thereof, we hypothesize that:

H3: There is a positive relationship between Creativity and Entrepreneurial Success among Rural Women.

17.2.4 OPENNESS TO EXPERIENCE AND ENTREPRENEURIAL SUCCESS

As personality trait, openness to experience makes a person more open to learning (Christensen et al., 2019). Openness to experience relates to innovation and creativity (Wartiovaara et al., 2019). Entrepreneurs need to have a learning mindset so that they can respond to the changes on the environment soon, as they occur. Agile leadership demands this openness to experience. We postulate that agile leaders need to be more open to the environment and be dynamic in their outlook. We test this for women entrepreneurs in rural areas in our study. So we hypothesize that:

H4: There is a relationship between Openness to Experience and Entrepreneurial Success among Rural Women.

17.2.5 EXTRAVERSION AND ENTREPRENEURIAL SUCCESS

Extraverts are individuals with an outgoing personality and are high on integration with others. Entrepreneurs need to be better interactive with others with openness for communication (Gao et al., 2020). Leaders are found high on extraversion (Silverthorne, 2001). Extraversion is rated high among entrepreneurs (İrengün 2015; Antoncic et al., 2015). This leads us to assume that in rural India, women entrepreneurs are high on extraversion, leading to entrepreneurial success. Thus we hypothesize that,

H5: There is a relationship between Extraversion and Entrepreneurial Success among Rural Women.

17.2.6 INNOVATION AND ENTREPRENEURIAL SUCCESS

Innovation refers to the production or adoption, absorption, and utilization of a value-added originality in economic and social aspects. Entrepreneurship requires innovation. Entrepreneurship in agile business looks for innovation. Innovation helps a business to thrive as customers are always on the lookout for innovative products. Agile entrepreneurs consider about farm duties from a different mindset, and they think differently. Among the various qualities of agile entrepreneurs, it is asserted that these people can learn to become more agile in their thoughts without losing the focal point required to be successful in business. Thus besides the internal factors that lead to entrepreneurial success, we seek to ask whether innovation in business can led to entrepreneurial success intensity among rural entrepreneurs. Thus we hypothesize that:

H6: Innovation acts as a moderator between internal qualities and entrepreneurial success among rural women entrepreneurs.

17.3 METHODOLOGY

17.3.1 SCALES USED

The scale used in the study contains 36 statements with a 5-point Likert scale. The scales have a reliability value of 0.89. Both positive and negative statements were included in the scale. The content and face validity of the scale had been verified by a group of three specialists. Items with an acceptance rate of 85% or above were kept in the research. Only a few statements were changed as a result of expert feedback. The final scale has 36 statements on it. Cronbach's alpha of 0.87 indicated high reliability. Other than the above scale, there were few semi-structured items involving categorical rating of responses in a dichotomous manner for the external factors or environmental variables. These items involved questions on whether the women entrepreneurs were successful/unsuccessful in terms of economic performance.

17.3.2 SAMPLE

Data was collected in June 2019 for conducting the current research. The researchers met with the rural women entrepreneurs in person, conducted focused group discussions, and subsequently developed a questionnaire. Data was collected from the subjects on a person-to-person basis. A total of 355 filled-in responses were collected from women rural entrepreneurs who were in business for than five years. The ages of the entrepreneurs were between 28-45 years. All entrepreneurs belonged to rural India covering East, West, North, and South rural areas in India. The data collection through interview was followed by a semi-structured interview as well.

17.4 RESULTS AND DISCUSSION

Entrepreneurs and entrepreneurship have been recognized as potentially significant factors required leading growth and employment (Kneller, and McGowan 2011). The present study attempted to understand the factors responsible for entrepreneurial success and later understand the role of innovation in businesses led by rural Indian women. The results in Table 1 depict those.

TABLE 17.1 Regression Coefficients

Predictors of entrepreneurial success	Model 1	Model 2
Proactivity	0.069**	
Openness to Experience	0.072**	
Leadership	0.012^{ns}	
Creativity	0.61***	
Extraversion	0.820***	
R^2 Determinants of Entrepreneurial Success	0.61***	
R^2 Innovation X Determinants of Entrepreneurial Success		0.72***
ΔR^2		0.11

The present study finds that there are certain intrinsic factors that lead to entrepreneurial success among rural women entrepreneurs. Also there is a role of innovation in success of rural entrepreneurship. This finding is in line with the studies of Trimi and Berbegal-Mirabent (2012; Lounsbury et al. 2019). To add here, some earlier studies have also found in general that women in rural areas frequently become pioneers in the entrepreneurial initiatives and such finding indicates gendered results (McGehee et al., 2007). This study adds that rural women become successful in entrepreneurship because some specific internal qualities, which when added with innovation, become more intense.

Small enterprises have slowly but surely captured a noteworthy place in mainstream business by offering employment opportunities and creating income opportunities for the aspirer. Innovation and entrepreneurship are active and complete procedures that may not be contained in the realm of the operational activities of a novel task. However, a need for greater research remains to be conducted to assess the real relation between innovation and entrepreneurship, as it is imperative that the synergy between the two functions foster organizational growth. Similarly, numerous authors have identified continual service innovation as the most effective method for businesses to achieve long-term success and organic development. Given the growing amount of research on the influence of entrepreneurs on small company success, relatively lesser number of this study has focused on small business service enterprises. (Glancey et al., 1998). Several researchers have suggested that service firms should develop a customer orientation mindset for attaining higher success (Agarwal et al., 2003; Appiah-Adu and Singh, 1998).

In today's global context characterized by rapid change and uncertainty, there is a dynamic relation between knowledge, innovation, and entrepreneurial success. The present study has confirmed that it is potentially feasible to recognize a set of variables that can be referred to as 'entrepreneurship dynamics. This set of variables can be effectively used and added to the list of researches that aim to distinguish between successful and unsuccessful women entrepreneurs (Brockhaus, 1980; Karimi et al., 2012; Bates, 2005). The key goal of the current research was to detect whether internal that is intrinsic (psychological) variables are important indicators to define a woman rural entrepreneur. The study established that few intrinsic factors play important roles in deciding firm success for a women entrepreneur. The relevant determiners for women entrepreneurs are found to be Proactivity, Openness to Experience, Creativity, and Extraversion. The regression coefficients are mentioned in Table 1. Leadership has not been found to be significant in this study as an indicator of rural entrepreneurial success. However, it has been found that innovation is a moderator of the internal factors for entrepreneurial success. This has consonance with previous researches (Bessant and Tidd, 2007; Hagedoorn, 1996). It is true that women are making momentous strides as leaders in business today and as entrepreneurs; still a significant number of them encounter a plethora of challenges that stop them from scaling their ventures to greater heights or even stop them from making the first move. Women entrepreneurs still need to regulate their negative emotions and maximize the positive ones, while meeting venture capitalists for financial support (Ganguly et al., 2019). If an identification of such entrepreneurial minded women is done, some form of assistance could be provided to the aspiring and existing women entrepreneurs to unleash their fullest potentials and reach the heights; it would not only help those women entrepreneurs to succeed in their businesses but also create a huge number of jobs in the market where many "solopreneur" women would engage employees in their business. Even business management programmes should incorporate added emphasis on entrepreneurial practical training and provide assistance for startups. If there is a robust assurance to empower women entrepreneurs to maneuver at their best, women with already high intrinsic qualities will do much better in business, and it will definitely make rooms for entrepreneurs to prosper and also construct stronger communities and enhance the local as well as national economies. Thus, it can be asserted that women entrepreneurship is a combination of psychological as well as environmental factors._Psychological factors like optimism, employee trust lead to satisfaction of the women entrepreneurs (Chakraborty et al., 2019). Earlier life experiences that include socioeconomic background, parenting styles and role models, scholastic achievements, social and emotional skills along with a positive self-concept play a significant role in being predictors of entrepreneurship (Schoon and Duckworth 2012). The current findings point to a combined influence of other social factors as well as individual intentions in deciding choice of occupation and its implementation. Earlier studies too ascertain that a woman's entrepreneurial motivation generates from "**pull**" **factors** (Buttner and Moore, 1997; Kirkwood, 2009).

17.5 CONCLUSION

The present study concludes that agile entrepreneurship related to rural Indian women has been found to be majorly predicted by certain internal factors of the entrepreneurs, and that innovation has a moderating effect on entrepreneurial success. The major limitation of the study is that we emphasize the exploratory nature of the study to conduct data collection and to know behaviors and approach of these women rural entrepreneurs; later, an in-depth study will be attractive, maybe conducted in the future as Indian rural entrepreneurs' case study.

KEYWORDS

- **rural entrepreneurship**
- **agile entrepreneurship**
- **innovation**
- **success factors**
- **women entrepreneurship**

REFERENCES

Agarwal, S.; Krishna Erramilli, M.; Dev, C. S. Market Orientation and Performance in Service Firms: Role of Innovation. *J. Serv. Market.* **2003**, *17* (1), 68–82.

Amore, M. D.; Garofalo, O.; Martin-Sanchez, V. Failing to Learn from Failure: How Optimism Impedes Entrepreneurial Innovation. *Organ. Sci.* **2021**, *32* (4), 940–964.

Antoncic, B.; Bratkovic Kregar, T.; Singh, G.; DeNoble, A. F. The Big Five Personality–Entrepreneurship Relationship: Evidence from Slovenia. *J. Small Bus. Manag.* **2015**, *53* (3), 819–841.

Appiah-Adu, K.; Singh, S. Customer Orientation and Performance: A Study of SMEs. *Manag. Decis.* **1998**, *36* (6), 385–394.

Basit, A.; Sze Wong, Z. H.; Sethumadhavan, S. Entrepreneurial Success: Key Challenges Faced by Malaysian Women Entrepreneurs in 21st Century. *Int. J. Bus. Manag.* **2020**, *15* (9), 122.

Bates, T. Analysis of Young, Small Firms that have Closed: Delineating Successful from Unsuccessful Closures. *J. Bus. Vent.* **2005**, *20* (3), 343–358.

Bessant, J.; Tidd, J. *Innovation and Entrepreneurship*. John Wiley & Sons, 2007.

Birley, S.; Moss, C.; Saunders, P. Do Women Entrepreneurs Require Different Training?. *Am. J. Small Bus.* **1987**, *12* (1), 27–36.

Botha, M. Measuring the Effectiveness of the Women Entrepreneurship Programme, as a Training Intervention, on Potential, Start-Up and Established Women Entrepreneurs in South Africa Doctoral Dissertation, University of Pretoriam, 2006.

Brockhaus, R. H. Psychological and Environmental Factors Which Distinguish the Successful from the Unsuccessful Entrepreneur: A Longitudinal Study. *Acad. Manag. Proc.* **1980**, *1980* (1), 368–372.

Buttner, E. H.; Moore, D. P. Women's Organizational Exodus to Entrepreneurship: Self-Reported Motivations and Correlates with Success. *J. Small Bus. Manag.* **1997**, *35* (1), 34.

Chakraborty T, Ganguly, M.; Natarajan, A. Predicting Entrepreneurial Satisfaction: The Role of Non-Financial Incentive Factors and Quality of Life among Women Digital Entrepreneurs. *J. Glob. Bus. Adv.* **2019**, *12* (3), 328–355.

Chakraborty, T.; Gupta, D.; Chatterjee, B. " Strong, Yet We Are Vulnerable": Role of Psychological Factors and Financial Affluence on Women's Entrepreneurial Success. *J. Glob. Bus. Adv.* **2018**, *11* (6), 726–752.

Chatterjee, N.; Das, N.; Srivastava, N. K. A Structural Model Assessing Key Factors Affecting Women's Entrepreneurial Success: Evidence from India. *J. Entrepren. Emerg. Econ.* **2019**.

Christensen, A. P.; Cotter, K. N.; Silvia, P. J. Reopening Openness to Experience: A Network Analysis of Four Openness to Experience Inventories. *J. Personal. Assessm.* **2019**, *101* (6), 574–588.

Gaddefors, J.; Anderson, A. R. Romancing the Rural: Reconceptualizing Rural Entrepreneurship as Engagement with Context (s). *Int. J. Entrepren. Innov.* **2019**, *20* (3), 159–169.

Ganguly, M.; Chakraborty, T.; Mishra, N. The Changing Face of Today's Agile Women Entrepreneurs: A Study of Women Entrepreneurs' Emotional Regulation while Meeting Investors. *Int. J. Manag. Technol. Eng.* **2019**, *IX* (V), 1855–1863

Gao, Y.; Zhang, D.; Ma, H.; Du, X. Exploring Creative Entrepreneurs' IEO: Extraversion, Neuroticism and Creativity. *Front. Psychol.* **2020**, *11*, 2170.

Glancey, K.; Greig, M.; Pettigrew, M. Entrepreneurial Dynamics in Small Business Service Firms. *Int. J. Entrepren. Behav. Res.* **1998**, *4* (3), 249–268.

Hachana, R.; Berraies, S.; Ftiti, Z. Identifying Personality Traits Associated With Entrepreneurial Success: Does Gender Matter?. *J. Innov. Econ. Manag.* **2018**, (3), 169–193.

Hagedoorn, J. Innovation and Entrepreneurship: Schumpeter Revisited. *Ind. Corp. Change* **1996**, *5* (3), 883–896.

Harper-Anderson, E. Intersections of Partnership and Leadership in Entrepreneurial Ecosystems: Comparing Three US Regions. *Econ. Dev. Quart.* **2018**, *32* (2), 119–134.

Hora, S.; Lemoine, G. J.; Xu, N.; Shalley, C. E. Unlocking and Closing the Gender Gap in Creative Performance: A Multilevel Model. *J. Organ. Behav.* **2021**, *42* (3), 297–312.

İrengün, O.; Arıkboğa, Ş. The Effect of Personality Traits On Social Entrepreneurship Intentions: A Field Research. *Proc.-Soc. Behav. Sci.* **2015**, *195*, 1186–1195.

Karimi, P.; Kloshani, M.; Bakhshizadeh, A. A Comparative Study on Emotional Intelligence and Cognitive Between Successful and Unsuccessful Entrepreneurs. *Manag. Sci. Lett.* **2012**, *2* (6), 2071–2076.

Kirkwood, J. Motivational Factors in a Push-Pull Theory of Entrepreneurship. *Gender Manag.* **2009**, *24* (5), 346–364.

Kneller, R.; McGowan, D.. *Entrepreneurship Dynamics, Market Size and Fiscal Policy* (No. 11003), 2011.

Ko, S.; Butler, J. E. Creativity: A Key Link to Entrepreneurial Behavior. *Bus. Horiz.* **2007**, *50* (5), 365–372.

Korreck, S. Women Entrepreneurs in India: What is Holding them Back?. *Observ. Res. Found. Issue Brief* **2019**, *Forthcoming.*

Kulkarni, S. R. An Empirical Study of the Barriers for the Development of Women Entrepreneurs in Pune Region. Unpublished Ph.D. Thesis, University of Pune: Pune, India, 2010.

Kumbhar, V. M. Some Critical Issues of Women Entrepreneurship in Rural India. *Eur. Acad. Res.* **2013**, *1* (2).

Kusmintarti, A.; Asdani, A.; Riwajanti, N. I. The Relationship Between Creativity, Entrepreneurial Attitude and Entrepreneurial Intention (Case Study on the Students of State Polytechnic Malang). *Int. J. Trade Glob. Markets* **2017**, *10* (1), 28–36.

Li, C.; Murad, M.; Shahzad, F.; Khan, M. A. S.; Ashraf, S. F.; Dogbe, C. S. K. Entrepreneurial Passion to Entrepreneurial Behavior: Role of Entrepreneurial Alertness, Entrepreneurial Self-Efficacy and Proactive Personality. *Front. Psychol.* **2020**, *11*, 1611.

Lounsbury, M.; Cornelissen, J.; Granqvist, N.; Grodal, S. Culture, Innovation and Entrepreneurship. *Innovation* **2019**, *21* (1), 1–12.

McElwee, G. and Smith, R. Researching Rural Enterprise. In *Handbook of Research on Entrepreneurship*; Fayolle, A. Ed.; Edward Elgar: Cheltenham, 2014; pp. 307–334.

McGehee, Kim, K.; Jennings, G.R. Gender and Motivation for Agri-Tourism Entrepreneurship. *Tour. Manag.* **2007**, *28* (1), 280–289.

Ndofirepi, T. M.; Rambe, P.; Dzansi, D. Y. The Relationship Among Technological Creativity, Self-Efficacy and Entrepreneurial Intentions of Selected South African University of Technology Students. *Acta Commercii* **2018**, *18* (1), 1–14.

Newman, A.; Neesham, C.; Manville, G.; Tse, H. H. Examining the Influence of Servant and Entrepreneurial Leadership on the Work Outcomes of Employees in Social Enterprises. *Int. J. Hum. Resour. Manag.* **2018**, *29* (20), 2905–2926.

Patel, B.; Chavda, K. Rural Entrepreneurship in India: Challenge and Problems. *Int. J. Adv. Res. Comput. Sci. Manag. Stud.* **2013**, *1* (2).

Sá, E.; Casais, B.; Silva, J. Local Development Through Rural Entrepreneurship, from the Triple Helix Perspective: The Case of a Peripheral Region in Northern Portugal. *Int. J. Entrepren. Behav. Res.* **2019**.

Schoon, I.; Duckworth, K. Who Becomes an Entrepreneur? Early Life Experiences as Predictors of Entrepreneurship. *Dev. Psychol.* **2012**, *48* (6), 1719

Sharma Y. Women Entrepreneurship in India. *IOSR J. Bus. Manag.* **2013**, *15* (3), 09–14

Silverthorne, C. Leadership Effectiveness and Personality: A Cross Cultural Evaluation. *Personal. Individ. Differ.* **2001**, *30* (2), 303–309.

Trimi, S.; Berbegal-Mirabent, J. Business Model Innovation in Entrepreneurship. *Int. Entrepren. Manag. J.* **2012**, *8* (4), 449–465.

Tsolakidis, P.; Mylonas, N.; Petridou, E. The Impact of Imitation Strategies, Managerial and Entrepreneurial Skills on Startups' Entrepreneurial Innovation. *Economies* **2020**, *8* (4), 81.

Wartiovaara, M.; Lahti, T.; Wincent, J. The Role of Inspiration in Entrepreneurship: Theory and the Future Research Agenda. *J. Bus. Res.* **2019**, *101*, 548–554.

Zhou, M.; Zhou, Y.; Zhang, J.; Obschonka, M.; Silbereisen, R. K. Person–City Personality Fit and Entrepreneurial Success: An Explorative Study in China. *Int. J. Psychol.* **2019**, *54* (2), 155–163.

CHAPTER 18

Linking Agile Methodology and Transformational Leadership to Project Success through Interplay of Team Building

SOHAIB JALIL MALIK[1], WAJID SHAKEEL AHMED[1*],
TAHIR MUMTAZ AWAN[1], NAVEEN FAROOQ[2], and MYRA KHALID[1]

[1]Department of Management Sciences, COMSATS University Islamabad, Islamabad, Pakistan

[2]Lecturer, Social Sciences Department, Fatima Jinnah University, Rawalpindi, Pakistan

**Corresponding author. E-mail: wajid_shakeel@comsats.edu.pk*

ABSTRACT

The success of the given work is centered on the impact of agile methodology and transformational leadership, mediated by team building in software houses projects. Deductive approach was adopted to accomplish this aim. Data was collected from different software houses and the sample size of the study was approximately 218 respondents. Data was collected with the help of questionnaires from 218 respondents through the convenience sampling technique. The structural equation modeling in Smart-PLS was used to analyze the collected data in order to observe the relationships. The outcomes between the relationship of project success and transformational leadership similarly demonstrate the mediating effect of team building. Agile methodology was accepted and considered to be directly associated with the

Agile Leadership for Industry 4.0: An Indispensable Approach for the Digital Era, Tanusree Chakraborty, Tahir Mumtaz Awan, Ashok Natarajan, and Muhammad Kamran (Eds.)

project's success. This methodology aligns with the preceding studies and the reasons for it is that adopting agile principles will help the project team to accomplish the project objectives and goals which leads to the success of projects. This study primarily focuses upon empirical investigation of team building, agile methodology, and transformational leadership, unique in its implication and scope, especially in the context of Pakistan.

18.1 INTRODUCTION

Continuous progressions in project management practices are currently interested in projects which are running in organizations in different levels, like individual, organizational, and temporary level (Drouin et al., 2018), for investigating to enhance the success ratio of projects (Bond-Barnard et al., 2018). The rapid change in the global environment gives more opportunities and challenges to manage projects successfully (Keeys and Huemann, 2017). The success of a project is highly dependent on consumption of time, scope, and the assigned budget (Raziq et al., 2018). Success of a project is presented as successful just when it by chance performs as per given requirements (Fernando et al., 2018). Henceforth, project manager job can be viewed as exceptionally basic while characterizing and moulding the ideal focuses on execution of project (Yu et al., 2018).

Project management writing requires increasingly experimental investigations on transformational leadership as 80% of project failures happen because of ineffectively created administration (Maqbool et al., 2017). Project Management Institute's (PMI) recent investigation of worldwide project management patterns uncover the wastage of one million dollar every 20 seconds and two trillion dollars consistently because of inadequately performing or potentially failure projects (Project Management Institute, 2018). Significant purposes behind the declining proportions of successful software projects feature the issues of missed cut-off times are of 75%, surpassing arranged spending plans are of 55%, poor correspondence is of 40%, and failure to meet project-explicit necessities of client are of 37% (Kazmi et al., 2016). These various project difficulties interest for transformational authority approach that drives project management greatness, empowers superior groups, and enhances chances to project success (Irfan and Hassan, 2017).

Pakistan is a developing country and takes a few attributes, which are characteristics of other developing nations, especially in Central and South Asia. It is surrounded by two financial organizations, namely, ECO, which takes about ten states from West Asia as well as Central Asia, and SAARC with countries of sub-continent of South Asia, expanding Distant East boundaries (ul Hassan et al., 2014). The software industry of Pakistan has been growing gradually over the past few years. In 1997, Pakistan's entire software trades were $20 million and reached $50 million till 1998 (Hassan, 1998). Previously, in 1993, total exports were not as much of 100,000. In 1996, Pakistan Software Export Board (PSEB) registered 82 software organizations; meanwhile, it has been expanded to more than 120 till March 1997. ul Hassan et al. (2014) emphasized that there are large number of factors that may affect the progress of software industry with financial conditions and government strategy.

Agile is a methodology for developing software that depends on time-frame system just as iterative procedure. Time-frame implies a software engineer appointed with explicit time span; within that time-frame, you need to create and discharge the software with their legitimate usefulness. The time period might be roughly about a month to finish every cycle of the module. It might be much longer in synchronous with higher consumer loyalty (Vijayasarathy and Turk, 2008). This time span will be increased until the software doesn't wind up finished. Agile methodology has been created to plan the software without upgrading, modifying, and refactoring (Adelyar and Norta, 2016; Nagwani and Singh, 2009). This methodology encourages us to comprehend what the client really wants, on the grounds that the dialogs that between the client and designer is about the same project or not (Edeki, 2015). It helps to identify with the module of the software that is required by the client that must be accessible in the item which is under creation stage or has been created. This, in turn, implies that the client sees whatever they had proposed for an item is conveyed or not, and in the event that it isn't the same as prior talk then engineer will apply at least one emphasis to finish their necessity or usefulness appropriately.

Pakistan's software industry has shown positive development trends throughout the previous four years. Over the course of 2016–2017, exports from Pakistan's software industry totaled 3.3 billion dollars, which increased to 5 billion dollars and is anticipated to rise to 6 billion dollars in future (Chung, 2018). The local software market is export oriented; but it is reliant on the imports of the latest services and technologies. Major worldwide brands of software from various nations such as Germany, Spain, USA, United Kingdom, and China officially have been established a durable manifestation in Pakistan ("Pakistan—Computer Software," 2018). Kudos to excellent efforts and increased motivation and incentive for IT, the software sector has seen tremendous development, for example, in Lahore, the Pakistan

Government supported the IT sector, Arfa Kareem Tower with some great inducements (Jamal, 2018), for instance, 100% owner of foreign property, 100% return on capital and earnings, excluding tax on all IT exportations up to June 2019, and investment property tax holiday up to June 2024.

Riaz et al. (2018) observed that implementation of agile software development approaches has turned into an integral part in software development in expansive and developed firms as well as in small firms, apart from the presence of confusion that agile strategies are substantial for expansive scale projects and established firms. The ratio of project failure in software industry is more than project success and number of projects are still going to increase day by day; it is very crucial to understand the cause of failure in projects and to maximize the project success rate (Johnson, 2018). In the rapidly changing environment, project outcomes can be better by adopting agile methodology (Dhir et al., 2019). Additionally, the effect of transformational leadership depends upon the process in which teams were built and it will eventually lead to success of the project (Gupta et al., 2019). Therefore, this research will try to fill a gap in previous studies and examine the transformation of leadership relationship with project success via team building and examine the connection between agile approach with project success in project-based organizations. Organizations may look at their projects using the model and make the required modifications to improve performance.

Since the influence on project success of transformation leadership in the context of an agile environment is neglected in the paradigm of software industry, the following research questions have been developed to address the issue:

RQ1: Does transformation leadership have a significant impact on project success through team building in software houses?;

RQ2: Is there a positive significant contribution of adopting agile methodology on the project success in software houses?

18.2 LITERATURE REVIEW

18.2.1 CONCEPTUALIZING AGILE METHODOLOGY IN RELATION TO STAKEHOLDER THEORY

Agile is a method in software development used in developing software. It is a modern management project management technique that helps the managers as well as employees to do work according to the project in the development and acquisition of software. As far back as 1958, a researcher named Koontz pointed out that "no effective manager makes a plan and then proceeds to put it into effect regardless of what events occur" (Koontz, 1959). Typically, this situation of deviation is very frequently identified in project management. Deviation is the phenomenon that occurs which may affect the real project positively or negatively that is it deviates the managers in carrying out the actual project (Hällgren and Maaninen-Olsson, 2005). Furthermore, solution for this problem of deviation is not always available in the initial plan, so that one has to plan after its occurrence. Meaning that using the standardized methodologies can't provide exact solution to this problem but somehow it has the ability to provide a way for solution. In today's environment of software development projects, the companies are not using the traditional ways of planning such as planning of front-end process as it needs more improvisation to overcome the deviation problem which can help to be do projects in iterative form. As most of the activities are done in designing phase, project managers and developers faces problems in designing as the mismatch occurs between the desired plan and the real design due to confronting critical issues (Van Der Westhuizen and Fitzgerald, 2005). Agile methods focus on the development as well as addressing management aspects from coding, testing and project delivery to people, processes, and techniques. The manifesto of Agile development concentrate on four fundamental values—customer cooperation during contractual negotiations; response to change in accordance with an established procedure; processes and tools interactions; software work on complete documentation, etc. (Dingsøyr et al., 2008).

18.2.2 CONCEPTUALIZING PROJECT SUCCESS IN RELATION TO AGENCY THEORY

Project success is a broad term on which the researchers are not agree upon yet (Pinto and Slevin, 1988). Traditionally, it was measured by fulfilling the triple constraints criteria quality, budget, and cost. Quality denotes the functionality of project whether it is working in accordance to the requirements (Atkinson, 1999; Kerzner, 2003). But now cost, budget, and quality are related to the project management as it ends with delivery of project to the customer. Scholars have measured success by the influence on the organization instead of satisfying triple constrains criteria. Sometimes the client or customer feel highly satisfied even if the real objectives are not fulfilled but sometime even fulfilment of real objectives does not satisfy the client which means that measurement of project success is not straight forward (Thomas and Mengel, 2008). The triple constrains criteria has role in determining the project success, but stakeholder or client satisfaction is considered as part of success as well. Other criteria of success are to meet planned goals, user benefits, client benefit, and overall project success (Dvir et al., 2003). These are highly correlated in making project as successful and creating win-win situation for all the stakeholders. The project success corresponds to motivating expectations on project quality, cost, scope, time, and diverse parties concerned, according to the Project Management Institute (PMI) (Ika, 2012). In terms of project success, the criterion remained important from the 1960s to 1980s, the "iron triangle" (quality, time and cost), and various different additional factors such as organization and project team benefits, team size, and strategic organization (Suyanto and Nurhaeni, 2017). Many of the time quality of project is not achieved which is related to the expertise and competence of the team as it is related to technical performance, achievement of objectives, and specifications of project.

18.2.3 CONCEPTUALIZING TRANSFORMATIONAL LEADERSHIP IN RELATION TO LEADERSHIP THEORY

There is a lack of empirical studies in the context of project management even if the issue of leadership is studied for several decades (Tyssen et al., 2014). The Full range leadership theory is the most widely accepted theory in leadership theory and explores transformational, transactional, and laissez-faire (Sohmen, 2013). The focus in this study is on transformation leading, as research has highlighted the significant impact on project-based organizations (Gundersen et al., 2012). In published literature, there is by all accounts wide trade off on the four measurements which expresses leadership for transformation: aspirational influence, inspiring motivation, intellectual stimulation, and individualizing consideration. In the aspirational influence, leaders consider themselves to be good examples to their supporters and their supporters are trusting, respecting, and increasing in value. Leaders with extraordinary perfect impact are dependable and prepared to go out on a limb (Voon et al., 2011). Inspirational motivation is shown when leaders convey an alluring and motivating vision of their wards and convey them with testing undertakings and expanded desires. Intellectual stimulation raises to the role of leaders in persuading the uniqueness, oddity, development, and innovation of their adherents by clarifying speculations and settling longstanding issues by novel approach. They usually push their followers to try new methods, means, and traps for solving the old issues. Personalized thought include the job of the pioneer to pay uncommon thought on necessities of every supporter by going about as a guide for accomplishing the desired goals (Bass and Riggio, 2006).

18.2.4 CONCEPTUALIZING TEAM BUILDING IN RELATION TO LEADERSHIP THEORY

Team building is considered as a central part of human resources management (HRM) in project-based organizations that concentrates on practices of HRM (Dubow et al., 2017; Turner and Lloyd-Walker, 2008). Klein et al. (2009) define team building as "formal and casual (informal) team-level mediations that focus on enhancing social connections and clarifying jobs, as well as taking care of undertaking and relational difficulties that impact team working." There is an accord in the writing that there are four unmistakable methodologies likewise that may be combined. Goal-setting, illuminating jobs, creating relational relations, and utilizing critical thinking strategies are example of these methodologies (Klein et al., 2009; Salas et al., 1999). Each of the team-building exercises is shown later. Setting goals: This approach entails defining the project's overall goals and explicit objectives for each member of the team, as well as describing subtasks and creating schedules. Individuals on the team who are involved in goal-setting are expected to become involved in activity planning in order to identify ways to achieve those goals. Goal-setting mediation, together with performance estimation and input, has been successfully linked in companies, according to studies (Salas et al., 1999).

18.3 CONCEPTUAL FRAMEWORK

Research model is consisting of two independent variables, one dependent variable and one mediator. Independent variables are Agile Methodology and Transformational leadership, dependent variable is Project Success, whereas team building as a Mediator. This conceptual framework can be visualized in Figure 18.1.

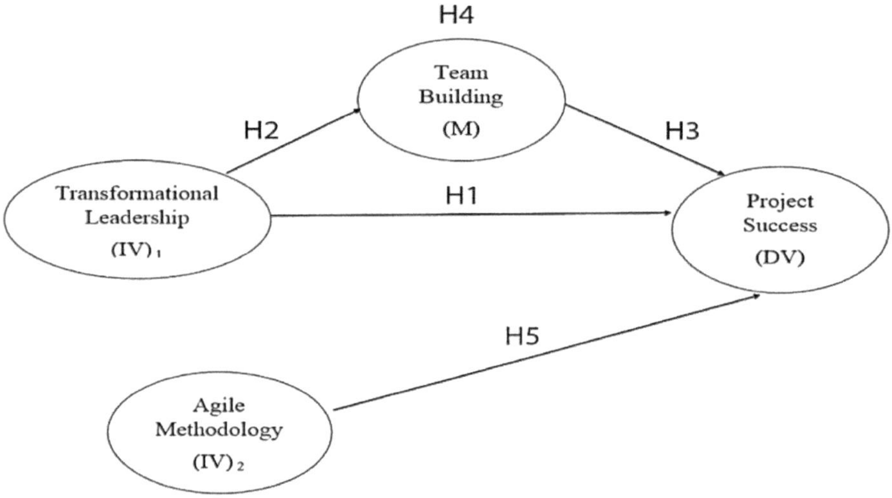

FIGURE 18.1 Conceptual framework.

18.3.1 STUDY HYPOTHESES

H1: There is a positive relationship existed between Transformational Leadership and Project Success.

H2: There is a positive relationship existed between Transformational Leadership and Team Building.

H3: There is a positive relationship existed between Team Building and Project Success.

H4: There is an influential mediating impact of team building among Transformational Leadership and Project Success.

H5: There is a significant association between Agile Methodology and Project Success.

18.3.2 DATASET AND RESEARCH METHODOLOGY

Significant and reframing issues, as well as generating hypotheses or suggesting clarification, are all part of research. To put it another way, research is the gathering and analyzing of data (Mukherjee, 2017). Data is collected using a variety of approaches in quantitative research, including surveys, questionnaires, and rating scales (Rohankar et al., 2015). In this study, data has been collected through questionnaires. Smart-PLS and SPSS software has been used to analyze the data in this study. Instead of constructing several statistics and regression models, the researcher may use Smart-PLS to run realistic models. The researcher can use Smart-PLS to define the link between the variables and estimate, decide, and retrieve their model as a graph.

18.3.2.1 SAMPLE SELECTION

Data was collected by convenience sample technique. A convenience sample is one that is obtained from a group of persons who are simple to contact or approach. This strategy aids in the collection of relevant data and information that would otherwise be impossible to get using probability sampling procedures, which need more formal access to population lists.

The software industry of Islamabad and Rawalpindi, Pakistan, has been chosen as the study's target population. There are 1114 registered software houses in Pakistan, with 245 registered software houses in Islamabad and just 80 registered software houses in Rawalpindi ("PSEB Registered Companies," 2019). So, there are a total of (245 + 80) 325 Registered Software Houses in Islamabad and Rawalpindi. The overall sample size for this study was 218 people (Sekaran and Bougie, 2016). Participants are important project professionals who are project managers or above and have helped lead the projects to improved performance. They were asked to compare and contrast the results of their respective teams. They were asked to compare the success of their respective companies over a three-year period or longer.

18.3.2.2 DATA SCREENING

The data was reviewed once it was collected using descriptive analysis. Outliers and variance are reported in order to gain a basic overview of the variables utilised in the study. It was then assured that there were no missing values in the data and if there was any missing value then it has been supplanted. The values of skewness and kurtosis of the data variables were used to check for non-normality in the data with values ranges between –2.5 and 2.5 (Gravetter and Wallnau, 2007).

18.3.2.3 RELIABILITY AND VALIDITY

Through Cronbach alpha, reliability analysis was used to evaluate the consistency of each concept's measurements. It indicates how strongly the dataset's elements are linked to one another. The Cronbach alpha value near to 1 indicated the highest measure consistency. Cronbach alpha must be greater than or equal to 0.7 as a rule of thumb for reliability tests (Jenkins and Taber, 1977). Validity is the extent to which a measurement values express to the variable they are proposed to. Tests that are valid are reliable too but tests that are reliable aren't valid all the times. More precisely, validity applies to both structure and techniques of the research. In data collection, validity implies that results correctly signify the phenomenon that is claiming to measure (Herbert et al., 1989).

18.3.2.4 FACTOR ANALYSIS

The approach utilized to decrease the data dimensionality was Exploratory Factor Analysis (EFA). It is the multivariate statistical correlational analyses that could be utilized to check the validity of items. This method is used to conclude the number of variables underlying one general variable (Ong and Puteh, 2017). EFA with component rotated with Promax with Kaiser Normalization have been applied to ensure the factor leading in their respective latent variables (Cureton and Mulaik, 1975). The measuring model allows the identification of the link between the variables and their items. The hypothesis of project success has been verified by the confirmatory factor analysis. The CFA is confirmatory method which is theory focused. Thus, analysis is determined by the connection between the observed and unobserved variables. Hypothesized model was used to determine the population variance matrix in CFA and compare the estimated matrix with observed variance matrix. In other words, CFA is used to minimize the different between observed and estimated matrices (Schreiber, 2010).

The suitable sample size was 218 of Islamabad and Rawalpindi software companies. In software houses, a total of 280 surveys were issued. About 250 questionnaires from various software houses have been returned. 218 questionnaires were used among these 250 respondents, of whom 206 were male and 12 were female (see Table 18.1). This results in a 78% response rate.

18.4 RESULTS

18.4.1 DESCRIPTIVE ANALYSIS

The responses were gender-based in analysis separated into two groups. 93.6% contributed to the male group while 6/4% contributed to the female group. In the last several years, the respondents were invited to share their experience. The responders with a 2–5-year experience supplied 63.2%, the greatest percentage of experience in sample data. The respondents were split into three categories based on analytical designations. Class1 (Team leaders) contributed 63.6%, class 2 (Developer) contributed 21.4%, and class 3 (Managers) contributed 15%, which has been shown in the Table 18.2.

TABLE 18.1 Descriptive Analysis of Respondents.

			Software houses		
Gender	**RWP & ISL**	**Experience**	**RWP & ISL**	**Designation**	**RWP & ISL**
Female	12	<2 years	20	Team leader	140
Male	206	2–5 years	139	Developer	47
		>5 years	61	Manager	33
Total	218		218		218

**RWP stands for Rawalpindi; ISL stands for Islamabad.*

18.4.2 STATISTICAL TOOL ANALYSIS

On the basis of selected sample, the internal consistency and reliability of each variable were tested. Researchers discovered a thorough and exact approach, Cronbach's alpha for measurement of internal consistency. The alpha value in Cronbach ranges from 0 to 1 and enhances the internal consistency as the coefficient value approaches 1. Cronbach Alpha value should be larger than 0.7, more specifically scaled as $\alpha > 0.9$ Excellent, $\alpha > 0.8$ excellent, $\alpha > 0.7$ acceptable, $\alpha > 0.6$ doubtable, $\alpha > 0.5$ bad, and $\alpha < 0.5$ not acceptable, according to George and Mallery (2003). The Cronbach alpha value for the whole scale was higher than 0.7, as seen from Table 18.3. This shows that it is quite reliable generally.

The measure of Kaiser-Meyer-Olkin (KMO) should exceed .70 and is insufficient, if it is below .50. The KMO test informs us if every factor predicts or not sufficient items. It is .897 here. So, it's good. The Bartlett test seems important (i.e., a value of less than .05) and this implies that the variables are significantly linked to offer a reasonable basis for the factor analysis. The relevant values of loading factors, composite reliability, and average variance are also well within the limits specified (see Table 18.3).

TABLE 18.2 Factor Loadings of Principal Components of Independent, Dependent, and Mediating Variables.

Variables	Factor				CR	AVE
	1	2	3	4		
Agile methodology					0.909	0.790
AM1			0.733			
AM2			0.925			
AM3			0.821			
AM4			0.882			
Transformational leadership					0.789	0.527
TL1				0.687		
TL2				0.758		
TL3				0.676		
TL4				0.748		
TL5				0.517		
Team building					0.939	0.731
TB1	0.819					
TB2	0.786					
TB3	0.913					
TB4	0.896					
TB5	0.858					
TB6	0.766					
TB7	0.722					
Project success					0.886	0.641
PS1		0.679				
PS2		0.854				
PS3		0.958				
PS4		0.736				
PS5		0.718				
PS6		0.387				

*CR is composite reliability; AVE means average variance explained.

18.4.2.1 T-VALUE ANALYSIS

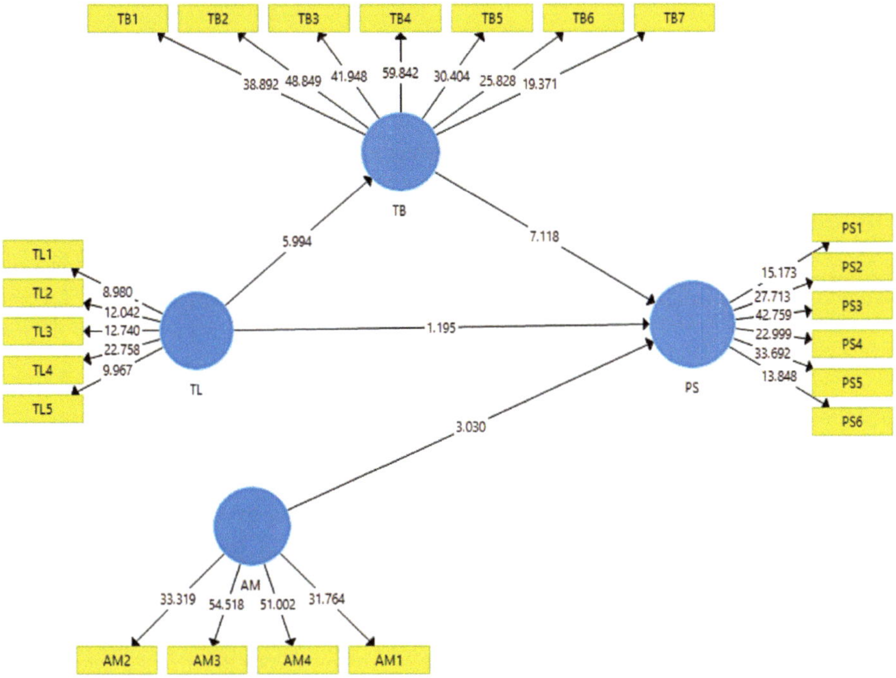

FIGURE 18.2 *t*-values for the structural equation model.

All *t*-values are above from 1.96, so all values seemed good. The *t*-values for TLTB is 5.994, for TB→PS is 7.118, for TL→PS is 1.195, and for AM→PS is 3.30.

18.4.2.2 PATH COEFFICIENTS (BOOTSTRAPPING)

The influence of transformational leadership on project success is mediated through team building. Results indicated that the success of project was a major predictor of team building positivity with 0.419, $t = 7.607$, $P < .005$, and that team building positivity was a significant predictor of Transformational leadership with 0.326, $t = 5.408$, $P < .005$. These results support the mediational hypothesis. The success of the project was no longer a major indicator of transformational leadership after controlling for the mediator with 0.067, t = 1.130, non-significant, consistent with full mediation, (see Table 18.4).

TABLE 18.3 Path Coefficients.

	Original sample (O)	Sample mean (M)	Standard deviation (STDEV)	*t*-statistics	*P* values
AM -> PS	0.210	0.208	0.067	3.124	0.002
TB -> PS	0.419	0.422	0.055	7.607	0.000
TL -> PS	0.067	0.075	0.059	1.130	0.259
TL -> TB	0.326	0.330	0.060	5.408	0.000

Note: AM, agile methodology; TL, transformational leadership; TB, team building; PS, project success.

18.4.2.3 TOTAL AND INDIRECT EFFECT (MEDIATION)

The study shows total indirect and total effect significant with 0.137 having t = 4.132 (P < 0.005) and total effect significance at 1% for all variables (Table 18.5). In path coefficients, direct path is no longer significant but mediational path is significant so we are having full mediation.

TABLE 18.4 Indirect Effects.

	Original sample (O)	Sample mean (M)	Standard deviation (STDEV)	t-statistics	*P* values
TL -> TB -> PS	0.137	0.140	0.033	4.132	0.000

Note: TL, transformational leadership; TB, team building; PS, project success.

TABLE 18.5 Total Effect.

	Original sample (O)	Sample mean (M)	Standard deviation (STDEV)	*t*-statistics	*P* values
AM -> PS	0.210	0.208	0.067	3.124	0.002
TB -> PS	0.419	0.422	0.055	7.607	0.000
TL -> PS	0.204	0.215	0.061	3.356	0.001
TL -> TB	0.326	0.330	0.060	5.408	0.000

Note: AM, agile methodology; TL, transformational leadership; TB, team building; PS, project success.

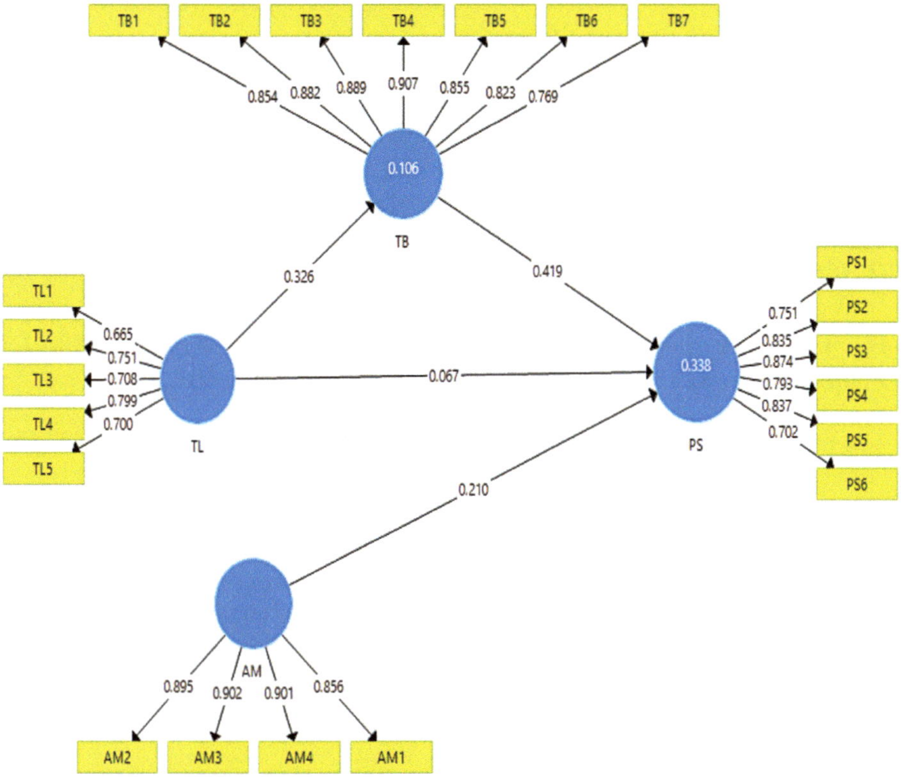

FIGURE 18.3 Factor loadings and path coefficients.

Figure 18.3 shows that factor loadings with each construct are greater than 0.5 which is good and path coefficients are for TL→PS, TL→TB, TB→PS, and AM→PS are 0.067, 0.326, 0.419, and 0.210, respectively.

18.4.3 HYPOTHESES RESULTS

H_1: There is a positive relationship between Transformational Leadership and Project Success.

The analysis carried out shows that the transformational leadership is positively linked to the success of project, which is demonstrated by the regression value (0.067) of the variable known as transformational leadership, which demonstrates the positive relationship between transformation leadership and success of project. Hence, null hypothesis for H_1 is rejected.

H_2: There is a positive relationship between Transformational Leadership and Team Building.

The second hypothesis postulated that, "There is a positive relationship existed between Transformational Leadership and Team Building." This is demonstrated by the regression value (0.326) of the variable called transformational leadership, which shows the positive relationship between transformational leadership and team building. Hence, null hypothesis for **H_2** is rejected.

H_3: There is a positive relationship existed between Team Building and Project Success.

The third hypothesis postulated that, "There is a positive relationship existed between Team Building and Project Success." The analyses carried out for this show that team building is positively and is strongly linked to the success of the project. The regression value (0.419) for the variable called team building is obvious, showing positive association between team building and the success of the project. Hence, null hypothesis for **H_3** is rejected.

H_4: There is an influential mediating impact of team building among Transformational Leadership and Project Success.

The fourth hypothesis postulated that, "There is an influential mediating impact of team building among Transformational Leadership and Project Success." The analysis being done for this demonstrates that the team building plays a **full mediation** role among the transformational leadership and the project success. When we control mediating variable, the direct path that is TL→PS become nonsignificant which results into full mediation. Hence, null hypothesis for **H_4** is rejected.

H_5: There is a significant association between Agile Methodology and Project Success.

The analysis done showed that agile methodology is positive, and is strongly associated with the success of the project, which was demonstrated by the value of regression (0.210) for variable called agile methodology, and that demonstrate the positive association between agile methodology and the success of the project. Hence, null hypothesis for **H_5** is rejected.

18.5 DISCUSSION

In project-based firm, agile methodology plays a significant role towards project success. Researchers have requested investigation of internal and external aspects of companies that may guide the link between transformational leadership and project success to achieve the inconsistent result management. ("Expressive-Level Comments and Social Media: Challenging a Model of Ethical Reflection," 2018). Transformational leadership contributes to and encourages team building and learning principles, such as cooperation or transparency. Transformational leadership may be seen as a key management approach for supporting project success (Jensen et al., 2019); a "difficult to understand" chaotic environment that demands a flexible organization to adapt swiftly and effectively to remain competitive (Sihotang et al., 2016).

Performance measurement is essential for coordination, support and leadership management, and decision-making and project success (Smith and Bititci, 2017). Organizations with transformational leadership can be more receptive to advancement and this is important because of the complexity of learning when a company enters a foreign market (Ferns, 2018). Team building techniques enhance the corporate knowledge of a company in the collection, sharing, and distribution of the market and business data necessary to successfully turn it into a market and enterprise (Mapalala et al., 2018). Agile methodology refers to the extremely diverse procedure that operates and strongly affect the project success (Schoenberg et al., 2013). In the presence of team building enterprises that pioneer creation and launch of new goods, new technologies or inventions managed by companies that exercise transformational leadership frequently perform better. (Lee and Chen, 2009).

Team building is both a procedure as well as outcome (Huang et al., 2019). Unlearning companies can enable teams to leverage existing data and develop new knowledge, elements that are important concerns for innovation implementation. (Cegarra-Navarro et al., 2016). Moreover, the success of the project offers important input on team building process efficiency and effectiveness. (Zid et al., 2020). Organizations are working nowadays in an extremely focused and internationally challenging field where agile approach plays its leading role. Agile approach creates enormous inventive thinking, challenges employees' mentality, and paves the door for new international opportunities and enhances the success of a project.

18.6 CONCLUSION AND RECOMMENDATIONS

Pakistan's software sector is progressing slightly. Taking into account the results of this study, certain proposals are relevant and significant to the software industry stakeholders. These crucial relationships with project stakeholders must be thought through by project managers. Before interacting with their group, they should evaluate every element of the company. Further future research may examine the effect on successful projects and the success of the projects of any transformational leadership and agile methodology characteristics. The employees' confidence must be strengthened by offering them the chance to think and act in order to implement their own learning and innovation. In future investigations, team performance as an exchange mediator between agile methodology and project success can increase the scope of this study. The creative ideas that employees develop can assist the team achieve its duties and goals and finally complete its project successfully and boost its company performance. This scenario involves transformational leadership as it increases the team building work side to achieve learning and innovation. The influence and the impact of culture were not taken into account.

However, culture and workplace both can have a major impact on the success of the project. A more worldwide investigation needs to be carried out to demonstrate diverse outcomes in the software business in order to have a better image of the study. Pakistan's software industry is thriving extremely quickly. Considering the study's outcomes, certain techniques are relevant and significant for the software industry's literature. Since this study is a part of project management along with software business, it also contributes in the literature of both industries as well.

Future studies concentrated in this professional group can duplicate a study research approach for greater sample sizes and more gender equilibrium. They might incorporate the effect on successfully completed projects of every transformative management and agile methods dimension. The team-building has an immense influence on the team and makes it more interesting and inventive for the team to engage in the project. Agile approach is very important in this circumstance since it increases the work side of organizational initiatives for improved performance and delivery.

KEYWORDS

- **transformational leadership**
- **team building**
- **agile methodology**
- **project success**
- **software projects**

REFERENCES

Adelyar, S. H.; Norta, A. In *Towards a Secure Agile Software Development Process*, Paper presented at the 2016 10th International Conference on the Quality of Information and Communications Technology (QUATIC); 2016.

Ambler, S. W. Agile Testing Strategies. *Dr. Dobb's Journal, January*. **2007.**

Atkinson, R. Project Management: Cost, Time and Quality, Two Best Guesses and a Phenomenon, Its Time to Accept other Success Criteria. *Int. J. Project Manag.* **1999,** *17* (6), 337–342.

Ayuso, S.; Navarrete-Báez, F. E. How Does Entrepreneurial and International Orientation Influence SMEs' Commitment to Sustainable Development? Empirical Evidence from Spain and Mexico. *Corp. Soc. Responsib. Environ. Manag.* **2018,** *25* (1), 80–94.

Bass, B. M.; Riggio, R. E. *Transformational Leadership*; Psychology Press, 2006.

Beukers, M. W. Project Management of Life-Science Research Projects: Project Characteristics, Challenges and Training Needs. *Drug Disc. Today* **2011,** *16* (3–4), 93–98.

Bond-Barnard, T. J.; Fletcher, L.; Steyn, H. Linking Trust and Collaboration in Project Teams to Project Management Success. *Int. J. Manag. Projects Bus.* **2018,** *11* (2), 432–457.

Cegarra-Navarro, J.-G.; Soto-Acosta, P.; Wensley, A. K. Structured Knowledge Processes and Firm Performance: The Role of Organizational Agility. *J. Busi. Res.* **2016,** *69* (5), 1544–1549.

Chung, C.-P. What are the Strategic and Economic Implications for South Asia of China's Maritime Silk Road initiative? *Pacific Rev.* **2018,** *31* (3), 315–332.

Cureton, E. E.; Mulaik, S. A. The Weighted Varimax Rotation and the Promax Rotation. *Psychometrika,* **1975,** *40* (2), 183–195.

Dhir, S.; Kumar, D.; Singh, V. Success and Failure Factors that Impact on Project Implementation using Agile Software Development Methodology. In *Software Engineering*; Springer, 2019; pp. 647–654.

Dingsøyr, T.; Dybå, T.; Abrahamsson, P. In *A Preliminary Roadmap for Empirical Research on Agile Software Development*; Paper Presented at the Agile 2008 Conference, 2008.

Drouin, N.; Müller, R.; Sankaran, S.; Vaagaasar, A. L. Balancing Vertical and Horizontal Leadership in Projects. *Int. J. Manag. Proj. Bus.* 2018.

Dubow, E. F.; Aber, J. L.; Betancourt, T. S.; Cummings, E. M.; Huesmann, L. R. Conducting Longitudinal, Process-Oriented Research with Conflict-Affected Youth: Solving the Inevitable Challenges. *Dev. Psychopathol.* **2017,** *29* (1), 85–92.

Dvir, D.; Raz, T.; Shenhar, A. J. An Empirical Analysis of the Relationship between Project Planning and Project Success. *Int. J. Project Manag.* **2003,** *21* (2), 89–95.

Edeki, C. Agile Software Development Methodology. *Eur. J. Math. Comput. Sci.* **2015,** *2*(1).

Fernando, Y.; Walters, T.; Ismail, M. N.; Seo, Y. W.; Kaimasu, M. Managing Project Success Using Project Risk and Green Supply Chain Management. *Int. J. Manag. Projects Bus.* **2018,** *11* (4).

Ferns, S. Collaboration, Cooperation and Consultation: Work-Integrated Learning Partnerships for Enhancing Graduate Employability. Doctoral Thesis, University of Western Australia, Perth, Australia, 2018.

George, D.; Mallery, M. *Using SPSS for Windows Step by Step: A Simple Guide and Reference*; 2003.

Gravetter, F.; Wallnau, L. *Statistics for the Behavioral Sciences*. Thomson Learning Inc.: Belmont, CA, 2007.

Gundersen, G.; Hellesøy, B. T.; Raeder, S. Leading International Project Teams: The Effectiveness of Transformational Leadership in Dynamic Work Environments. *J. Leaders. Organiz. Stud.* **2012,** *19* (1), 46–57.

Gupta, V. K.; Niranjan, S.; Markin, E. Entrepreneurial Orientation and Firm Performance: the Mediating Role of Generative and Acquisitive Learning Through Customer Relationships. *Rev. Manag. Sci.* **2019,** 1–25.

Hällgren, M.; Maaninen-Olsson, E. Deviations, Ambiguity and Uncertainty in A Project-Intensive Organization. *Proj. Manag. J.* **2005**, *36* (3), 17–26.

Hassan, S. Z. A Framework for It Industry Development: A Case Study of Pakistan. *J. Glob. Inform. Technol. Manag.* **1998,** *1* (4), 38–55.

Herbert, W.; Seliger, H. W.; Shohamy, E. G.; Shohamy, E. *Second Language Research Methods*; Oxford University Press, 1989.

Huang, H.; Zuo, L.; Wang, G.; Qin, L. *Entrepreneurial Network Pathways to Venture Growth: An fsQCA Analysis*, Paper Presented at the International Conference on Innovation and Entrepreneurship, 2019.

Ika, L. A. Project Management for Development in Africa: Why Projects are Failing and What can be Done about It. *Proj. Manag. J.* **2012,** *43* (4), 27–41.

Irfan, M.; Hassan, M. The Effect of Project Success on Corporate Reputation of the Public Sector Organizations in Pakistan. *Int. J. Econ. Manag.* **2017,** *11.*

Jamal, A. C. Coworking Spaces in Mid-Sized Cities: A Partner in Downtown Economic Development. *Environ. Plann. A: Econ. Space* **2018,** *50* (4), 773–788.

Jenkins, G. D.; Taber, T. D. A Monte Carlo Study of Factors Affecting Three Indices of Composite Scale Reliability. *J. Appl. Psychol.* **1977,** *62* (4), 392.

Jensen, U. T.; Andersen, L. B.; Bro, L. L.; Bøllingtoft, A.; Eriksen, T. L. M.; Holten, A.-L.; ... Salomonsen, H. H. Conceptualizing and Measuring Transformational and Transactional Leadership. *Admin. Soc.* **2019,** *51* (1), 3–33.

Johnson, J. *CHAOS Report: Decision Latency Theory: It is all About the Interval*; Lulu. com, 2018.

Kazmi, S. S.; Rafi, A.; Hassan, M. Impact of Benefit Realization Management on Two-Dimensional Model of Project Success: Evidence from Pakistani Telecom Industry. *Int. Rev. Manag. Mark.* **2016,** *6* (4S), 32–37.

Keeys, L. A.; Huemann, M. Project Benefits Co-Creation: Shaping Sustainable Development Benefits. *International Journal of Project Management,* **2017**, *35*(6), 1196–1212.

Kerzner, H. *Advanced Project Management: Best Practices on implementation*; John Wiley & Sons, 2003.

Klein, C.; DiazGranados, D.; Salas, E.; Le, H.; Burke, C. S.; Lyons, R.; Goodwin, G. F. Does Team Building Work? *Small Group Res.* **2009,** *40* (2), 181–222.

Koontz, H. Management Control: A Suggested Formulation of Principles. *Calif. Manag. Rev.* **1959,** *1* (2), 47–55.

Lee, R. P.; Chen, Q. The Immediate Impact of New Product Introductions on Stock Price: the Role of Firm Resources and Size. *J. Prod. Innov. Manag.* **2009,** *26* (1), 97–107.

Mapalala, M. J.; West, G. R.; Winston, B. Examining the Relationship between Entrepreneurial Orientation and Organizational Performance: The Moderating Role of Organizational Learning. *Int. Leaders. J.* **2018,** *15* (3), 40–62.

Maqbool, R.; Sudong, Y.; Manzoor, N.; Rashid, Y. The Impact of Emotional Intelligence, Project Managers' Competencies, and Transformational Leadership on Project Success: An Empirical perspective. *Proj. Manag. J.* **2017,** *48* (3), 58–75.

Mukherjee, M. A Review of Research Design. **2017**.

Nagwani, N. K.; Singh, P. An Agile Methodology based Model for Change-Oriented Software Engineering. *Int. J. Recent Trends Eng.* **2009,** *1* (1), 128.

Ong, M. H. A.; Puteh, F. Quantitative Data Analysis: Choosing between SPSS, PLS, and AMOS in Social Science Research. *Int. Interdisc. J. Sci. Res.* **2017,** *3* (1), 14–25.

Pinto, J. K.; Slevin, D. P. Critical Success Factors in Effective Project Implementation. In *Project Management Handbook*; Cleland, D. I., King, W. R., Eds.; Vol. 479; 1988; pp. 167–190.

Raziq, M. M.; Borini, F. M.; Malik, O. F.; Ahmad, M.; Shabaz, M. Leadership Styles, Goal Clarity, and Project Success. *Leadersh. Organ. Dev. J.* **2018,** *39* (5).

Riaz, M. N.; Mahboob, A.; Buriro, A. Social Success Factors Affecting Implementation of Agile Software Development Methodologies in Software Industry of Pakistan: An Empirical Study. *Int. J. Adv. Comp. Sci. Appl.* **2018,** *9*, 94–98.

Rohankar, R.; Katti, C.; Kumar, S. Comparison of Energy Efficient Data Collection Techniques in Wireless Sensor Network. *Procedia Comp. Sci.* **2015,** *57*, 146–151.

Salas, E.; DiazGranados, D.; Weaver, S. J.; King, H. Does Team Training Work? Principles for Health Care. *Acad. Emer. Med.* **2008,** *15* (11), 1002–1009.

Salas, E.; Rozell, D.; Mullen, B.; Driskell, J. E. The Effect of Team Building on Performance: An Integration. *Small Group Res.* **1999,** *30* (3), 309–329.

Schoenberg, R.; Collier, N.; Bowman, C. Strategies for Business Turnaround and Recovery: A Review and Synthesis. *Eur. Bus. Rev.* **2013**.

Schreiber, S. J. Interactive Effects of Temporal Correlations, Spatial Heterogeneity and Dispersal on Population Persistence. *Proc. R. Soc. B: Biol. Sci.* **2010,** *277* (1689), 1907–1914.

Sekaran, U.; Bougie, R. *Research Methods for Business: A Skill Building Approach*; John Wiley & Sons, 2016.

Sihotang, J.; Kartini, D.; Rufaidah, P. Environmental Turbulence, Entrepreneurial Orientation and Business Unit Performance: Effects on Dynamic Capabilitie and Strategic Alliance Formation and its Role to Build Sustainable Competitive Advantage. *Int. J. Econ. Comm. Manag.* **2016,** *317*.

Smith, M.; Bititci, U. S. Interplay between Performance Measurement and Management, Employee Engagement and Performance. *Int. J. Oper. Prod. Manag.* **2017,** *37*, 1207–1228.

Sohmen, V. S. Leadership and Teamwork: Two Sides of the Same Coin. *J. IT Econ. Dev.* **2013,** *4* (2), 1–18.

Suyanto, U. Y.; Nurhaeni, I. D. A. Transformational Leadership Roles in Improving Employees' Creativity in E-Commerce-Based Firms in Indonesia. *KnE Soc. Sci.* **2017,** *2* (4), 259–265.

Thomas, J.; Mengel, T. Preparing Project Managers to Deal with Complexity–Advanced Project Management Education. *Int. J. Proj. Manag.* **2008,** *26* (3), 304–315.

Turner, R.; Lloyd-Walker, B. Emotional Intelligence (EI) Capabilities Training: Can it Develop EI in Project Teams? *Int. J. Manag. Proj. Bus.* **2008,** *1*, 512–534.

Tyssen, A. K.; Wald, A.; Spieth, P. The Challenge of Transactional and Transformational Leadership in Projects. *Int. J. Proj. Manag.* **2014,** *32* (3), 365–375.

ul Hassan, I.; Qadri, S.; Bashir, R. N.; Saleem, R. M.; Naeem, M. Issues of Implementation of CMMI in Pakistan Software Industry. *Int. J. Innov. Appl. Stud.* **2014,** *9* (2), 547.

Van Der Westhuizen, D.; Fitzgerald, E. P. In *Defining and Measuring Project Success*, Paper presented at the Proceedings of the European Conference on IS Management, Leadership and Goverance, 2005.

Vijayasarathy, L.; Turk, D. Agile Software Development: A Survey of Early Adopters. *J. Inform. Technol. Manag.* **2008,** *19* (2), 1–8.

Voon, M. L.; Lo, M. C.; Ngui, K. S.; Ayob, N. B. The Influence of Leadership Styles on Employees' Job Satisfaction in Public Sector Organizations in Malaysia. *Int. J. Bus. Manag. Soc. Sci.* **2011,** *2* (1), 24–32.

Yu, W.-D.; Cheng, S.-T.; Ho, W.-C.; Chang, Y.-H. Measuring the Sustainability of Construction Projects Throughout their Lifecycle: A Taiwan Lesson. *Sustainability* **2018,** *10* (5), 1523.

Yukl, G. *Leadership in Organizations*, 9th ed.; Pearson Education India, 1981.

Zid, C.; Kasim, N.; Soomro, A. R. Effective Project Management Approach to Attain Project Success, based on Cost-Time-Quality. *Int. J. Proj. Organ. Manag.* **2020,** *12* (2), 149–163.

CHAPTER 19

Job Satisfaction among School Teachers: Probing into the Nature of Agility in Government and Private Secondary School

ANUP BAUL* and MIHIR KUMAR SHOME

Department of Management & Humanities, NIT, Arunachal Pradesh, India

**Corresponding author. E-mail: anupmax.max@gmail.com*

ABSTRACT

This study discusses the job contentment level between government and private secondary school teachers in Tripura, India. Job satisfaction refers to the positive or favorable attitude toward one's work. A satisfactory attitude encourages a teacher to work efficiently while dissatisfaction leads to disappointment of instructive objectives. A questionnaire developed by the Minnesota University to assess the level of satisfaction (MSQ) was used as an instrument to analyze the collected data. Mean scores of 20 facets were calculated and a *t*-test was applied to compare the job contentment level between government teachers and private school teachers and male and female teachers. The responses were traced on a 5-point Likert scale. A total number of 160 teachers were taken into consideration as respondents for the study from 20 schools altogether. The schools were categorized into two groups; 10 government or government-aided schools and remaining 10 from private-owned schools. Altogether 80 respondents were from government schools and the remaining 80 respondents were from private schools. The

Agile Leadership for Industry 4.0: An Indispensable Approach for the Digital Era, Tanusree Chakraborty, Tahir Mumtaz Awan, Ashok Natarajan, and Muhammad Kamran (Eds.)

study was conducted in the west region of Tripura. The result shows that there is a huge contrast in the mentality of the educators toward job contentment on different elements. However, it was also found that that there are no noteworthy disparity in job contentment level among male versus female teachers working at secondary level schools and no critical contrast between teachers working in the government and the private sector toward job satisfaction level.

19.1 INTRODUCTION

Education prepares a child for a productive, progressive, refined, and cultured being where the teacher dwells in the central position of the whole education system. Hence, it is essential to provide the utmost facilities to a teacher. Teaching includes substantial responsibility, which might result in a feeling of frustration and dissatisfaction toward the assigned role. The educators would be interested to show their understudies adequately when they are happy with their positions. If they are happy with their work, they can perform well and address the difficulties in the field of education. Velmurugan's 2016 study (as cited in Sasi Kumar and P. Balasubramani, 2019) has pointed out that the authoritative improvement has hands-off, hands-on satisfaction of workers. So, every organization should be familiar with the term *job satisfaction.* In keeping with the said issue, this study was conducted to trace the level of job contentment among the secondary school teachers in Tripura, a state in India.

The Indian education system has always been staunchly favoring the fullest progress of a student and state as well. In view of this context, satisfied teachers in the school can give better education to students in contrast to dissatisfied teachers. Teachers can revolutionize the overall society, the economic well-being of the student, and the country. It is an immense load for the school management to provide qualitative education due to the factors leading against it. Teachers' job satisfaction involves performing their responsibilities with utmost commitment while getting fulfillment from their work. The significance of job contentment lies in the way that it is firmly connected to the execution and efficiency of an individual and is influenced by various variables; to name a few are promotion, recognition and relation with management-influenced teachers, and positivity; so such factors should be addressed in order to improve the educational system in the country. Bamundo and

Kuppleman's 1980 study (as cited in Bhatt, 2018) inspected the impacts of seven factors on job satisfaction and found that instruction and pay decidedly and unequivocally directed occupation satisfaction and life accomplishment relationship.

Since 1957, academic studies began to assess various diagnostic tools to measure and evaluate work adjustment, which was embodied in the work "Theory of Work Adjustment." This work stresses on the vocational needs and abilities of an employee; the tools were built to serve as a manual for the said purpose (David et al., 1967).

Teachers working in schools feel affected on grounds of promotion, recognition, and in relation with management (Chauhan et al., 2017), and this leads to job discontentment. Drawing in and retaining great instructors are essential and are the challenges of the instructive foundations of educational institutions.

19.1.1 JOB CONTENTMENT IN AGILE ORGANIZATION

Job contentment has consistently been a blaze point of conversation among the specialists and researchers since long. It has turned into a significant piece of every organization nowadays. The education system likewise has been changed into an association. Hence, estimating job contentment has turned into a great focal point of consideration for analysts to make it dynamic and effective. It is significant to offer teachers with the best facilities to make them feel satisfied about the status of their job, which would lead to better performance of a teacher. Job satisfaction is only possible if the job fulfils the fundamental requirements of instructors in terms of pay and better status, as clarified by Newstrom (2007) (cited in Rajendran & Veerasekaran, 2013) that job satisfaction is a bunch of great or negative inclination and feelings with which worker see their work and the administrators should be ready about worker work-fulfillment level.

Given the importance of job satisfaction among teachers, estimating their job satisfaction has become a vital factor for analysts and public policy officials. The educational management nowadays aims to give educators the most extreme offices so that they could be happy with their work and thereby drive a superior exhibition from an instructor. Occupational fulfilment is conceivable if the work satisfies the essential necessities of the instructors in terms of pay and better status.

19.1.2 ARE EDUCATIONAL SYSTEMS AGILE ENOUGH?

Rajendran and Veerasekaran (2013) find that there was a massive gap of job contentment among teachers. Hence, it is utmost necessary to beat the issue and provide job satisfaction to educators. Today, the education system has been allotted in the development strategy and cognizant endeavors have made toward the monstrous extension of instructive organizations in the state. A few commissions in the nation have expressed worry about the quality, work fulfilment, and inspiration of the educators. Educators with professional degrees, advanced strategies/policies, incentives, and remunerations couldn't stop the deteriorating education standards, creating an alarming situation in the country. Job satisfaction among educators has been considered as a crucial factor in the significance of the training framework and it communicates the degree of the match between the worker's assumptions and work fulfillment.

19.2 REVIEW OF LITERATURE

Bhat (2018) carried out a study on job fulfillment among high school teachers and found that there is no major disparity in the attitude of male and female teachers working in high school, in terms of job satisfaction. Further he added that there is no significant difference between high school teachers working in government schools and those working in the private sector in terms of job satisfaction, but there is major cooperation impact of sex with the sort of organization/school on work fulfillment.

Choudhury and Demor (2019) investigated the job satisfaction of English teachers at the secondary level in the city of Vadodara, Gujarat, and showed that there was no significant relationship between overall teacher's job contentment level and number of years in teaching. However, there was a moderate negative relationship between the satisfaction with operating procedures and the years of teaching experience. Further they added that, with the increase of teaching experience, teachers were less satisfied with the load of teaching in their work.

Chauhan et al. (2017) accomplished a study about job liking among higher secondary school teachers in Meerut, India, and found that most of the school teachers in Meerut had better job satisfaction in terms of salary and working conditions, which increased their morale and assisted them in accomplishing their assigned tasks. Further, they added that there were other factors, like promotion, recognition, and relation with management, that influenced teachers positively.

Achanta et al. (2014) found that there is a dissimilarity in the intensity of job contentment between both the genders. Kumar et al. (2019) stresses that the acknowledgment for extra work taken, clearly laid out advancement strategies in term of promotion, and salary enabled the teachers to work as per the way of thinking/vision/mission/technique of the administration. Rajendran and Veerasekaran (2013) conducted a study and found that there was a significant difference in work fulfillment among male and female teachers working at the secondary school level, though no critical distinction was secured between urban and rural teachers in their attitude toward work fulfillment. Raymond (2018), in his review presumed that the exploration members in his review experienced low inspiration and low cleanliness as of previous instructors and workers of Oklahoma City Public Schools. However, Gihar and Mishra (2018) investigated a case study about the job contentment among teachers working at secondary level at Bareilly district of UP and found that there are no significant gender, locality, and work experience differences toward job contentment. Nigama et al. (2018) studied job contentment level among the school teachers and found that there is no significant difference in the height of contentment irrespective of sex.

19.3 STATEMENT OF THE PROBLEM

The overall effectiveness of an organization depends upon the job satisfaction of the employees of that organization and work fulfillment level among teachers in the educational field is considered to be vital in reducing the negative factors, such as growth in dropout rates, mental disorder among the teachers and students, decrease in enrolment ratio of students due to lack of interest, etc. Dissatisfied teachers cannot utilize their full potential in their assigned role leading to disastrous condition of the whole education system. There are many factors responsible in this regard such as wage or salary, working hours, environment, scope for career development, emotional bonding, etc. Several studies have been conducted to assess the level of satisfaction among the employees of various organizations or institutions with varied findings. Hence, this study focused to assess the job contentment level among teachers working in both government and private schools in Tripura.

19.4 SIGNIFICANCE OF THE STUDY

The study is important because it would help to sort out and examine the factors affecting job satisfaction among high school teachers in Tripura. Work-related satisfaction is the most significant viewpoint for any work and for teachers it is very important since teaching is the most rousing vocation on the sphere. The education segment in Tripura has always neglected various issues, like job security. For instances are the termination of 10,323 teachers from various teaching posts contrary to the provision of NCT guidelines of recruitment, 1993; case no. 2868 at Tripura high court (2020); fixed pay scale for 5 years in fresh recruitment; contractual recruitment by private vendors; muddled transfer policy; huge gap with the central pay scale; increased dropout rates at the secondary school level (Dropout Drive, 2010–11), etc. Many social organizations took steps to influence the higher authorities for some corrective course and to save this decent profession from menaces such as dissatisfaction. The outcomes would likewise assist with understanding the components liable for work pressure and change among instructors. The study might be useful in drawing attention of the authorities to some of the exertion of the instructors. It would assist them to understand different components of change and discover available resources to adapt to changes. The study would also help all the stakeholders for better strategies in order to reduce work feelings of anxiety and to develop work fulfillment in teachers.

19.5 METHODS OF STUDY

The sample for the study comprised of the high school teachers working in the west locale of Tripura. A sample of 160 teachers consisting of 10 private school and 10 government schools were taken into consideration and from each category 80 teachers were taken with a ratio of 50% male and remaining 50% female teachers. Hence, altogether 160 numbers of teachers were taken for the study.

19.5.1 OBJECTIVES OF THE STUDY

The objectives of this study are listed as follows:

- To recognize the issues affecting job contentment of teachers in Tripura at the secondary level.

- To draw a comparative study between private and government school teacher's attitude toward job satisfaction in Tripura.
- To identify if gender is a factor affecting the level of job satisfaction of teachers in Tripura at the secondary level.

19.5.2 TOOLS USED FOR THE STUDY

Research instruments are the tools used for any study and for this study the instrument used to collect data was a questionnaire framed by the University of Minnesota. This tool is widely used for analyzing job satisfaction among employees of various organizations including teachers. The questionnaire consists of 20 facets of job satisfaction with choice based on a 5-point Likert scale. Mean score of their responses were calculated and ANOVA and *t*-test were applied. The responses were documented on a 5-point Likert scale. For data analysis or calculation, computational software like MS Excel was used. The segment factors like sex and school type were also taken from the respondents. The tool is available in two outlines: a long form and a short form. This study uses the short form with 20 factors.

19.5.2.1 STATISTICS USED

As per the nature of the data, mean, standard deviation, degree of freedom, and *t*-test were calculated at the background of the study.

19.5.3 HYPOTHESIS

Ho: There is no major disparity in the level of job contentment among secondary school teachers of Tripura.

H1: There is a major inconsistency in the level of job contentment between government and private school teachers in Tripura.

H2: There is a major disparity in the level of job contentment between male and female teachers at secondary level in Tripura.

19.5.4 DATA ANALYSIS AND INTERPRETATION

TABLE 19.1 Overall Satisfaction Level Scores from 160 Participants from 20 Schools.

Scale	Total score	Mean	*p* Value is	*t*-test value	Level of significance
Very dissatisfied	**481**	24.05	**.000243**	**12.399777**	**The result is significant at $p < .05$.**
Dissatisfied	1417	70.85			
Neutral	**299**	14.95			
Satisfied	687	34.35			
Very satisfied	**316**	15.8			

Table 19.1 indicates that there is a major disparity in the attitude of secondary school teachers toward job satisfaction in Tripura. Hence, we can reject the null hypothesis as there is a significant difference at $p < .05$ in the level of job contentment among the teachers of Tripura on various facets such as policies, support, supervision, freedom, etc.

TABLE 19.2 Analysis of Sample Data by Gender (Male vs. Female).

Gender	Group	Mean	p Value	f-Ratio value	Level of significance	
Male teachers	N 1 = 80	29.05			The result	
Female teachers	N 2 = 80	20.06	0.90	0.012	is not significant at $p < .05$	
ANOVA: Single factor						
Summary						
Groups	*Count*	*Sum*	*Average*	*Variance*		
Column 1	5	1550	310	29961.5		
Column 2	5	1650	330	89408.5		
ANOVA						
Source of variation	*SS*	*df*	*MS*	*f*	*p* Value	*f* Crit
Between groups	1000	1	1000	0.01	0.90	5.32
Within groups	477480	8	59685			
Total	478480	9				

The *f*-ratio value is 0.01676. The *p*-value is .900205. The outcome is not considerable at .05 level of significance.

Table 19.2 reflects that there is no noteworthy variation in the attitude of male versus female secondary school teachers toward job satisfaction in Tripura at $p < .05$; drawing a similarity with the study of Gihar and Mishra (2018) that there is no major disparity in respect to gender, locality, and experiences toward job satisfaction among secondary school teachers.

TABLE 19.3 Analysis of Sample Data by Type of School (Government vs. Private).

Gender	**Group**	**Mean**	***p* Value**	***f*-Ratio value**	**Level of significance**
Teachers in government schools	N 1 = 80	29.05	**0.91**	**0.02**	The result is not significant at $p < .05$
Teachers in private schools	N 2 = 80	20.06			

ANOVA: Single factor

Summary

Groups	**Count**	**Sum**	**Average**	**Variance**		
Column 1	5	1558	311.6	36541.3		
Column 2	5	1644	328.8	77189.2		
ANOVA						
Source of variation	*SS*	*df*	*MS*	*f*	*p* Value	*f* Crit
Between groups	739.6	1	739.6	0.02	0.91	5.32
Within groups	454922	8	56865.25			
Total	455661.6	9				

The *f*-ratio value is 0.01301. The *p*-value is .912013. The result is not significant at $p < .05$.

Table 19.3 shows that there is no noteworthy variation in the attitude of government versus private secondary school teachers toward job satisfaction in Tripura at $p < .05$; drawing a similarity with the study of Bhat (2018) that there is no significant difference in the attitude of job satisfaction between high school teachers working in the government and private sector in job satisfaction.

19.6 RESULTS AND FINDINGS

To recognize the issues affecting job contentment of teachers in Tripura at secondary level.

TABLE 19.4 Tabulation of the Number of Responses Toward Each Facet of All Teachers.

Sl. no	Facets	On my present job, this is how I feel about	Very dissatisfied	Dissatisfied	Neutral	Satisfied	Very satisfied
01	Ability Utilization	Being able to keep busy all the time	12	89	19	32	8
02	Achievement	The chance to work alone on the job	26	48	29	45	12
03	Activity	The chance to do different things from time to time	29	52	30	27	22
04	Advancement	The chance to be "somebody" in the community	16	34	22	69	19
05	Authority	The way my boss handles his/her workers	28	96	0	22	14
06	Education Policy	The competence of my supervisor in making decision	16	45	11	59	29
07	Compensation	Being able to do things that don't go against my conscience	34	26	56	26	18
08	Co-worker	The way my job provides for steady employment	39	59	4	46	12
09	Creativity	The chance to do things for other people	18	58	22	62	0
10	Independence	The chance to tell people what to do	36	89	0	33	2
11	Moral Value	The chance to do something that makes use of my abilities	23	98	0	17	22
12	Recognition	The way company policies are put into practice	42	87	7	8	16
13	Responsibility	My pay and the amount of work I do	12	102	0	33	13
14	Security	The chance for advancement on this job	33	82	13	13	19
15	Social Service	The freedom to use my own judgement	38	79	18	20	5
16	Social Status	The chance to try my own methods of doing the job	12	68	22	38	20
17	Supervision Human Relation	The working condition	18	69	4	52	17
18	Supervision Technical	The way my co-workers get along with each other	10	78	18	32	22

TABLE 19.4 *(Continued)*

Sl. no	Facets	On my present job, this is how I feel about	Very dissatisfied	Dissatisfied	Neutral	Satisfied	Very satisfied
19	Variety	The praise I get for doing a good job	22	93	4	29	12
20	Working Condition	The feeling of accomplishment I get from the job	17	65	20	24	34
	Total responses obtained from each category		481	1417	299	687	316
	Mean		24.05	70.85	14.95	34.35	15.8
	Percentage		15.03	44.28	9.34	21.46	9.87

TABLE 19.5 The Level of Attitude of All Secondary Teachers Toward Job Satisfaction in Tripura.

All respondents together (N = 160)				
Sl no.	**Scale**	**No. of responses N = 160 * 20 facets = 3200 responses**	**Mean**	**Percentage % response**
01	Very dissatisfied	481	24.05	15.03
02	Dissatisfied	1417	70.85	44.28
03	Neutral	299	14.95	9.34
04	Satisfied	687	34.35	21.46
05	Very satisfied	316	15.8	9.87
Total				**100 %**

Table 19.5 shows that the dominant part of reactions is toward disappointment with a mean estimation of 70.85 while the satisfactory reactions, with a mean estimation of 34.35. However, 9.34% teachers are with neutral response and 15.31% teachers are very dissatisfied in contrast to very satisfied with 9.87% responses toward it.

To identify if gender is a factor affecting the level of job satisfaction of teachers in Tripura at secondary level.

TABLE 19.6 Attitude, by Gender (Male versus Female Teachers) Toward Job Satisfaction.

Categories	**No. of male teachers response**		**No. of female teachers response**		**Total**
	N = 80	%	N = 80	%	
Very dissatisfied	183	**11.43**	298	**18.62**	481
Dissatisfied	580	**36.25**	837	**52.31**	1417
Neutral	187	**11.68**	112	**07**	299
Satisfied	388	**24.25**	299	**18.68**	687
Very satisfied	212	**13.25**	104	**6.5**	316

Table 19.6 shows that dominant part of reactions is toward disappointment in both the cases of male and female teachers.

To draw a comparative study between private and government school teacher's attitude toward job satisfaction in Tripura.

Table 19.7 shows that the dominant part of reactions is toward disappointment in both the cases of government and private teachers.

TABLE 19.7 The Level of Attitude of Government versus Private School Teachers Toward Job Satisfaction.

Categories	No. of government teachers response		No. of private teachers response		Total
	N = 80	**%**	**N = 80**	**%**	
Very dissatisfied	218	**13.62**	263	**16.43**	481
Dissatisfied	606	**37.87**	811	**50.68**	1417
Neutral	157	**9.81**	142	**8.87**	299
Satisfied	402	**25.12**	287	**17.93**	687
Very satisfied	175	**10.93**	141	**8.81**	316

19.7 DISCUSSION

The motivation behind this examination was to investigate and look at the extent of job contentment of school teachers in Tripura working at secondary level. The finding of this exploration shows that the teachers as a whole are not happy regarding the various elements of the work, for example, approaches, support, oversight, opportunity, and so on. However, they are marginally happy concerning their independency, acknowledgment, pay, climate, and so forth; though there is no significant contrast in the mentality of job satisfaction between male versus female teachers and working in government versus private-owned schools. Based on the conclusion of this investigation, it is suggested that options might be furnished with better compensation bundle, quick advancement, openings for innovative work, acknowledgment by the division, dynamic cooperation in approach definition, and better working condition. Moreover, it may be stated that such study might be undertaken for additional examination at different samples.

TABLE 19.8 Tabulation of the Level of Satisfaction of All the Teachers, Using the Scales Used in the Study.

Level of satisfaction/ scales	Items no.	Mean	Remark
Intrinsic satisfaction	1,2,3.4,7,8,9,10, 11,15,16,20.	27.375	27% (approx) Teachers are satisfied on internal factors.
Extrinsic satisfaction	5,6,12,13,14,19.	17.166667	17% (approx) Teachers are satisfied on external factors.
General satisfaction	All	25.075	25% (approx) Teachers are satisfied on general factors.

On the basis of MSQ recommendations (p. 4), the short form of the questionnaire consists of three scales of satisfaction level and as per Table 19.8 it may be concluded that only 27% teachers are satisfied on internal factors and 17% (approximately) teachers are satisfied on external factors, while the overall general satisfaction level is 25%.

19.8 CONCLUSION

The reason for the review was to discover the degree of occupational fulfillment among the teachers of Tripura working at secondary schools and furthermore to find out if gender and type of school (government or privately owned) play a role affecting the same. The findings of the research shows that only 27% teachers are satisfied on internal factors such as independency, recognition, salary, environment, etc. and 17% are satisfied based on external factors such as policies, support, supervision, freedom, etc. Overall, 25% teachers are satisfied in general, while 75% of the teachers offered a dissatisfactory response.

The dissatisfaction toward their job is due to lack of independence among the government teachers from doing creative activity; in case of recognition, no moral support is given; discriminatory salary package; relation with the co workers; and an unsuitable environment.

The results also show that there is no huge contrast in the extent of job contentment between male and female teachers at secondary level in Tripura and no significant difference between high school teachers working in government and private sector in terms of job satisfaction.

KEYWORDS

- **job satisfaction**
- **probing**
- **agility**
- **teacher**
- **gender**
- **school**
- **attitude**

REFERENCES

Achanta, S.; Reddy, D. V. Job Satisfaction among Primary School Teachers in Krishna District. *Indian J. Res. Paripex*, **2014,** *3* (12), 95–96.

Bamundo, P. J.; Kuppleman, R. F. The Moderating Effects of Occupation, Age and Urbanisation on the Relationship between Job Satisfaction. *J. VocT. Behav.* **1980,** *17* (1), 106–123.

Bhat, A. A. Job Satisfaction among High School Teachers. *Int. J. Indian Psychol.* **2018,** *6* (1).

Chauhan, V.; Chauhan, S.; Chauhan, S. S. Job Satisfaction among Senior Secondary School Teachers: Case Study of Meerut Region. *Int. J. Bus. Gen. Manag.* **2017,** *6* (5), 39–48.

Choudhury, P.; Demor, D. A Study on Job Satisfaction of English Teachers at Secondary School Level of Vadodara City. *IERJ*, **2019,** *5* (1).

Gihar, S.; Mishra, D. K. Job Satisfaction among Secondary School Teachers: A Study. *Sch. Res. J. Humanit. Sci. Engl. Lang.* **2018,** *6* (29).

Newstrom, J. W. *Organizational Behavior*, 12th ed.; Tata McGraw-Hill Publishing Company Limited: New Delhi. 2007; pp. 204–205.

Nigama, K.; Selvabaskar, S.; Surulivel, S. T.; Alamelu, R.; Joice, U. D. Job Satisfaction among School Teachers. *Int. J. Pure Appl. Math.* **2018,** *119* (7), 2645–2655.

Rajendran, R, D. R.; Veerasekaran. A Study of Job Satisfaction of Secondary School Teachers. *Int. Glob. Res. Anal.* **2013,** *2* (6), 1–2.

Raymond, C. J. A Case Study Exploring Teacher Job Satisfaction and Teacher Retention Issues in a Large Urban Oklahoma School District. Master;s Thesis. 1 February, Retrieved December 14, 2020, 2018.

Sasi K. M.; Balasubramani, D. P. A Study on the Job Satisfaction of Primary School Teachers in Purasaiwakkam Area. *Int. J. Eng. Manag. Res.* **2019,** *9* (4). doi:https://doi.org/10.31033/ijemr

Index

D

E

F

J

T